Smooth Dynamical Systems

Pure and Applied Mathematics

A Series of Monographs and Textbooks

Editors **Samuel Eilenberg and Hyman Bass**

Columbia University, New York

RECENT TITLES

I. MARTIN ISAACS. Character Theory of Finite Groups

JAMES R. BROWN. Ergodic Theory and Topological Dynamics

C. TRUESDELL. A First Course in Rational Continuum Mechanics: Volume 1, General Concepts

GEORGE GRATZER. General Lattice Theory

K. D. STROYAN AND W. A. J. LUXEMBURG. Introduction to the Theory of Infinitesimals

B. M. PUTTASWAMAIAH AND JOHN D. DIXON. Modular Representations of Finite Groups

MELVYN BERGER. Nonlinearity and Functional Analysis: Lectures on Nonlinear Problems in Mathematical Analysis

CHARALAMBOS D. ALIPRANTIS AND OWEN BURKINSHAW. Locally Solid Riesz Spaces

JAN MIKUSINSKI. The Bochner Integral

THOMAS JECH. Set Theory

CARL L. DEVITO. Functional Analysis

MICHIEL HAZEWINKEL. Formal Groups and Applications

SIGURDUR HELGASON. Differential Geometry, Lie Groups, and Symmetric Spaces

ROBERT B. BURCKEL. An Introduction to Classical Complex Analysis: Volume 1

JOSEPH J. ROTMAN. An Introduction to Homological Algebra

C. TRUESDELL AND R. G. MUNCASTER. Fundamentals of Maxwell's Kinetic Theory of a Simple Monatomic Gas: Treated as a Branch of Rational Mechanics

BARRY SIMON. Functional Integration and Quantum Physics

GRZEGORZ ROZENBERG AND ARTO SALOMAA. The Mathematical Theory of L Systems.

DAVID KINDERLEHRER and GUIDO STAMPACCHIA. An Introduction to Variational Inequalities and Their Applications.

H. SEIFERT AND W. THRELFALL. A Textbook of Topology; H. SEIFERT. Topology of 3-Dimensional Fibered Spaces

IN PREPARATION

LOUIS HALLE ROWEN. Polynominal Identities in Ring Theory

ROBERT B. BURCKEL. An Introduction to Classical Complex Analysis: Volume 2

DRAGOS M. CVETKOVIC, MICHAEL DOOB, AND HORST SACHS. Spectra of Graphs

DONALD W. KAHN. Introduction to Global Analysis

EDWARD PRUGOVECKI. Quantum Mechanics in Hilbert Space

ROBERT M. YOUNG. An Introduction to Nonharmonic Fourier Series

Smooth Dynamical Systems

M. C. IRWIN

The University of Liverpool
Department of Pure Mathematics

1980

ACADEMIC PRESS

A Subsidiary of Harcourt Brace Jovanovich, Publishers
London   New York   Toronto   Sydney   San Francisco

ACADEMIC PRESS, INC. (LONDON) LTD.
24/28 Oval Road, London NW1 7DX

United States Edition published by
ACADEMIC PRESS, INC.
111 Fifth Avenue, New York, New York 10003

British Library Cataloguing in Publication Data

Irwin, M C
    Smooth dynamical systems. — (Pure
    and applied mathematics).
    1. Differential equations
    I. Title   II. Series
    515'.35   QA371   80–40031

    ISBN 0-12-374450-4

PRINTED BY J. W. ARROWSMITH LTD., BRISTOL, ENGLAND

CONTENTS

Preface

In 1966 I began teaching a third year undergraduate course in the geometric theory of differential equations. This had previously been given by my friend and colleague Stewart Robertson (now of Southampton University). We both felt that no modern text book really covered the course, and we decided to collaborate in writing one. We had in mind something very simple, with plenty of pictures and examples and with clean proofs of some nice geometric results like the Poincaré–Bendixson theorem, the Poincaré–Hopf theorem and Liapunov's direct method. Unfortunately, over the years, this book stubbornly refused to materialize in a publishable form. I am afraid that I was mainly responsible for this. I became increasingly interested in detailed proofs and in presenting a coherent development of the basic theory, and, as a result, we lost momentum. Eventually two really excellent introductions (Arnold [1] and Hirsch and Smale [1]) appeared, and it is to these that one would now turn for an undergraduate course book. The point of this piece of history is to emphasize the very considerable contribution that Professor Robertson has made to the present book, for this has developed out of our original project. I am very happy to have the opportunity of thanking him both for this and also for his help and encouragement in my early years at Liverpool.

The book that has finally appeared is, I suppose, mainly for postgraduates, although, naturally, I should like to foist parts of it upon undergraduates as well. I hope that it will be useful in filling the gap that still exists between the above-mentioned text books and the research literature. In the first six chapters, I have given a rather doctrinaire introduction to the subject, influenced by the quest for generic behaviour that has dominated research in recent years. I have tried to give rigorous proofs and to sort out answers to questions that crop up naturally in the course of the development. On the other hand, in Chapter 7, which deals with some aspects of the rich flowering of the subject that has taken place in the last twenty-odd years, I have gone in for informal sketches of the proofs of selected theorems. Of course, the choice of results surveyed is very much a function of my own interests and, particularly, competence. This explains, for example, my failure to say anything much about ergodic theory or Hamiltonian systems.

I have tried to make the book reasonably self-contained. I have presupposed a grounding in several-variable differential calculus and a certain amount of elementary point set topology. Very occasionally results from algebraic topology are quoted, but they are of the sort that one happily takes on trust. Otherwise, the basic material (or, at least, enough of it to get by with) is contained in various slag heaps, labelled Appendix, that appear at the end of chapters and at the end of the book. For example, there is a long appendix on the theory of smooth manifolds, since one of the aims of the book is to help students to make the transition to the global theory on manifolds. The appendix establishes the point of view taken in the book and assembles all the relevant apparatus. Its later pages are an attempt to alleviate the condition of the student who shares my congenital inability to grasp the concept of affine connection. To make room for such luxuries, I have, with regret, omitted some attractive topics from the book. In particular, the large body of theory special to two dimensions is already well treated in text books, and I did not feel that I could contribute anything new. Similarly, there is not much emphasis on modelling applications of the theory, except in the introduction. I feel more guilty about ducking transversality theory, and this is, in part, due to a lack of steam. However, after a gestation period that would turn an Alpine black salamander green with envy, it must now be time to stand and deliver.

When working my way into the subject, I found that the books by Coddington and Levinson [1], Hurewicz [1], Lefschetz [1], Nemitskiĭ and Stepanov [1] and, at a later stage, Abraham [1] and Abraham and Robbin [1] were especially helpful. I should like to express my gratitude to my colleague at Liverpool, Bill Newns, who at an early stage read several of the chapters with great care and insight. I am also indebted to Plinio Moreira, who found many errors in a more recent version of the text, and to Andy du Plessis for helpful comments on several points. Finally, a special thank-you to Jean Owen, who typed the whole manuscript beautifully and is still as friendly as ever.

Introduction

In the late nineteenth century, Henri Poincaré created a new branch of mathematics by publishing his famous memoir (Poincaré [1]) on the qualitative theory of ordinary differential equations. Since then, differential topology, one of the principal modern developments of the differential calculus, has provided the proper setting for this theory. The subject has a strong appeal, for it is one of the main areas of cross-fertilization between pure mathematics and the applied sciences. Ordinary differential equations crop up in many different scientific contexts, and the qualitative theory often gives a major insight into the physical realities of the situation. In the opposite direction, substantial portions of many branches of pure mathematics can be traced back, directly or indirectly, to this source.

Suppose that we are studying a process that evolves with time, and that we wish to model it mathematically. The possible states of the system in which the process is taking place may often be represented by points of a differentiable manifold, which is known as the *state space* of the model. For example, if the system is a single particle constrained to move in a straight line, then we may take Euclidean space $\mathbf{R}^2$ as the state space. The point $(x, y) \in \mathbf{R}^2$ represents the state of the particle situated x units along the straight line from a given point in a given direction moving with a speed of y units in that direction. The state space of a model may be finite dimensional, as in the above case, or it may be infinite dimensional. For example, in fluid dynamics we have the velocity of the fluid at infinitely many different points to take into account and so the state space is infinite dimensional. It may happen that all past and future states of the system during the process are completely determined by its state at any one particular instant. In this case we say that the process is *deterministic*. The processes modelled in classical Newtonian mechanics are deterministic; those modelled in quantum mechanics are not.

In the deterministic context, it is often the case that the processes that can take place in the system are all governed by a smooth vector field on the

1

state space. In classical mechanics, for example, the vector field involved is just another way of describing the *equations of motion* that govern all possible motions of the system. We can be more explicit as to what we mean by a vector field governing a process. As the process develops with time, the point representing the state of the system moves along a curve in the state space. The velocity of this moving point at any position x on the curve is a tangent vector to the state space based at x. The process is *governed* by the vector field if this tangent vector is the value of the vector field at x, for all x on the curve.

In the *qualitative* (or *geometric*) *theory*, we study smooth vector fields on differentiable manifolds, focusing our attention on the collection of parametrized curves on the manifold that have the tangency property described above. Our hope is that any outstanding geometrical feature of the curve system will correspond to a significant physical phenomenon when the vector field is part of a good mathematical model for a physical situation. This seems reasonable enough, and it is borne out in practice. We complete this motivational introduction by examining some familiar examples in elementary mechanics from this viewpoint. The remainder of the book is more concerned with the mathematical theory of the subject than with its modelling applications.

I. THE SIMPLE PENDULUM

Consider a particle P of mass m units fixed to one end of a rod of length l units and of negligible mass, the other end Q of the rod being fixed. The rod is free to rotate about Q without friction or air resistance in a given vertical plane through Q. The problem is to study the motion of P under gravity. The mechanical system that we have described is known as the *simple pendulum* and is already a mathematical idealization of a real life pendulum. For simplicity we may as well take $m = l = 1$, since we can always modify our units to produce this end. The first stage of our modelling procedure is completed by the assumption that gravity exerts a constant force on P of g units/sec^2 vertically downwards.

We now wish to find a state space for the simple pendulum. This is usually done by regarding the rotation of PQ about Q as being positive in one direction and negative in the other, and measuring

(i) the angular displacement θ radians of PQ from the downwards vertical through Q, and

(ii) the angular velocity ω radians/sec of PQ (see Figure 0.1).

We can then take $\mathbf{R}^2$ as the state space, with coordinates (θ, ω).

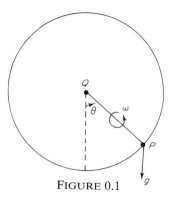

FIGURE 0.1

The equation of motion for the pendulum is

(0.2)
$$\theta'' = -g \sin \theta,$$

where $\theta'' = d^2\theta/dt^2$. Using the definition of ω, we can replace this by the pair of first order equations

(0.3)
$$\theta' = \omega,$$
$$\omega' = -g \sin \theta.$$

A solution of (0.3) is a curve (called an *integral curve*) in the (θ, ω) plane parametrized by t. If the parametrized coordinates of the curve are $(\theta(t), \omega(t))$ then the tangent vector to the curve at time t is $(\omega(t), -g \sin \theta(t))$, based at the point $(\theta(t), \omega(t))$. We get various integral curves corresponding to various initial values of θ and ω at time $t = 0$, and these curves form the so-called *phase portrait* of the model. It can be shown that the phase portrait looks like Figure 0.4. One can easily distinguish five

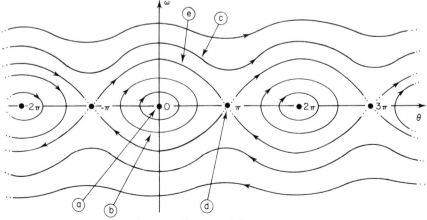

FIGURE 0.4

types of integral curves by their dissimilar appearances. They can be interpreted as follows:

(a) the pendulum hangs vertically downwards and is permanently at rest,
(b) the pendulum swings between two positions of instantaneous rest equally inclined to the vertical,
(c) the pendulum continually rotates in the same direction and is never at rest,
(d) the pendulum stands vertically upwards and is permanently at rest,
(e) the limiting case between (b) and (c), when the pendulum takes an infinitely long time to swing from one upright position to another.

The phase portrait in Figure 0.4 has certain unsatisfactory features. Firstly, the pendulum has only two equilibrium positions, one *stable* (downwards) and one *unstable* (upwards). However, to each of these there correspond infinitely many point curves in the phase portrait. Secondly, solutions of type (c) are periodic motions of the pendulum but appear as nonperiodic curves in the phase portrait. The fact of the matter is that unless we have some very compelling reason to do otherwise we ought to regard $\theta = \theta_0$ and $\theta = \theta_0 + 2\pi$ as giving the same position of the pendulum, since there is no way of instantaneously distinguishing between them. That is to say, the *configuration space*, which is the differentiable manifold representing the spatial positions of the elements of the mechanical system, is really a circle rather than a straight line. To obtain a state space that faithfully describes the system, we replace the first factor $\mathbf{R}$ of $\mathbf{R}^2 = \mathbf{R} \times \mathbf{R}$ by the circle S^1, which is the real numbers reduced modulo 2π. Keeping θ and ω as our parameters, we obtain the phase portrait on the cylinder $S^1 \times \mathbf{R}$ shown in Figure 0.5.

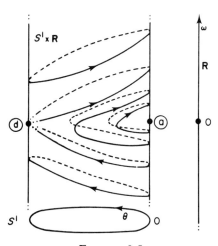

FIGURE 0.5

Consider now the *kinetic energy* T and the *potential energy* V of the pendulum, given by $T(\theta, \omega) = \frac{1}{2}\omega^2$ and $V(\theta, \omega) = g(1 - \cos\theta)$. Writing $E = T + V$ for the *total energy* of the pendulum, we find that equations (0.3) imply that $E' = 0$. That is to say E is constant on any integral curve. In view of this fact, the mechanical system is said to be *conservative* or *Hamiltonian*. In fact, in this example, the phase portrait is most easily constructed by determining the level curves (contours) of E. A pleasant way of picturing the role of E (due to E. C. Zeeman) is to represent the state space cylinder $S^1 \times \mathbf{R}$ as a bent tube in Euclidean 3-space and to interpret E as height. This is illustrated in Figure 0.6. The two arms of the tube contain solutions

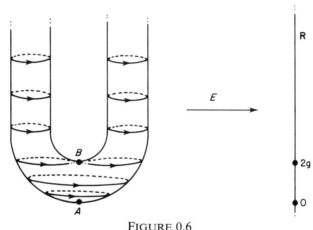

FIGURE 0.6

corresponding to rotations of the pendulum in opposite directions with the same energy E, with $E > 2g$, the potential energy of the unstable equilibrium.

The stability properties of individual solutions are apparent from the above picture. In particular, any integral curve through a point that is close to the stable equilibrium position A remains close to A at all times. On the other hand, there are points arbitrarily close to the unstable equilibrium position B such that integral curves through them depart from a given small neighbourhood of B. Note that the energy function E attains its absolute minimum at A and is stationary at B. In fact it has a saddle point at B.

II. A DISSIPATIVE SYSTEM

The conservation of the energy E in the above example was due to the absence of air resistance and of friction at the pivot Q. We now take these

forces into account, assuming for simplicity that they are directly propor-
tional to the angular velocity. Thus we replace equation (0.2) by

(0.7) $\theta'' = -g \sin \theta - a\theta'$

for some positive constant a, and (0.3) becomes

(0.8)
$$\theta' = \omega,$$
$$\omega' = -g \sin \theta - a\omega.$$

We now find that $E' = -a\omega^2$ is negative whenever $\omega \neq 0$. Thus the energy is
dissipated along any integral curve, and the system is therefore said to be
dissipative. If, as before, we represent E as a height function, the inequality
$E' < 0$ implies that the integral curves cross the (horizontal) contours of E
downwards, as shown in Figure 0.9.

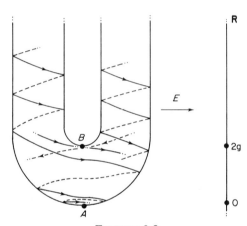

FIGURE 0.9

The reader may care to sketch dissipative versions of Figures 0.4 and 0.5.
Notice that the stable equilibrium is now *asymptotically stable*, in that nearby
solutions tend towards A as time goes by. We still have the unstable
equilibrium B and four strange solutions that either tend towards or away
from B. In practice we would not expect to be able to realize any of these
solutions, since we could not hope to satisfy the precise initial conditions
needed, rather than some nearby ones which do not have the required effect.
(One can, in fact, sometimes stand a pendulum on its end, but our model is a
poor one in this respect, since it does not take "limiting friction" into
account.)

A comparison of the systems of equations (0.3) and (0.8) gives some hint
of what is involved in the important notion of *structural stability*. Roughly

speaking, a system is structurally stable if the phase portrait remains qualitatively the same when the system is modified by any sufficiently small perturbation of the right-hand sides. By *qualitatively* (or *topologically*) *the same*, we mean that some homeomorphism of the state space maps integral curves of the one onto integral curves of the other. The existence of systems (0.8) shows that the system (0.3) is not structural stable, since the constant a can be as small as we like. To distinguish between the systems (0.3) and (0.8), we observe that most solutions of the former are periodic, whereas the only periodic solutions of the latter are the equilibria. (Obviously this last properly holds in general for any dissipative system, since E decreases along integral curves.) The systems (0.8) are themselves structurally stable, but we do not attempt to prove this fact.

III. THE SPHERICAL PENDULUM

In the case of the simple pendulum, it is desirable, but not essential, to use a state space other than Euclidean space. With more complicated mechanical systems, the need for non-Euclidean state spaces is more urgent; it is often impossible to study them globally using only Euclidean state spaces. We need other spaces on which systems of differential equations can be globally defined, and this is one reason for studying the theory of differentiable manifolds.

Consider, for example, the *spherical pendulum*, which we get from the simple pendulum by removing the restriction that PQ moves in a given plane through Q. Thus P is constrained to lie on a sphere of radius 1 which we may as well take to be the unit sphere $S^2 = \{(x, y, z): x^2 + y^2 + z^2 = 1\}$ in Euclidean 3-space. We use Euler angles θ and ϕ to parametrize S^2, as in Figure 0.10.

The motion of P is then governed by the second order equations

(0.11)
$$\theta'' = \sin\theta\cos\theta(\phi')^2 + g\sin\theta,$$
$$\phi'' = -2(\cot\theta)\theta'\phi',$$

which we replace by the equivalent system of four first order equations

(0.12) $\theta' = \lambda,$

$\phi' = \mu,$

$\lambda' = \mu^2\sin\theta\cos\theta + g\sin\theta,$

$\mu' = -2\lambda\mu\cot\theta.$

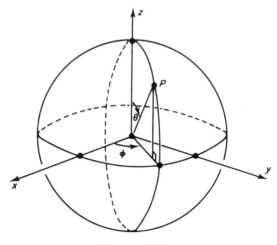

FIGURE 0.10

However, since the parametrization of S^2 by θ and ϕ is not even locally one-to-one at the two poles $(0, 0, \pm 1)$, it is wrong to expect that the four numbers $(\theta, \phi, \lambda, \mu)$ can be used without restriction to parametrize the state space of the system as $\mathbf{R}^4$. They *can* be employed with restrictions (for example $0 < \theta < \pi$, $0 < \phi < 2\pi$) but they do not then give the whole state space. In fact, the state of the system is determined by the position of P on the sphere, together with its velocity, which is specified by a point in the 2-dimensional plane tangent to S^2 at P. The state space is not homeomorphic to $\mathbf{R}^4$, nor even to the product $S^2 \times \mathbf{R}^2$ of the sphere with a plane, but is the tangent bundle TS^2 of S^2. This is the set of all planes tangent to S^2 and it is an example of a non-trivial vector bundle. *Locally*, TS^2 is topologically indistinguishable from $\mathbf{R}^4$, and we can use the four variables θ, ϕ, λ and μ as local coordinates in TS^2, provided that (θ, ϕ) does not represent the north or south pole of S^2.

The system is conservative, so again $E' = 0$ along integral curves, where the energy E is now a real function on TS^2 which, in terms of the above local coordinates, has the form

$$E(\theta, \phi, \lambda, \mu) = \tfrac{1}{2}(\lambda^2 + \mu^2 \sin^2 \theta) + g(1 + \cos \theta).$$

Thus every solution is contained in a contour of E. The contour $E = 0$ is again a single point at which E attains its absolute minimum, corresponding to the pendulum hanging vertically downwards in a position of stable equilibrium. The contour $E = 2g$ again contains the other equilibrium point, where the pendulum stands vertically upright in unstable equilibrium. At this point E is stationary but not minimal. The reader who is acquainted with

Morse theory (see Hirsch [1] and Milnor [3]) will know that for $0 < c < 2g$ the contour $E^{-1}(c)$ is homeomorphic to S^3, the unit sphere in $\mathbf{R}^4$. In any case, it is not hard to see this by visualizing how the contour is situated in TS^2. For $c > 2g$, $E^{-1}(c)$ intersects each tangent plane to S^2 in a circle, and thus can be deformed to the unit circle bundle in TS^2. This can be identified with the topological group $SO(3)$ of orthogonal 3×3 matrices, for (the position vector of) a point of S^2 and a unit tangent vector at this point determine a right-handed orthonormal basis of $\mathbf{R}^3$. Moreover, rather less obviously (see, for example, Proposition 7.12.7 of Husemoller [1]), $SO(3)$ is homeomorphic to real projective space $\mathbf{RP}^3$.

The spherical pendulum is, as a mechanical system, symmetrical about the vertical axis l through the point of suspension Q. By this we mean that any possible motion of the pendulum gives another possible motion if we rotate the whole motion about l through some angle k, and that, similarly, we get another possible motion if we reflect it in any plane containing l. This symmetry shows itself in the equations (0.12), for they are unaltered if we replace ϕ by $\phi + k$ or if we replace ϕ and μ by $-\phi$ and $-\mu$. We say that the orthogonal group $O(2)$ acts on the system as a group of symmetries about the axis l. Symmetry of this sort is quite common in mechanical systems, and it can reveal important features of the phase portrait. In this case, for any c with $0 < c < 2g$, the 3-sphere $E^{-1}(c)$ is partitioned into a family of tori, together with two exceptional circles. The picture that we have in mind is Figure 0.13 rotated about the vertical straight line m. This decomposes $\mathbf{R}^3$ into a family of tori, together with a circle (through p and q) and the line m. Compactifying with a "point at ∞" (see the appendix to Chapter 2) turns $\mathbf{R}^3$ into a topological 3-sphere and the line m into another (topological) circle.

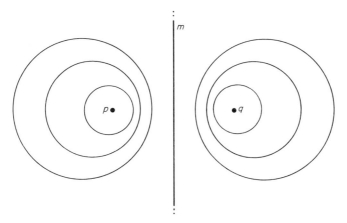

FIGURE 0.13

The submanifolds of this partition are each generated by a single integral curve under the action of $SO(2)$ (i.e. rotations only, not reflections). They are the intersections of $E^{-1}(c)$ with the contours of the angular momentum function on TS^2. The two exceptional circles correspond to the pendulum bob P revolving in a horizontal circle in the two possible directions. Halfway between them comes a torus corresponding to simple pendulum motions in the various planes through l. It is a useful exercise to investigate the similar decomposition of $E^{-1}(c)$ for $c > 2g$, and to see what happens at the critical case $c = 2g$.

IV. VECTOR FIELDS AND DYNAMICAL SYSTEMS

In all the above examples, the dynamical state of the system is represented by a point of the state space, which is the tangent bundle ($S^1 \times \mathbf{R}$ or TS^2) of the configuration space (S^1 or S^2). The equation of motion yields a vector field on the state space. Its integral curves give the possible motions of the mechanical system.

A useful way of visualizing a vector field v on an arbitrary manifold X is to imagine a (compressible) fluid flowing on X. We suppose that the velocity of the fluid at each point x of X is independent of time and equal to the value $v(x)$ of the vector field. In this case the integral curves of v are precisely the paths followed by particles of fluid. Now let $\phi(t, x)$ be the point of X reached at time t by a particle of fluid that leaves the point x at time 0. We can make some rather obvious comments. Firstly $\phi(0, x)$ is always x. Secondly, since velocity is independent of time, $\phi(s, y)$ is the point reached at time $s + t$ by a particle starting at the point y at time t. If we put $y = \phi(t, x)$, so the particle started from the point x at time 0, we deduce that $\phi(s, \phi(t, x)) = \phi(s + t, x)$. Finally we would expect the smoothness of ϕ, regarded as a function of t and x, to be influenced by the smoothness of v.

The map ϕ may not be defined on the whole of the space $\mathbf{R} \times X$, because particles may very well flow off X in a finite time. However, if ϕ is a well defined smooth map from $\mathbf{R} \times X$ to X with the above properties, we call it, in line with the above analogy, a smooth *flow* on X; otherwise we call it a smooth *partial flow* on X. It is said to be the *integral flow* of v or the *dynamical system* given by v.

Smooth vector fields and smooth flows on differentiable manifolds are the main objects of study in this book. If $\phi: \mathbf{R} \times X \to X$ is a smooth flow on X, then, for any $t \in \mathbf{R}$, we may define a map $\phi^t: X \to X$ by $\phi^t(x) = \phi(t, x)$, and this is clearly a diffeomorphism, with inverse ϕ^{-t}. If we put $f = \phi^a$ for some $a \in \mathbf{R}$, we have, by induction, that $\phi(na, x) = f^n(x)$ for all integers n. Thus, if

a is small and non-zero, we often get a good picture of the properties of ϕ by studying the iterates f^n of the single map f (just as real events can be described reasonably well by the successive stills of a motion picture). The theory of *discrete dynamical systems* or *discrete flows*, as the study of iterates of a single homeomorphism is called, resembles the theory of flows in many parts, and is sometimes rather easier. We carry the two theories side by side throughout the book, and use the term *dynamical systems* to cover both theories.

CHAPTER 1

Some Simple Examples

The theory of smooth dynamical systems can be thought of as an outcrop of a more general theory known as *topological dynamics*. Topological dynamics deals with continuous actions of any topological group G on a topological space X. Smooth dynamical systems are smooth actions of the group $\mathbf{R}$ or $\mathbf{Z}$ on a differentiable manifold X. Naturally, adding extra structure in this way revolutionizes the subject, just as it does when one makes the transition from point-set topology to differential calculus. However, in our two opening chapters we are, for the most part, laying basic foundations in which differentiability does not play a significant role. Thus these chapters have a flavour of topological dynamics. We begin by illustrating a few fundamental definitions with some simple examples. We sidestep a detailed discussion of *group action* in the text, since this would slow us down at the outset. We do, however, return to this point in the appendix to this chapter.

I. FLOWS AND HOMEOMORPHISMS

Let G denote either the additive topological group $\mathbf{R}$ of real numbers or the additive (discrete) topological group $\mathbf{Z}$ of integers. A *dynamical system* on a topological space X is a continuous map $\phi: G \times X \to X$ such that, for all $x \in X$ and, for all $s, t \in G$,

(1.1)
$$\phi(s + t, x) = \phi(s, \phi(t, x))$$
$$\phi(0, x) = x.$$

(1.2) Examples. For any X, the *trivial* dynamical system is defined by $\phi(t, x) = x$. For $X = \mathbf{R}$, $\phi(t, x) = e^t x$ defines a C^ω (analytic) dynamical system on X.

The space X is called the *phase space* of ϕ. If X is a differentiable manifold and ϕ is a C^r map, for $r \geqslant 0$, then we call ϕ a C^r dynamical system. *Throughout the book the adjective "smooth" means "C^r for some $r \geqslant 1$".*

Let ϕ be a dynamical system on X. Given $t \in G$, we define the partial map $\phi^t : X \to X$ by $\phi^t(x) = \phi(t, x)$. If $G = \mathbf{R}$, we sometimes call ϕ^t the *time t map* of ϕ. Similarly, given $x \in X$, we define the partial map $\phi_x : G \to X$ by $\phi_x(t) = \phi(t, x)$. Note that if ϕ is C^r, then so are ϕ^t and ϕ_x. Equation (1.1) may be written as

$$\phi^{s+t} = \phi^s \phi^t$$

(1.3)

$$\phi^0 = id.$$

Throughout the book juxtaposition of maps denotes composition and id denotes the relevant identity map, here the identity on X.

(1.4) Proposition. *For all $t \in G$, ϕ^t is a homeomorphism. If ϕ is C^r then ϕ is a C^r diffeomorphism.*

Proof. By equations (1.3) $\phi^t \phi^{-t} = \phi^{-t} \phi^t = \phi^0 = id$. That is, ϕ^t is invertible, and its inverse is ϕ^{-t}, which is C^r when ϕ is C^r. $\qquad\qquad\square$

For brevity we sometimes denote $\phi(t, x)$ by $t \cdot x$ when the context makes it obvious which dynamical system is under discussion. With this convention, equations (1.1) and (1.3) become

$$(s + t) \cdot x = s \cdot (t \cdot x)$$

(1.5)

$$0 \cdot x = x.$$

If $G = \mathbf{R}$ then the dynamical system ϕ is called a *flow* on X, or a *one-parameter group of homeomorphisms* of X. If $G = \mathbf{Z}$, then ϕ is completely determined by the homeomorphism ϕ^1, and it is usual to talk in terms of the homeomorphism rather than the dynamical system ϕ (sometimes called a *discrete dynamical system* or *discrete flow*) that it generates.

II. ORBITS

Let ϕ be a dynamical system on X. We define a relation $\sim$ on X by putting $x \sim y$ if and only if there exists $t \in G$ such that $\phi^t(x) = y$.

(1.6) Proposition. *The relation* $\sim$ *is an equivalence relation*   □

The equivalence classes of $\sim$ are called *orbits* of ϕ, or of the homeomorphism ϕ^1 in the case $G = \mathbf{Z}$. For each $x \in X$, the equivalence class containing x is called the *orbit through* x. It is the image of the partial map $\phi_x : G \to X$. We denote it $G \cdot x$ when it is clear to which dynamical system we refer. Proposition 1.6 implies that two orbits either coincide or are disjoint. We denote the quotient space $X/\sim$ by X/ϕ, and call it the *orbit space* of ϕ. The quotient map, which takes x to its equivalence class, is denoted $\gamma_\phi : X \to X/\phi$, or just $\gamma : X \to X/\phi$ when no ambiguity can occur. As usual, we give X/ϕ the finest topology with respect to which γ is continuous (that is, a subset U of X/ϕ is open in X/ϕ if and only if $\gamma^{-1}(U)$ is open in X).

III. EXAMPLES OF DYNAMICAL SYSTEMS

(1.7) Every orbit of the trivial dynamical system $t \cdot x = x$ is a singleton $\{x\}$. (From now on we shall usually denote both the point and the subset ambiguously by x.)

(1.8) If $G = \mathbf{R}$ the non-trivial flow $t \cdot x = e^t x$ of Example 1.2 has three orbits, namely the origin and the positive and negative half lines. In Figure 1.8 the arrows on the orbits indicate the *orientations* induced on them by the flow. That is to say, they give the direction in which $t \cdot x$ moves as t increases.

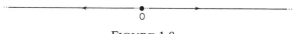
O

FIGURE 1.8

(1.9) For all $t, x \in \mathbf{R}$, put $t \cdot x = x + t$. This flow has only one orbit, $\mathbf{R}$ itself.

FIGURE 1.9

(1.10) For all $t, x \in \mathbf{R}$, put $t \cdot x = (x^{1/3} + t)^3$. Again this flow has $\mathbf{R}$ as its only orbit.

The preceding three examples are flows on the space $\mathbf{R}$ of real numbers with its standard topology. We now give a rather pathological example on $\mathbf{R}^\flat$, which is the set $\mathbf{R}$ with the indiscrete topology, in which the only open sets are $\mathbf{R}$ itself and the empty set.

(1.11) Regard **R** as a vector space over the field **Q** of rational numbers, and, using the axiom of choice, extend 1 to a basis $\mathscr{B}$ of **R**. Let $T: \mathbf{R} \to \mathbf{R}$ be the unique **Q**-linear map such that $T(1) = 1$ and $T(t) = 0$ for all other $t \in \mathscr{B}$. Put $t \cdot x = x + T(t)$ for all $t \in \mathbf{R}$ and all $x \in \mathbf{R}^{b}$. Then the orbit through x of this flow on $\mathbf{R}^{b}$ is the set $\{x + q : q \in \mathbf{Q}\}$.

Let S^{1} denote the circle $\mathbf{R}/\mathbf{Z}$ (see Example 1.3 of Appendix A) and let $[x] \in S^{1}$ denote the equivalence class of $x \in \mathbf{R}$.

(1.12) *Rotation.* Define a diffeomorphism $f: S^{1} \to S^{1}$ by $f([x]) = [x + \theta]$, for some fixed $[\theta] \in S^{1}$. Then each orbit of f consists of s points if θ is rational, $\theta = r/s$ with r, s coprime integers and $s > 0$. If, on the other hand, θ is irrational, then every orbit of f is dense in S^{1}, for the orbit through $[0]$ is the infinite cyclic subgroup of S^{1} generated by $[\theta]$.

(1.13) *Rotation flows.* For any $\theta \in \mathbf{R}$, put $t \cdot [x] = [x + \theta t]$. If $\theta = 0$ we have the trivial flow on S^{1}. Otherwise we have the single orbit S^{1}. If we embed S^{1} in the plane by the standard embedding $[x] \mapsto (\cos 2\pi x, \sin 2\pi x)$, then the rotation is anti-clockwise if θ is positive, and clockwise if θ is negative. We call θ the *speed* of the flow.

The next four examples describe flows on the plane $\mathbf{R}^{2}$. It is sometimes convenient to identify $\mathbf{R}^{2}$ with the complex line **C**, since the two are indistinguishable as topological spaces.

FIGURE 1.13 $(\theta > 0)$

(1.14) For all $t \in \mathbf{R}$ and for all $(x, y) \in \mathbf{R}^{2}$, put $t \cdot (x, y) = (x e^{t}, y e^{t})$. The origin is the only point orbit, and all other orbits are open rays issuing from the origin.

(1.15) If the formula giving the previous example is changed slightly to $t \cdot (x, y) = (x e^{t}, y e^{-t})$, the *phase portrait* (i.e. the partition of the phase space into orbits) is radically altered, since the new flow has only two orbits beginning at the origin. The picture (without arrows) is familiar to anyone who has sketched contours of a real valued function on $\mathbf{R}^{2}$. It is associated with a *saddle-point* of the function. We shall investigate later the connection between contours and flows (see Example 3.3).

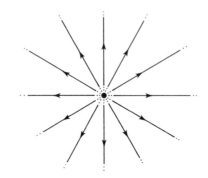

FIGURE 1.14

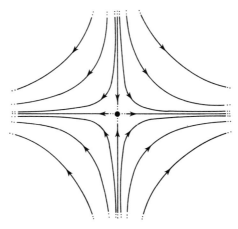

FIGURE 1.15

(1.16) For all $t \in \mathbf{R}$ and $z = x + iy \in \mathbf{C}$, put $t \cdot z = z\,e^{it}$. The origin is a point orbit, and the other orbits are all circles with centre the origin.

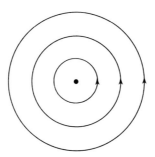

FIGURE 1.16

(1.17) For all $t \in \mathbf{R}$ and $z \in \mathbf{C}$, put $t \cdot z = z\, e^{(i-1)t}$. The origin is a point orbit, and all other orbits spiral in towards it.

FIGURE 1.17

(1.18) Exercise. Sketch the phase portraits of the flows on $\mathbf{R}^2$ given by

(i) $t \cdot (x, y) = (x, tx + y)$,

(ii) $t \cdot (x, y) = (x\, e^{-t}, (tx + y)\, e^{-t})$.

(1.19) Exercise. Describe the orbit spaces of the dynamical systems (1.7)–(1.17). Which are Hausdorff topological spaces?

IV. CONSTRUCTING SYSTEMS

There are various ways of constructing new dynamical systems from given ones.

(1.20) *Product.* Let $\phi: G \times X \to X$ and $\psi: G \times Y \to Y$ be dynamical systems. The *product* of the two systems is the dynamical system on $X \times Y$ defined, for all $t \in G$ and $(x, y) \in X \times Y$, by

$$t \cdot (x, y) = (t \cdot x, t \cdot y).$$

We shall usually denote the product system by $\phi \times \psi$, although this is strictly speaking an abuse of notation (the cartesian product of the maps ϕ and ψ has domain $G \times X \times G \times Y$, not $G \times X \times Y$, and takes (s, x, t, y) to $(s \cdot x, t \cdot y)$). Note that the product of two discrete dynamical systems is the system corresponding to the cartesian product of their generating homeomorphisms.

One can regard this construction either as a way of building up complicated examples from simple ones, or, perhaps more importantly, as a way of
decomposing complicated examples into simple ones. For instance Example
1.14 is the product of two copies of Example 1.8, and Example 1.15 is the
product of Example 1.8 and the flow $t \cdot x = x e^{-t}$ on $\mathbf{R}$ (see Figure 1.20).

FIGURE 1.20

(1.21) Example. Let ϕ be the flow of Example 1.8 and let ψ be the rotation
flow of Example 1.13 (for some $\theta > 0$, say). Then $\phi \times \psi$ is a flow on the
circular cylinder $\mathbf{R} \times S^1$. The circle $\{0\} \times S^1$ is an orbit, and all other orbits
spiral away from it.

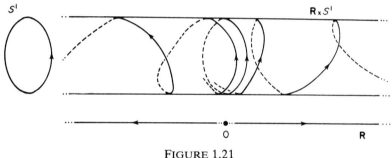

FIGURE 1.21

(1.22) Embedding. Let $\phi : \mathbf{R} \times X \to X$ be any flow. For any $t \in \mathbf{R}$, $\phi^t : X \to X$
is a homeomorphism which generates a discrete dynamical system
$\psi : \mathbf{Z} \times X \to X$. We say that ϕ^t (or ψ) is *embedded* in ϕ. For example, any
rotation of S^1 (Example 1.12) is embedded in any non-trivial rotation flow
on S^1 (Example 1.13).

It is not true that every homeomorphism of every topological space X can
be embedded in a flow on X. For example, when X is a pair of points (with
the discrete topology) there is only one flow on X, namely the trivial flow, so
only the identity homeomorphism of X is embeddable (in a flow on X).
However there is another homeomorphism of X, namely the one which
interchanges the two points. More generally, any embeddable homeomorphism of X is isotopic to the identity, since ϕ itself supplies an isotopy from
ϕ^t to ϕ^0. Thus, for example, the homeomorphism $f : \mathbf{R} \to \mathbf{R}$ defined by
$f(x) = -x$ is not embeddable.

Summing up, every flow on X yields a homeomorphism of X (in fact,
many homeomorphisms), but the reverse is not usually true. This is rather

unfortunate, since we wish to develop the theories of flows and homeomorphisms side by side, and we need some way of associating with a given homeomorphism f a flow with similar properties. We can do this provided we allow the flow to be on a larger space Y. Thus we have the space X embedded in the space Y and the homeomorphism of X embedded in a flow ϕ on Y. The standard, and most economical, way of constructing the space Y and the flow ϕ is known as *suspension* (warning: this is different from the construction of the same name in algebraic topology).

(1.23) *Suspension.* Let $f: X \to X$ be a homeomorphism (generating a discrete dynamical system ψ). Let $\sim$ be the equivalence relation defined on $\mathbf{R} \times X$ by $(u, x) \sim (v, y)$ if and only if $u = v + m$ for some $m \in \mathbf{Z}$ and $y = f^m(x)$. Then there is a flow $\phi: \mathbf{R} \times Y \to Y$ on $Y = (\mathbf{R} \times X)/\sim$ defined by $\phi(t, [u, x]) = [u + t, x]$ where $[u, x]$ denotes the equivalence class of $(u, x) \in \mathbf{R} \times X$. The flow ϕ is called the *suspension* of the homeomorphism f (or of ψ). For any $u \in R$, the restriction of ϕ^1 to any *cross section* $[u, X]$ with the obvious identification $[u, X] = X$ coincides with f.

For example, if $f: \mathbf{R} \to \mathbf{R}$ is defined by $f(x) = -x$, then the suspension is a flow on the open Möbius band, and all its orbits are topologically circles (see Figure 1.23).

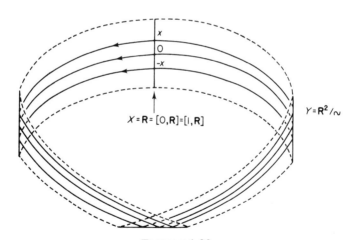

FIGURE 1.23

(1.24) Example. A *rational flow* ϕ on the torus $T^2 = S^1 \times S^1$ is given by suspending a rational rotation of S^1 (Example 1.12). Each orbit of ϕ is topologically a circle. The two diagrams in Figure 1.24 illustrate one orbit of ϕ for a rotation $\theta = \frac{2}{3}$. The first shows the torus cut open. To glue it together again, we need to identify the top and bottom edge AB and then the two

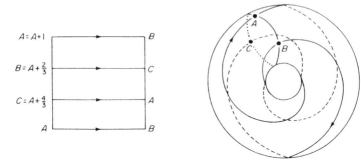

FIGURE 1.24

ends of the cylinder as indicated by the lettering. We then get the second picture, which portrays T^2 embedded as an "anchor ring" or "tyre inner tube" in $\mathbf{R}^3$. Each orbit of ϕ winds twice round "the air inside the tube" and three times round "the axis of the wheel".

(1.25) Example. An *irrational flow* ϕ on T^2 is defined by suspending an irrational rotation of S^1. In this case, ϕ has neither point orbits nor circle orbits. Every orbit of ϕ is the image of $\mathbf{R}$ under a continuous injection. However, none of these embeddings is a topological embedding; in fact, every orbit of ϕ is dense in T^2.

Rational and irrational flows on T^2 may be given another definition as the product of two rotation flows on S^1 (Example 1.13). One gets a rational flow when the ratio of the speeds of the two factors is rational, and an irrational flow when it is irrational. We discuss the relation between systems given by the two different constructions in Chapter 2 (Example 2.9). We also show there that all rational flows are essentially the same, but that there are infinitely many different types of irrational flow. (This is initially rather startling; looking at the pictures one might almost expect the reverse to be the case.) For the moment we just draw the important conclusion that *the phase portrait of a product flow is not uniquely determined by the phase portraits of its factors*, since the phase portraits of rational and irrational flows are completely different topologically, but come from factors with identical phase portraits.

(1.26) Induced systems. Let $\phi: G \times X \to X$ be a dynamical system on X, let $\alpha: G \to G$ be a continuous automorphism of the additive group G, and let $h: X \to Y$ be a homeomorphism. Then $\psi = h\phi(\alpha \times h)^{-1}$ is a dynamical system on Y. We call ψ the dynamical system *induced from* ϕ *by the pair* (α, h) (or *by* h, if $\alpha = id$). For example, if $h = id: X \to X$ and $\alpha = -id: \mathbf{R} \to \mathbf{R}$, we obtain ϕ^-, the *reverse flow* of ϕ, given by $\phi^-(t, x) = \phi(-t, x)$. Intuitively in ϕ^- points move along the orbits of ϕ with the same speed but in the opposite direction.

Notice that the continuous automorphisms of **R** are exactly the linear maps $t \mapsto ct$ where c is a non zero real constant. One talks of *speeding up* (or *slowing down*) the flow *by the factor c* when one changes to the flow induced by (α, id). The continuous automorphisms of **Z** are just id and $-id$, so in the discrete case the homeomorphism induced from f is just hfh^{-1} or its inverse.

(1.27) Example. Let $h: \mathbf{R} \times S^1 \to \mathbf{R}^2 \backslash \{0\}$ be the homeomorphism defined for all $x \in \mathbf{R}$ and all $z \in S^1$, by $h(x, z) = e^x z$, and let ζ be the flow on $\mathbf{R} \times S^1$ defined in Example 1.21. Then h induces a flow on $\mathbf{R}^2 \backslash \{0\}$ which we may extend to a flow on $\mathbf{R}^2$ by making the origin a point orbit. The phase portrait of this flow is illustrated in Figure 1.27.

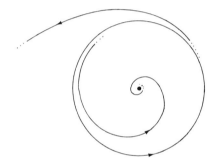

FIGURE 1.27

The construction of induced systems is not in itself a particularly interesting process, since one usually regards the induced systems as being equivalent to the original ones. However the following generalization produces genuinely new systems.

(1.28) *Quotient systems.* Let $\phi: G \times X \to X$ be a dynamical system on X and let $\sim$ be an equivalence relation on X such that, for all $t \in G$ and all $x, y \in X$, $\phi(t, x) \sim \phi(t, y)$ if (and hence only if) $x \sim y$. Then ϕ induces a dynamical system ψ, called the *quotient* system, on the quotient space $X/\sim$ by $\psi(t, [x]) = [\phi(t, x)]$, where $t \in \mathbf{R}$ and $[x]$ is the equivalence class of $x \in X$.

(1.29) Example. If f and g are *commuting homeomorphisms* of X (i.e. $fg = gf$) then f takes orbits of g onto orbits of g and hence induces a homeomorphism of the orbit space of g. This is an example of a quotient system, where ϕ is the discrete dynamical system generated by f and $\sim$ is the equivalence relation giving orbits of g as equivalence classes. Similarly if ϕ and ψ are *commuting flows* on X (i.e. $\phi^s \psi^t = \psi^t \phi^s$ for all $s, t \in \mathbf{R}$) then ϕ

induces a quotient flow on the orbit space X/ψ. Notice that such a pair of commuting flows on X is equivalent to an action of $\mathbf{R}^2$ on X (define $(s, t) \cdot x$ to be $\phi^s(\psi^t(x))$).

(1.30) Example. Define an equivalence relation $\sim$ on $\mathbf{R}^n$ by $x \sim y$ if and only if $x - y \in \mathbf{Z}^n$. The quotient $\mathbf{R}^n/\sim$ is the n-dimensional torus T^n, the cartesian product of n copies of the circle S^1. Let f be a linear automorphism of $\mathbf{R}^n$ whose matrix A (with respect to the standard basis of $\mathbf{R}^n$) is in $GL_n(\mathbf{Z})$. That is to say, A has integer entries and $\det A = \pm 1$ (or equivalently, A^{-1} also has integer entries). Then f maps $\mathbf{Z}^n$ onto itself, and hence f and f^{-1} preserve the relation $\sim$. Thus f induces a homeomorphism (in fact, a diffeomorphism) of T^n. Suppose further that A has no eigenvalues on the unit circle in the Argand diagram $\mathbf{C}$. Then we call the induced homeomorphism of T^h a *hyperbolic toral automorphism*. Thus, for example, we use the term if

$$A = \begin{bmatrix} 2 & 1 \\ 1 & 1 \end{bmatrix} \quad \text{but not if } A = \begin{bmatrix} 1 & 0 \\ 0 & -1 \end{bmatrix}.$$

Considering for a moment the particular case

$$A = \begin{bmatrix} 2 & 1 \\ 1 & 1 \end{bmatrix},$$

we observe that the phase portrait of f is qualitatively the same as that of the time 1 map, ϕ^1 say, of the flow in Example 1.15. Thus the saddle point picture in Figure 1.15 is still a good illustration for it, with the reservation that the hyperbolae are merely unions of certain orbits, and do not have any special significance. The straight lines, on the other hand, are very significant; they are the loci of points whose iterates under positive or negative iterates of f tend to the fixed point 0. The important difference between f and ϕ^1 is that for f these lines have irrational slope, and so the corresponding curves on T^2 for the induced toral automorphism wrap themselves densely around T^2. This rather complicated behaviour is typical of hyperbolic toral automorphisms.

Another remarkable fact about any hyperbolic toral automorphism $g : T^n \to T^n$ is that both its periodic and non-periodic point sets are dense in T^n. Of course, a point $x \in T^n$ is *periodic* if $g^r(x) = x$ for some $r > 0$.

(1.31) Theorem. *The periodic set of the hyperbolic toral automorphism $g : T^n \to T^n$ is precisely $\mathbf{Q}^n/\sim$, where $\mathbf{Q}$ is the rational numbers.*

Proof. First suppose that $x \in \mathbf{Q}^n$, and let m be the L.C.M. of the denominators of the coordinates of x. Since A, the matrix of f, has integer entries, the coordinates of $f^r(x)$ $(r > 0)$ are integer combinations of those of x, and hence their denominators have L.C.M. dividing m. There are only finitely many points $[y]$ of T^n corresponding to points $y \in \mathbf{Q}^n$ whose coordinates satisfy the above condition. Thus there are only finitely many possible values for $[f^r(x)] = g^r([x])$, and hence $g^r([x]) = g^s([x])$ for some $r > s > 0$. Thus $g^{r-s}([x]) = [x]$, and $[x]$ is a periodic point of g.

Conversely, suppose that $x \in \mathbf{R}^n$ is such that $[x] \in T^n$ is a periodic point of g. Then $g^r([x]) = [x]$ for some $r > 0$, and hence $(f^r - id)(x) = y$, say, is in $\mathbf{Z}^n$. Now $f^r - id$ is a linear automorphism (since 1 is not an eigenvalue of f^r) with integer entries. Thus $(f^r - id)^{-1} = (\det (f^r - id))^{-1} \operatorname{adj} (f^r - id)$ has rational entries, and hence $x = (f^r - id)^{-1}(y)$ is in $\mathbf{Q}^n$.   □

Hyperbolic toral automorphisms are the simplest examples of Anosov diffeomorphisms (see Chapter 7) on compact manifolds. They first aroused interest as a counterexample (due to R. Thom) to the conjecture that structurally stable diffeomorphisms (in the sense of Chapter 7) have finite periodic sets.

(1.32) Exercise. Verify that induced and quotient systems are indeed dynamical systems.

(1.33) Exercise. Let ϕ and ψ be flows on $\mathbf{R}^2$ defined by $\phi(t, z) = z e^t$ and $\psi(t, z) = z e^{it}$. Describe (i) the flow that ψ induces on the orbit space of ϕ, and (ii) the flow that ϕ induces on the orbit space of ψ.

V. PROPERTIES OF ORBITS

A cursory examination of the preceding examples reveals at least five topologically different types of orbit in the case of flows, and infinitely many in the case of homeomorphisms. We now make some remarks which are directly concerned with sorting out orbits into topological types.

Let ϕ be a dynamical system on a topological space X. For each $x \in X$, the subset $G_x = \{t \in G : \phi(t, x) = x\}$ is a subgroup of G called the *stabilizer* (or *isotropy subgroup*) *of x*, or *of ϕ at x*.

(1.34) Proposition. *If X is a T_1 space, then for all $x \in X$, G_x is a closed subgroup of G.*

Proof. The subset $\{x\}$ is closed, and $G_x = \phi_x^{-1}(\{x\})$.   □

We must place some sort of separation condition on X if we require closed stabilizers. Example 1.11 shows that if X has a very coarse topology then $\mathbf{R}_x$ may be a rather unpleasant non-closed subgroup of $\mathbf{R}$. However one may get by with weaker conditions than in the above proposition.

(1.35) Exercise. Prove that Proposition 1.34 still holds with T_1 replaced by T_0.

The connection between the stabilizer of a point and the topology of its orbit is made as follows:

(1.36) Proposition. *For all $x \in X$, there is a continuous bijection*

$$\beta : G/G_x \to G \cdot x.$$

Proof. Let $\pi: G \to G/G_x$ be the quotient map. Define β by $\beta(\pi(t)) = t \cdot x$. To show that β is well defined, let $\pi(s) = \pi(t)$. Thus $s - t \in G_x$, whence $(s - t) \cdot x = x$ and so $s \cdot x = t \cdot x$.

Suppose now that $\beta(\pi(s)) = \beta(\pi(t))$, i.e. that $s \cdot x = t \cdot x$. Thus $s - t \in G_x$, and so $\pi(s) = \pi(t)$. This proves that β is injective. Since β is trivially surjective, we have shown that β is bijective.

Finally, let $A = U \cap (G \cdot x)$ be an open subset of $G \cdot x$, where U is open in X. Then $W = \phi_x^{-1}(U)$ is open in G. But $W = \pi^{-1}\beta^{-1}(U)$. Hence $\beta^{-1}(U)$ is open in G/G_x, by definition of the topology on G/G_x.   □

(1.37) Corollary. *If X is Hausdorff and G/G_x is compact, then the orbit $G \cdot x$ is homeomorphic to G/G_x*   □

To capitalize on this corollary, we must analyse the possible closed (by Proposition 1.34) subgroups of G.

In the case $G = \mathbf{Z}$, with ϕ generated by the homeomorphism f, G_x is either (i) the zero subgroup or (ii) isomorphic to $\mathbf{Z}$ itself and generated by some positive integer n. Case (ii) gives a compact quotient space, a set of n distinct points. Correspondingly the orbit is a set of n distinct points $\{x, f(x), \ldots, f^{n-1}(x)\}$, and x satisfies $f^n(x) = x$. We say that x is *periodic* of *period n* and write per $x = n$. If $n = 1$ we call x a *fixed point*.

One may easily prove that any closed subgroup H of $\mathbf{R}$ is either (i) the zero subgroup, (ii) $\mathbf{R}$ itself, or (iii) an infinite cyclic subgroup generated by some positive number t_0. (*Hint*: if $H \neq \{0\}$, inf $\{t \in H: t > 0\} \in H$.) Thus if $G = \mathbf{R}$ and X is Hausdorff, we have correspondingly the three possibilities that G/G_x is (i) homeomorphic to $\mathbf{R}$ itself (ii) a single point or (iii) homeomorphic to the circle. In case (ii) the orbit $G \cdot x$ is the single point x, which is called a *fixed point* of ϕ. In case (iii) $G \cdot x$ is homeomorphic to the circle S^1, and $\phi_x: \mathbf{R} \to X$ is periodic of period t_0. One calls $G \cdot x$ a *periodic orbit, circle orbit, cycle* or *closed orbit*. The last term is open to the criticism that point orbits are also closed in X, and orbits in case (i) may very well be closed in X as well (see Example 1.9 above). We shall show in Corollary 2.36 below that this never happens if X is compact). However, it is probably the commonest of the four terms, so we shall employ it whenever it does not create actual confusion. We call t_0 the *period* of the orbit $G \cdot x$, and write per $x = t_0$.

(1.38) Exercise. Let X be Hausdorff. Prove that the *fixed point set* Fix ϕ of a dynamical system ϕ on X is closed in X. Find, among the examples of this

chapter, counterexamples to the following statements:
  (i) The *periodic point set* Per f of a homeomorphism $f: X \to X$ is closed in
      X.
 (ii) The set of all periodic points with a given period n_0 of a homeomor-
      phism $f: X \to X$ is closed in X.
(iii) the set of all points on periodic orbits with a given period t_0 of a flow ϕ
      on X is closed in X.

(1.39) Exercise. Show that if p is a fixed point of a flow ϕ then, for any
neighbourhood U of p there exists a neighbourhood V of p such that
$\phi(t, x) \in U$ for all $x \in V$ and all $t \in [0, 1]$. Deduce that $\phi(t, x) \to p$ as $t \to \infty$ if
and only if $\phi(n, x) \to p$ as $n \to \infty$, where $t \in \mathbf{R}$ and $n \in \mathbf{Z}$. Prove, similarly, that
if $\mathbf{R} \cdot p$ is a closed orbit of a flow ϕ on a metric space then the distance
$d(\phi(t, x), \phi(t, p)) \to 0$ as $t \to \infty$ in $\mathbf{R}$ if and only if $\phi(n\tau, x) \to \phi(n\tau, p) = p$ as
$n \to \infty$ in $\mathbf{Z}$, where τ is the period of $\mathbf{R} \cdot p$.

(1.40) Exercise. We have shown that every orbit of a flow on a Hausdorff
space either is a point, is homeomorphic to the circle or is the image of $\mathbf{R}$
under a continuous bijection. Prove that these three possibilities are mutually exclusive.

   In case (i) the continuous bijection β of Proposition 1.36 is (identified
with) the map $\phi_x: \mathbf{R} \to \mathbf{R} \cdot x$. It may or may not be the case that this map is a
homeomorphism. It is in most of the examples given above, but it is not for
irrational flow on the torus (Example 1.25). It turns out that the orbit $\mathbf{R} \cdot x$ is
homeomorphic to $\mathbf{R}$ (or, equivalently, ϕ_x is a homeomorphism) if and only if
it is locally compact. For this result, see Theorem 2.51 below, or construct a
proof directly, in which case you will probably need the Baire category
theorem (see § 5.3 of Chapter 9 of Bourbaki [1]).

   There is, we should finally say, one rather obvious topological property
that is shared by orbits of flows on all topological spaces, Hausdorff or not.

(1.41) Proposition. *Every orbit of every flow is connected.*

Proof. For any flow ϕ on any space X and for all $x \in X$, ϕ_x is continuous, and
so preserves the connectedness of $\mathbf{R}$. $\square$

Appendix 1

I. GROUP ACTIONS

Let X be a topological space. The set Homeo X of all homeomorphisms of X (onto itself) acquires the structure of a group when one defines the group product of two such homeomorphisms f and g to be the composed map fg. The identity of Homeo X is the identity map $id: X \to X$ (sending each $x \in X$ to itself), and the group inverse of f is the inverse map f^{-1} (defined, for all $x, y \in X$, by $f^{-1}(y) = x$ if and only if $f(x) = y$). One would like to turn Homeo X into a *topological group* (that is, to give it a topology with respect to which the multiplication map $(f, g) \to fg$ and the inversion map $f \to f^{-1}$ are continuous). One would then define an *action* of a topological group G on X to be a continuous homomorphism $\alpha: G \to$ Homeo X. The triple (G, X, α) is known as a *topological transformation group*. *Topological dynamics* is the study of topological transformation groups.

The reader may already have encountered the *compact-open* (C.O.) *topology* as an example of a function space topology. A sub-basis for this topology for Homeo X is given by all subsets of the form $\{f \in$ Homeo $X : f(K) \subset U\}$, where K, U range respectively over all compact and all open subsets of X. Thus every open subset in the C.O. topology is a union of finite intersections of subsets of the given form. Homeo X is not always a topological group with respect to the C.O. topology. Group multiplication may fail to be continuous when X is not locally compact, and so may inversion even when X is locally compact. However X *is* a topological group with respect to the C.O. topology if X is either compact Hausdorff or locally connected, locally compact Hausdorff. Moreover the former condition suggests that by compactifying X we may obtain a suitable topology (called the Arens G-topology) on Homeo X for any locally compact Hausdorff X. The interested reader is referred to § 3 of Chapter 9 of Bourbaki [1] together with its associated exercises.

The connection between group actions and dynamical systems is fairly obvious. Let G be a topological group and let X be a topological space such that Homeo X is a topological group with respect to the C.O. topology. Any map $\alpha: G \to$ Homeo X determines a map $\phi: G \times X \to X$ given by $\phi(g, x) = \alpha(g)(x)$. The map α is a homomorphism if and only if, for all $x \in X$ and all $g, h \in G$,

(A.1) $\phi(g*h, x) = \phi(g, \phi(h, x))$,

and (redundantly)

(A.2) $\phi(e, x) = x$,

where $*$ is the group operation of G and e is the identity element of G. When $G = \mathbf{R}$ or $\mathbf{Z}$ we have the equations (1.1) in the definition of "dynamical system". Conversely, if $\phi \times X \to X$ is a continuous map satisfying (1) and (2) then the argument of Proposition 1.4 shows that ϕ determines a map $\alpha: G \to$ Homeo X by $\alpha(g) = \phi^g$, and α is a homomorphism, as above. To complete the picture we have to relate continuity of α and ϕ. It is straightforward to show that if ϕ is continuous then α is continuous. One checks that the inverse image of a sub-basic open set is open in G; this requires no extra conditions on G and X. Showing, conversely, that if α is continuous then ϕ is continuous requires a little more care. The obvious argument needs X to be locally compact (i.e. for all $x \in X$, every neighbourhood of x contains a compact neighbourhood of x). However this is not usually a serious restriction, since we are already assuming some such condition in order to make Homeo X a topological group.

Later in the book we restrict our attention to smooth manifolds X and smooth dynamical systems ϕ. Corresponding to such systems we would naturally expect a smooth version of the above theory of group actions. The analogue of a topological group in the smooth theory is a *Lie group*, which is a C^∞ manifold with a group structure such that composition and inversion are C^∞ maps. The group need not be abelian. A *(left) Lie transformation group* or *(left) action* of a Lie group G on a C^∞ manifold X is usually defined to be a C^∞ map $\phi: G \times X \to X$ satisfying the multiplicative version of (1.1)

$$\phi(gh, x) = \phi(g, \phi(h, x)), \qquad \phi(e, x) = x.$$

See, for example, Warner [1] or Brickell and Clark [1]. Another approach would be to define an action as a C^∞ homomorphism of G into Diff$^\infty X$, the space of all C^∞ diffeomorphisms of X, but it is not straightforward to give Diff$^\infty X$ a Lie group structure (see Leslie [1]).

There is a good deal of general theory of Lie group actions, especially for compact G. We shall not investigate this theory; we are quite happy to work with the particular groups $\mathbf{R}$ and $\mathbf{Z}$.

CHAPTER 2

Equivalent Systems

The *classification* of dynamical systems is one of the main goals of the subject. One begins by placing an equivalence relation upon the set of all dynamical systems. This relation should be a natural one, in the sense that it is based on qualitative resemblances of the systems. One then attempts to list the equivalence classes, distinguishing between them by numerical and algebraic *invariants* (quantities that are associated with all systems and that are equal for all systems in the same equivalence class). If one cannot achieve a classification of the set of all dynamical systems, one would at any rate like to classify a large (in some sense) subset of it. A good classification requires a careful choice of basic equivalence relation, with tractable invariants. There are several "obvious" relations to be considered and we discuss these in some detail.

The search for invariants involves a study not only of the structure of individual orbits of a system, but also of the topological relationships between orbits. This results in the formulation of such concepts as *limit set, minimal set* and *non-wandering set.*

I. TOPOLOGICAL CONJUGACY

Let $f: X \to X$ and $g: Y \to Y$ be homeomorphisms of topological spaces X and Y. A *topological conjugacy* from f to g is a homeomorphism $h: X \to Y$ such that $hf = gh$.

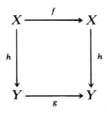

28

The homeomorphisms f and g (and the discrete flows that they generate) are said to be *topologically conjugate* if such a homeomorphism h exists. Trivially, topological conjugacy is an equivalence relation.

(2.1) Exercise. Show that a topological conjugacy maps orbits onto orbits, periodic points to periodic points, and preserves periods. Prove that if h is a topological conjugacy from f to g, and if $f^n(x) \to a$ as $n \to \infty$, then $g^n(y) \to b$ as $n \to \infty$, where $y = h(x)$ and $b = h(a)$.

(2.2) Remark. In this book we are mainly concerned with the situation where X and Y are differentiable manifolds and f and g are diffeomorphisms. In this context, it might seem natural to require the map h to be a diffeomorphism, rather than just a homeomorphism. This modification gives the notion of *differentiable conjugacy*. Differentiable conjugacy is a stronger relation than topological conjugacy and, correspondingly, there are, in general, many more equivalence classes with respect to it. This does not inevitably mean that a differentiable classification is harder to make than a topological one (see Exercise 2.3 below). However, we do find that, with differentiable conjugacy, *stable* diffeomorphisms (ones which stay in the same equivalence class when slightly perturbed) are very rare. We also have to class as non-equivalent diffeomorphisms which most people would feel are qualitatively the same (for example, the contractions $x \mapsto \frac{1}{2}x$ and $x \mapsto \frac{1}{3}x$ of the real line $\mathbf{R}$). For these reasons, topological conjugacy continues as the basic equivalence relation when we restrict ourselves to the differentiable category.

(2.3) Exercise. Prove, by differentiation at the origin, that if two linear automorphisms of $\mathbf{R}^n$ are differentiably conjugate then they are similar. (Thus the differentiable classification of $GL(\mathbf{R}^n)$ is the classical theory of (real) Jordan canonical form (see Chapter 4). In contrast, the topological classification is very hard, and not completely solved at the time of writing (see Kuiper and Robbin [1] and Chapter 4).) Prove, more generally, that if f and g are diffeomorphisms of a differentiable manifold X and if a differentiable conjugacy h from f to g takes a fixed point p to q then $T_p f$ is similar to $T_q g$.

II. HOMEOMORPHISMS OF THE CIRCLE

To get some feeling for the difficulties involved in the classification problem, we discuss briefly the situation when X is the circle S^1. The results here (due to Poincaré and Denjoy) are too complete to be typical of compact manifolds in general. As is usual in the one-dimensional case, the ordering of the real line plays a distinctive role. We sketch the ideas behind the theory, and leave some of the details as Exercise 2.4 below. A more complete account is given in Chapter 1 of Nitecki [1].

Let $f : S^1 \to S^1$ be a homeomorphism. We suppose that f *preserves orientation*, in that, when any point moves clockwise on S^1 then so does its image under f (see Exercise 2.4 and the Appendix to this chapter for more accurate definitions). We focus our attention on some point $x \in S^1$, and measure the average angle $\theta(n, x)$ turned through by this point in n successive applications of the map f. The reader will immediately, and correctly, protest that this average is not well defined, since the individual angles are all ambiguous by integer multiples of 2π. However, we shall see in the exercise that it is possible, by moving from S^1 to its covering space $\mathbf{R}$, to resolve the ambiguities well enough to overcome this objection. We now let n tend to ∞. For any $y \in S^1$, the sequence $\theta(n, x) - \theta(n, y)$ will tend to zero, since otherwise, intuitively, one point will overtake the other under successive applications of f. Similarly, but perhaps rather less clearly, it can be argued that $\theta(n, x)$ converges, since otherwise x will overtake itself. Thus we are led to believe in a *rotation number* $\rho(f)$, defined modulo 1, such that, for all $x \in S^1$, the average angle turned through by x under infinitely many successive applications of the map f is $2\pi\rho(f)$. It is clear that this number is an invariant of orientation preserving topological conjugacy, for, if g and h are orientation preserving homeomorphisms of S^1 such that $g = hfh^{-1}$, then $g^n = hf^nh^{-1}$. For large n, the single maps h and h^{-1} make a negligible contribution to the average rotation, so $\rho(g) = \rho(f)$.

If f has a periodic point x of period s, then $\rho(f)$ is rational, of form r/s. This is immediate, since f^s rotates x through an angle $2\pi r$ for some integer r. Rather less obviously, the converse also holds. Thus, for example, $\rho(f) = 0$ if and only if f has a fixed point. It is clear that the rotation number does not by itself classify homeomorphisms of S^1. For instance, given any closed subset E of S^1, one may construct a homeomorphism f of S^1 whose fixed point set is precisely E. The complement $S^1 \backslash E$ is essentially a countable collection of open real intervals and one merely defines f as an increasing homeomorphism on each such interval. (Moreover one can, if one is careful, construct f to be as smooth as one wishes.) Of course two homeomorphisms with non-homeomorphic fixed point sets are not topologically conjugate (see Exercise 2.1). A rather similar situation holds for $\rho(f)$ any rational number r/s, since then f^s has fixed points.

The case $\rho(f)$ irrational is in complete contrast. Here, provided the maps concerned are C^2, $\rho(f)$ *is* a complete set of invariants of orientation preserving topological conjugacy. That is to say, two C^2 diffeomorphisms f and g with irrational rotation numbers are topologically conjugate if and only if $\rho(f) = \rho(g)$ or $\rho(f) = 1 - \rho(g)$. If we allow maps which are not C^1 with bounded variation, we introduce new conjugacy classes for each irrational rotation number. These have been classified by Markley [1].

(2.4) Exercise. Let $S^1 = \mathbf{R}/\mathbf{Z}$ and let $p : \mathbf{R} \to S^1$ be the quotient map. Let $f : S^1 \to S^1$ be a homeomorphism. A *lifting* of f is a continuous map $\tilde{f} : \mathbf{R} \to \mathbf{R}$ such that $p\tilde{f} = fp$. Note that p maps every open interval of length 1 in $\mathbf{R}$ homeomorphically onto the complement of a point in S^1. Using this fact, obtain a lifting of f by first constructing it on $]0, 1[$ and then extending it to $\mathbf{R}$ in the only way possible. Prove that any two liftings of f differ by an integer constant function. We say that f is *orientation preserving* if its liftings are increasing functions, and *orientation reversing* if they are decreasing.

Now let f be orientation preserving, and let $\tilde{f}$ be a lifting of f. Prove that, for all $n > 0$, $\tilde{f}^n - id$ is periodic of period 1. Let α_n be the greatest integer $\leqslant \tilde{f}^n(0)$. Prove that, for all $m > 0$ and $x \in [0, 1]$, $m\alpha_n \leqslant \tilde{f}^{mn}(x) \leqslant m(\alpha_n + 2)$, and hence that

$$\left| \frac{1}{mn} \tilde{f}^{mn}(0) - \frac{1}{n} \tilde{f}^n(0) \right| \leqslant \frac{2}{n}.$$

Deduce that $(1/n)\tilde{f}^n(0)$ is a Cauchy sequence. Prove that, for all $x, y \in \mathbf{R}$, $|\tilde{f}^n(x) - \tilde{f}^n(y)| \leqslant |x - y| + 1$. Deduce that, as $n \to \infty$, $(1/n)\tilde{f}^n(x)$ converges to a limit which is independent of x. Since two liftings differ by an integer, the congruence class of this limit modulo 1 depends only on the underlying map f, and is called the *rotation number* $\rho(f)$ of f.

Prove that $\rho(f)$ is an invariant of orientation preserving topological conjugacy, and that if g is topologically conjugate to f by an orientation reversing conjugacy then $\rho(g) = 1 - \rho(f)$. Prove further that $\rho(f)$ is rational of form r/s if and only if f has a periodic point of period s (for necessity, show that f^s has a fixed point). Finally, suppose that $\rho(f)$ is irrational, and prove that, for all $x \in \mathbf{R}$, the map sending $n\rho(f) + m$ to $\tilde{f}^n(x) + m$ is an increasing map of the subset $\{m + n\rho(f) : m, n \in \mathbf{Z}\}$ of $\mathbf{R}$ into $\mathbf{R}$. Deduce that if the orbit of $p(x)$ is dense in S^1 then f is topologically conjugate to the rotation map $p(y) \mapsto p(y + \rho(f))$.

III. FLOW EQUIVALENCE AND TOPOLOGICAL EQUIVALENCE

Let ϕ and ψ be flows on topological spaces X and Y respectively. We say that $h : X \to Y$ is a *flow map from ϕ to ψ* if it is continuous and if there exists an increasing continuous homomorphism $\alpha : \mathbf{R} \to \mathbf{R}$ such that the diagram

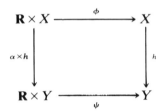

commutes. Recall that α is just multiplication by a positive constant. If, further, h is a homeomorphism we call the pair (α, h), or the single map h in case $\alpha = id$, a flow equivalence from ϕ to ψ. We say that ϕ is *flow equivalent* to ψ if there exists such a pair (α, h). Thus, in this case, ψ is the flow induced on Y from ϕ by (α, h) (see Example 1.26).

(2.5) Exercise. Let (α, h) be a flow equivalence from a flow ϕ on X to a flow ψ on Y. Show that h is a topological conjugacy from $f = \phi^1$ to $g = \psi^t$, where $t = \alpha(1)$.

Although the notion of flow equivalence seems very natural, it is rather too strong for the qualitative theory of flows. For example, it preserves ratios of periods of closed orbits, and flows may differ in this aspect and yet have a very similar appearance (in fact they may have identical phase portraits!). We now define a weaker equivalence relation which is usually regarded as the basic one in the subject. We say that $h : X \to Y$ is a *topological equivalence* from ϕ to ψ if it is a homeomorphism which maps each orbit of ϕ onto an orbit of ψ, and preserves orientation of orbits. Intuitively this last requirement means that h takes the direction of increasing t on each orbit of ϕ to the direction of increasing t on the corresponding orbit of ψ (see Figure 2.6). More precisely, *h preserves the orientation* of $\mathbf{R}.x$ if there exists an increasing homeomorphism $\alpha : \mathbf{R} \to \mathbf{R}$ such that, for all $t \in \mathbf{R}$, $h\phi(t, x) =$

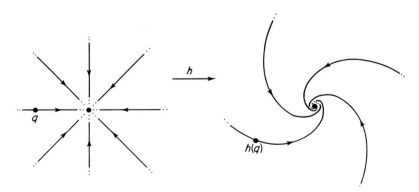

FIGURE 2.6

$\psi(\alpha(t), h(x))$. Similarly, h *reverses the orientation* of $\mathbf{R}.x$ if there exists a decreasing homeomorphism $\alpha : \mathbf{R} \to \mathbf{R}$ satisfying this equation. It is straight-forward to check that these properties are independent of the choice of x on the orbit, for if $x' \in \mathbf{R}.x$ then $\phi_x^{-1}\phi_{x'} : \mathbf{R} \to \mathbf{R}$ is a translation.

(2.7) Remarks. (a) Remark 2.2 made the point that it is undesirable to strengthen "topological" to "smooth" in the definition of conjugacy of diffeomorphisms. This applies equally to the definition of equivalence of smooth flows, for the same reasons.

(b) The above definitions of a homeomorphism preserving and reversing orientation of orbits are perfectly satisfactory, but they leave various questions unanswered. For example, can a homeomorphism of one orbit onto another fail to either preserve or reverse the orientation? We deal with this and similar points in the appendix to this chapter.

(c) The reader may have noticed that topological conjugacy is the immediate analogue of flow equivalence and wondered about the discrete analogue of topological equivalence of flows. We say that homeomorphisms f and g of a topological space X are *topologically equivalent* if there is a homeomorphism h of X taking orbits of f onto orbits of g. This relation has not been widely studied. It is different from topological conjugacy. For example, rotations of S^1 through $[\frac{1}{5}]$ and $[\frac{2}{5}]$ are topologically equivalent (with $h = id$), but, as we have seen above, they are not topologically conjugate. However, quite often topological equivalence of f and g is the same as topological conjugacy of f and either g or g^{-1}. This is certainly the case if the complement of Per f in X is path connected and dense in X, a result due to Kupka [2]. For let f and g be related by a homeomorphism h as above. Any two points x and y in $X\backslash$Per f may be joined by a con-tinuous curve $\gamma : I \mapsto X\backslash$Per f, where $I = [0, 1]$. If we denote by $C_n\{t \in I : g^n h\gamma(t) = hf\gamma(t)\}$ for all $n \in \mathbf{Z}$, then it is easy to see that the set C_n are mutually disjoint, closed and have union I. It follows from Lemma 2.49 below that $I = C_n$ for some $n \in \mathbf{Z}$, and hence that $g^n h = hf$ on $X\backslash$Per f. We deduce that $n = \pm 1$, since otherwise images of orbits of f under h are not whole orbits of g. Finally, by continuity, $g^n h = hf$ on X.

(2.8) Exercise. Prove that flow equivalence and topological equivalence are equivalence relations, and that the former implies the latter. Find a flow on $\mathbf{R}^2$ that is topologically equivalent but not flow equivalent to Example 1.16. Prove that Examples 1.15 and 1.16 are topologically equivalent to their reverse flows. Prove that any topological equivalence induces a homeo-morphism of orbit spaces.

(2.9) Example. *Rational and irrational flows on the torus.* Our task here is to compare the suspension and product constructions for rational and irrational flows given in Chapter 1, and to classify the flows obtained. Recall that the suspension of a rotation through $[\theta] \in S^1 = \mathbf{R}/\mathbf{Z}$ is a flow on the space $X = (\mathbf{R} \times S^1)/\sim$, where $\sim$ is defined by $(x, [y]) \sim (x', [y'])$ if and only if $x = x' + m$, $m \in \mathbf{Z}$, and $[y'] = [y + m\theta]$. The flow is defined by $t.[x, [y]] = [t + x, [y]]$, where the outer brackets denote equivalence classes with respect to $\sim$. It is not completely clear at first that X is the torus. We define a map h from X to $T^2 = S^1 \times S^1$ by $h([x, [y]]) = ([x], [y + \theta x])$. It is well defined since, all m and $n \in \mathbf{Z}$,

$$h([x - m, [y + n + m\theta]]) = ([x - m], [y + n + m\theta + \theta(x - m)]) = ([x], [y + \theta x]).$$

It is continuous since the quotient maps concerned are local homeomorphisms and h is induced by them from a linear isomorphism of $\mathbf{R}^2$. Moreover, h is a homeomorphism, since it has a well defined continuous inverse given by $h^{-1}([x], [y]) = [x, [y - \theta x]]$. Finally note that h is a flow equivalence from the suspension flow to the product of a pair of rotation flows with speeds 1 and θ. This is because

$$t.h([x, [y]]) = t.([x], [y + \theta x]) = ([x + t], [y + \theta x + \theta t]) = h([x + t, [y]])$$
$$= h(t.[x, [y]]).$$

Thus:

(2.10) Theorem. *For any $\theta \in \mathbf{R}$, the suspension of a rotation of S^1 through $[\theta]$ is flow equivalent to the produce of two rotation flows on S^1 with speeds 1 and θ.*   □

The maps h and h^{-1} are, in fact, C^ω, so the two constructions are equivalent in the strongest possible sense. For the rest of the section then, we may think in terms of product flows as we proceed towards a classification.

We begin with some trivial remarks. For any pair of flows ϕ and ψ, the product flow $\phi \times \psi$ is flow equivalent to $\psi \times \phi$, by the homeomorphism which interchanges factors. Thus if we consider a product of two rotation flows and if the product is non trivial we may assume that the first factor is non trivial. Moreover, by speeding up or slowing down both factors simultaneously, we may assume that the speeds of the factors are respectively 1 and θ. Note also that by Theorem 2.10 we get a flow equivalent flow if we replace θ by any $\theta' \in [\theta]$.

Before we go any further, we must say a few words about homeomorphisms of T^2, and introduce some notation. Let $\pi : \mathbf{R}^2 \to T^2$ be the quotient map $\pi(x, y) = ([x], [y])$, and let $h : T^2 \to T^2$ be any homeomorphism satisfying $h\pi(0, 0) = \pi(0, 0)$. Then h lifts to a unique continuous map $H : \mathbf{R}^2 \to \mathbf{R}^2$ such that $H(0, 0) = (0, 0)$ and the diagram commutes. This is because π is a *covering* (see Greenberg [1]); we have homeomorphisms of the form $(\pi|V)^{-1}h(\pi|U)$ for any open square U of side 1 in $\mathbf{R}^2$ and for any connected component V of the set $\pi^{-1}(h\pi(U))$, and continuity determines how we

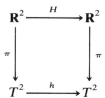

should stick some of these homeomorphisms together to form H. The map H is a homeomorphism, its inverse H^{-1} being the lifting of h^{-1}. It has the property that there exist integers p, q, r and s such that, for all $(x, y) \in \mathbf{R}^2$ and for all $(m, n) \in \mathbf{Z}^2$,

$$H(x + m, y + n) = H(x, y) + (pm + rn, qm + sn).$$

In particular, H maps $\mathbf{Z}^2$ linearly into itself with matrix

$$A = \begin{bmatrix} p & r \\ q & s \end{bmatrix}.$$

In fact, $H | \mathbf{Z}_2$ is just the map h_* (or $H_1(h)$): $H_1(T^2) \to H_1(T^2)$ of homology theory (see, for example, Greenberg [1]). It describes the fact that h wraps the circle $S^1 \times \{0\}$ p times round the first factor of $S^1 \times S^1$ and q times round the second factor, with corresponding numbers r and s for $\{0\} \times S^1$. Since H^{-1} also has an associated integer matrix, B say, and $HH^{-1} = id$, we deduce that B is the inverse of A, and so $A \in GL_2(\mathbf{Z})$. The matrix A defines a linear automorphism, denoted by L_h, of $\mathbf{R}^2$, and this covers a homeomorphism denoted by l_h, of T^2. Notice that if

$$\begin{bmatrix} p & r \\ q & s \end{bmatrix} \in GL_2(\mathbf{Z})$$

then p and q are coprime, because any common factor also divides the determinant of A, and this is ± 1. Conversely if p and q are coprime integers then there exist r and s such that

$$\begin{bmatrix} p & r \\ q & s \end{bmatrix} \in GL_2(\mathbf{Z}).$$

To see this, assume $0 \leqslant p \leqslant q$, since

$$\begin{bmatrix} p & r \\ q & s \end{bmatrix} \in GL_2(\mathbf{Z})$$

if

$$\begin{bmatrix} q & s \\ p & r \end{bmatrix} \quad \text{or} \quad \begin{bmatrix} -p & -r \\ -q & -s \end{bmatrix} \in GL_2(\mathbf{Z}).$$

We observe that

$$\begin{bmatrix} 0 & 1 \\ 1 & 0 \end{bmatrix} \quad \text{and} \quad \begin{bmatrix} 1 & 0 \\ 1 & 1 \end{bmatrix} \in GL_2(\mathbf{Z}),$$

and if $0 < p < q$, we may choose for s any integer with $s(q - p) \equiv 1 \bmod q$ and put $r = (1 + ps/q)$.

The situation as regards classification should now be somewhat clearer. The reason that the rational flows corresponding to $\theta = 0$ and $\theta = \frac{2}{3}$ look different is that we are not mentally allowing for homeomorphisms of T^2 that alter the homology of the generators $S^1 \times \{0\}$ and $\{0\} \times S^1$; we are letting our picture of T^2 as a submanifold of $\mathbf{R}^3$ dominate and limit our imagination. For each circle of rational slope q/p in T^2, there is a linear automorphism of $\mathbf{R}^2$ covering a homeomorphism of T^2 that maps the generator $S^1 \times \{0\}$ onto the circle. Thus the homeomorphism equates the $\theta = 0$ case to the $\theta = q/p$ case. On the other hand, since there are only countably many such homeomorphisms, they cannot equate the uncountable infinity of irrational flows. Of course there are uncountably many other homeomorphisms h of T^2 to consider as possible equivalences, but there is a countable restriction placed upon them by the associated maps l_h, and this turns out to be vital. We now fill in the details. We first prove:

(2.11) Theorem. *All rational flows on T^2 are flow equivalent.*

Proof. It is enough to give a flow equivalence from the product of two rotation flows with speeds 1 and θ to the product of two rotation flows with speeds p and q, where p and q are coprime integers with $p \neq 0$. Choose

$$A = \begin{bmatrix} p & r \\ q & s \end{bmatrix} \in GL_2(\mathbf{Z}),$$

and let $h : T^2 \rightarrow T^2$ be the homeomorphism covered by the linear automorphism of $\mathbf{R}^2$ with matrix A. Then h is a flow equivalence, since

$$h(t.([x], [y])) = h([x + t], y)$$
$$= ([px + pt + ry], [qx + qt + sy])$$
$$= t.([px + ry], [qx + sy])$$
$$\doteq t.h([x], [y]). \qquad \square$$

In classifying irrational flows it is, as we have seen above, sufficient to consider products of two rotation flows with speeds 1 and θ.

(2.12) Theorem. *For any irrational numbers α and β, the $\theta = \alpha$ and $\theta = \beta$ cases are topologically equivalent if and only if there exists $A \in GL_2(Z)$ such that $A\begin{bmatrix} 1 \\ \alpha \end{bmatrix} = \begin{bmatrix} 1 \\ \beta \end{bmatrix}$.*

Proof. Suppose such a matrix A exists. The corresponding linear automorphism of $\mathbf{R}^2$ takes lines of slope α to lines of slope β, and so induces a
homeomorphism h of T^2 taking orbits of the $\theta = \alpha$ flow to orbits of the $\theta = \beta$
flow. In fact, h is a flow equivalence, since

$$h(t.([x],[y]) = h([x+t],[y+\alpha t])$$
$$= h([x],[y]) + h([t],[\alpha t])$$
$$= h([x],[y]) + ([t],[\beta t])$$
$$= t.h([x],[y]).$$

Now suppose, conversely, that $h : T^2 \to T^2$ is a topological equivalence
from the $\theta = \alpha$ flow to the $\theta = \beta$ flow. We may suppose that $h([0],[0]) =$
$([0],[0])$, since any translation of $\mathbf{R}^2$ induces an orbit preserving
homeomorphism of T^2. Consider the homeomorphism $f = l_h^{-1}h$ of T^2. Let ϕ
be the flow on T^2 induced from the $\theta = \alpha$ flow by f. The orbits of ϕ lift to the
family of all lines of slope γ in $\mathbf{R}^2$, where $(1, \gamma) = l_h^{-1}(1, \beta)$. Let S be the
circle $\{0\} \times S^1$ in T^2, and let $T = f(S)$. The time 1 map of the $\theta = \alpha$ flow maps
S homeomorphically onto itself. Let g denote this homeomorphism of S.
Then g is covered by the translation $t \mapsto t + \alpha$ of $\mathbf{R}$, so the rotation number
$\rho(g)$ of g is α (see Exercise 2.4). Since f is a flow equivalence from the $\theta = \alpha$
flow to ϕ, $f|S$ conjugates g with the time 1 homeomorphism $\phi^1|T$. Notice
that $\phi^1|T$ is the *first return homeomorphism* of T with respect to the flow ϕ;
that is to say, for any $x \in T$, $\phi^1(x)$ may be described as $t.x$ for the first $t > 0$
for which $t.x \in T$.

Let $F : \mathbf{R}^2 \to \mathbf{R}^2$ be the lifting of f such that $F(0, 0) = (0, 0)$. Then F is a
homeomorphism satisfying

$$F(x+m, y+n) = F(x, y) + (m, n)$$

for all $(x, y) \in \mathbf{R}^2$ and all $(m, n) \in \mathbf{Z}^2$. Let $\mathbf{R}_m$ denote the line $\{m\} \times \mathbf{R}$ in
$\mathbf{R}^2$. Thus $\pi^{-1}(S) = \bigcup\{\mathbf{R}_m : m \in \mathbf{Z}\}$, and correspondingly $\pi^{-1}(T) =$
$\bigcup\{F(\mathbf{R}_m) : m \in \mathbf{Z}\}$. Each of these continuous arcs $F(\mathbf{R}_m)$ is the translate of
$F(\mathbf{R}_0)$ by the vector $(m, 0)$. Since F maps lines of slope α onto lines of slope
γ, and each line of slope α intersects $\mathbf{R}_0$ in a single point, each line of slope γ
intersects $F(\mathbf{R}_0)$ in a single point. Thus there is a continuous map $\lambda : \mathbf{R} \to \mathbf{R}$,
which is periodic of period 1, such that $F(\mathbf{R}_0) = \{(\lambda(t), t + \gamma\lambda(t)) : t \in \mathbf{R}\}$. This
enables us to define a homeomorphism, k say, from $S^1 = \mathbf{R}/\mathbf{Z}$ to T by
$k([t]) = ([\lambda(t)], [t + \gamma\lambda(t)])$. Now the portion of the orbit of ϕ between
$([\lambda(t)], [t + \gamma\lambda(t)])$ and the point of its next return to T lifts to the line
segment joining $(\lambda(t), t + \gamma\lambda(t))$ in $F(\mathbf{R}_0)$ to $(1 + \lambda(t + \gamma), t + \gamma + \lambda(t + \gamma))$
in $F(\mathbf{R}_1)$. This latter point projects to the same point of T as does
$(\lambda(t + \gamma), t + \gamma + \lambda(t + \gamma))$, and so k^{-1} conjugates $\phi^1|T$ with the map of S^1
taking $[t]$ to $[t + \gamma]$. Since this has rotation number γ and is conjugate to f, we

deduce that $[\gamma] = [\alpha]$. Hence

$$\begin{bmatrix} 1 \\ \gamma \end{bmatrix} = \begin{bmatrix} 1 & 0 \\ n & 1 \end{bmatrix} \begin{bmatrix} 1 \\ \alpha \end{bmatrix}$$

for some $n \in \mathbf{Z}$. But $(1, \beta) = l_h(1, \gamma)$, and the matrix of l_h is in $GL_2(\mathbf{Z})$. Therefore

$$\begin{bmatrix} 1 \\ \beta \end{bmatrix} = A \begin{bmatrix} 1 \\ \alpha \end{bmatrix}$$

for some $A \in GL_2(\mathbf{Z})$, as required. $\square$

IV. LOCAL EQUIVALENCE

It is often possible to modify the definition of a global property or relation by inserting the word "local" and paraphrasing it as "in some neighbourhood of a given point". Even if the new definition does not immediately make perfect sense, it often points quite clearly towards a sensible concept. In many situations we need to handle topological conjugacy, topological equivalence and flow equivalence in a local form. There is an element of doubt as to whether the adaptation process described above produces the most useful definitions of local equivalence for these relations. For example, it is arguable that, since the theory of discrete dynamical systems is essentially a study of homeomorphisms with respect to their orbit structure, we should in no circumstances call periodic points of different periodicities locally equivalent. Similarly, in the flow case, two points may have neighbourhoods on which the phase portraits are identical and may yet be distinguished by certain types of recurrence involving orbits that leave the said neighbourhoods but return at some later time. In spite of these objections we find the straightforwardly adapted definitions useful, with the reservation that the equivalence relations that they give are very weak for regular points (see the first section of Chapter 5 below; a point is *regular* if it is not a fixed point).

Let U and V be open subsets of topological spaces X and Y respectively, and let $f: X \to X$ and $g: Y \to Y$ be homeomorphisms. We say (by abuse of notation) that $f|U$ is *topologically conjugate* to $g|V$ if there is a homeomorphism $h: U \cup f(U) \to V \cup g(V)$ such that $h(U) = V$ and, for all $x \in U$, $hf(x) = gh(x)$. If $p \in X$ and $q \in Y$, we say that f is *topologically conjugate at p to g at q* if there exist open neighbourhoods U of p and V of q such that $f|U$ is topologically conjugate to $g|V$ by a conjugacy h taking p to q. We also say, by a further abuse, that $f|p$ is *topologically conjugate* to $g|q$, or even that p is *topologically conjugate* to q.

We now turn to flows. Let $\phi : D \to X$ be a continuous map, where X is a topological space and D is a neighbourhood of $\{0\} \times X$ in $\mathbf{R} \times X$. We write $t.x$ for $\phi(t, x)$ and D_x for the set $\{t \in \mathbf{R} : (t, x) \in D\}$. We say that ϕ is a *partial flow* on X if, for all $x \in X$,

  (i) D_x is an interval,
  (ii) $0.x = x$,
  (iii) for all $t \in D_x$ with $s \in D_{t.x}$, $(s + t).x = s.(t.x)$,
  (iv) for all $t \in D_x$, $D_{t.x} = \{s - t : s \in D_x\}$.

Thus, as the name suggests, ϕ is a flow that is not defined for all time. Condition (iv) implies that ϕ is maximal; it cannot be extended without offending the preceding conditions. We show this in the following proposition, which gives a few useful properties of partial flows.

(2.13) Proposition, *Let $\phi : D \to X$ be a partial flow on X. Then*
  (i) *D is open in $\mathbf{R} \times X$,*
  (ii) *if $D_x = \,]a, b[$ with $b < \infty$, ϕ cannot be extended to a continuous map of $D \cup \{(b, x)\}$ into X,*
  (iii) *if x is a fixed point or a periodic point of ϕ then $D_x = \mathbf{R}$.*

Proof. Let $(t, x) \in D$. Then for some neighbourhood U of $t.x$ and for some $\varepsilon > 0$, $]-\varepsilon, \varepsilon[\times U \subset D$, since D is an open neighbourhood of $\{0\} \times X$ in $\mathbf{R} \times X$. By continuity of ϕ, there is some neighbourhood V of x such that $t.y \in U$ for all $y \in V$. Since for all such y, $]-\varepsilon, \varepsilon[\subset D_{t.y}$ it follows by (iv) of the definition that $]t - \varepsilon, t + \varepsilon[\subset D_y$. Thus D contains $]t - \varepsilon, t + \varepsilon[\times V$, and hence D is open in $\mathbf{R} \times X$.

To see (ii), note that if ϕ could be extended by putting $b.x = y$, then $t.x$ would necessarily converge to y as $t \to b -$. But, as in the first part, this would imply that for some $\varepsilon > 0$, $]-\varepsilon, \varepsilon[\subset D_{t.x}$ for t near b, and thus that $]t - \varepsilon, t + \varepsilon[\subset D_x$. This contradicts the definition of b when t is larger than $b - \varepsilon$. In fact, we have shown that $(t.x)$ has no cluster points in X as $t \to b -$.

Part (iii) of the proposition is immediate from (iv) of the definition.   □

We may define flow equivalence and topological equivalence for partial flows in the obvious way. Thus if ψ is a partial flow on a topological space Y then a flow equivalence from ϕ to ψ is a pair (α, h) where $h : X \to Y$ is a homeomorphism, $\alpha : \mathbf{R} \to \mathbf{R}$ is multiplication by a positive constant, and $h(t.x) = \alpha(t).h(x)$ for all $(t, x) \in D$. A *topological equivalence* from ϕ to ψ is a homeomorphism $h : X \to Y$ that maps all orbits $D_x.x$ of ϕ onto orbits of ψ and preserves their orientation by increasing t.

(2.14) Proposition. *Flow equivalence and topological equivalence are equivalence relations on the set of all partial flows on topological spaces.*

The only non trivial point in the proof of this proposition is to establish the symmetry of flow equivalence. This requires the following lemma:

(2.15) Lemma. *Let* (α, h) *be a flow equivalence from* ϕ *to* ψ. *Then* $\alpha \times h$ *maps the domain* D *of* ϕ *homeomorphically onto the domain* E *of* ψ.

Proof. By the definition of flow equivalence $(\alpha \times h)(D) \subset E$, and it suffices to show that $(\alpha \times h)(D) = E$. But (α, h) induces a partial flow $\tilde{\psi} : (\alpha \times h)(D) \to Y$ on Y, and $\tilde{\psi} = \psi$ on $(\alpha \times h)(D)$. Since by Proposition 2.13 (ii) partial flows cannot be extended, the domain of $\tilde{\psi}$ equals the domain of ψ.   □

Now let ϕ be any flow (or, indeed, partial flow) on X and let U be an open subset of X. Then, for all $x \in X$, $\phi_x^{-1}(U)$ is open in $\mathbf{R}$ and hence a countable union of disjoint open intervals. Let D_x be the one containing 0. We call $\phi_x(D_x)$ the *orbit component of* ϕ *through* x (*in* U) (see Figure 2.16). We

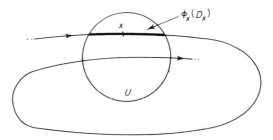

FIGURE 2.16

denote by D the subset $\bigcup_{x \in U} D_x \times \{x\}$ of $\mathbf{R} \times U$ (see Figure 2.17), and define a map $\phi|U : D \to U$ by $(\phi|U)(t, x) = \phi(t, x)$. Of course this is an abuse of notation, since U is not in the domain of ϕ. By the same abuse, we call $\phi|U$ the *restriction of the flow* ϕ *to the subset* U. It is very easy to check that $\phi|U$ is a partial flow on U. Thus if ψ is a flow on a topological space Y and V is an

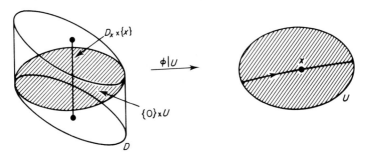

FIGURE 2.17

open subspace of Y, then to say that $\phi|U$ is *flow* or *topologically equivalent to* $\psi|V$ means that they are equivalent as partial flows. If $p \in X$ and $q \in Y$, we say that ϕ is equivalent at p to ψ at q if there exist open neighbourhoods U of p and V of q and an equivalence from $\phi|U$ to $\psi|V$ taking p to q. Once again, we write that $\phi|p$ *is equivalent to* $\psi|q$, or even that p *is equivalent to* q.

(2.18) Proposition. *Flow equivalence and topological equivalence are equivalence relations on* $\{(\phi, p) : \phi$ *is a flow,* $p \in$ *the phase space of* $\phi\}$. *If* $\phi|p$ *is flow equivalent to* $\psi|q$ *then* $\phi|p$ *is topologically equivalent to* $\psi|q$. $\qquad\square$

V. LIMIT SETS OF FLOWS

We now begin an investigation of properties of orbits of flows that are preserved under topological equivalence. An obvious example is the topological types of the orbits; for instance if the sets of fixed points of two flows have different cardinalities then we can say immediately that the flows are not topologically equivalent.

Let ϕ and ψ be flows on topological spaces X and Y. Suppose that h is a topological equivalence from ϕ to ψ. Then h maps the closure $\bar{\Gamma}$ in X of each orbit Γ of ϕ onto the closure $\overline{h(\Gamma)}$ of $h(\Gamma)$ in Y. Consequently, h maps the set $\bar{\Gamma}\backslash\Gamma$ onto the set $\overline{h(\Gamma)}\backslash h(\Gamma)$. Compare, for instance, Examples 1.14 and 1.15 which are distinguished topologically by the number of orbits Γ with $\bar{\Gamma}\backslash\Gamma$ empty (see Figure 2.19).

This observation takes no account of the fact that topological equivalence must preserve orientation of orbits. Example 1.14 is not topologically equivalent to its reverse flow, and the difference obviously lies in the way that the orbits begin at the origin in Example 1.14 but end there in

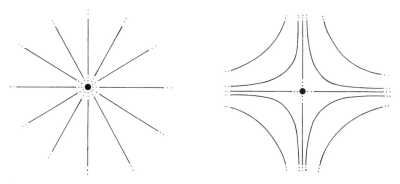

FIGURE 2.19

the reverse flow. In order to handle these differences, we must analyse closures of orbits more carefully, and pick out those parts corresponding to "large positive t" and "large negative t". These *limit sets* will be the main objects of study for the rest of this chapter.

Let I_t denote the closed half-line $[t, \infty[$. The ω-*set* (or ω-*limit set*) $\omega(x)$ of a point $x \in X$ (with respect to the flow ϕ) is defined by

(2.20)
$$\omega(x) = \bigcap_{t \in \mathbf{R}} \overline{\phi_x(I_t)}.$$

Intuitively, $\omega(x)$ is the subset of X that $t.x$ approaches as $t \to \infty$. For instance, in the saddle point picture of Example 1.15, if x is on either of the two vertical orbits then $t.x$ approaches the origin as $t \to \infty$, and this is reflected in the fact that $\omega(x) = \{0\}$. It follows immediately from the definition of flow that, for all $t \in \mathbf{R}$, $\omega(t.x) = \omega(x)$. Thus we may define the ω-set $\omega(\Gamma)$ of any orbit Γ of ϕ by $\omega(\Gamma) = \omega(x)$ for any $x \in \Gamma$. Notice that if Γ is a fixed point or periodic orbit then $\phi_x(I_t) = \Gamma$ for all $x \in \Gamma$ and $t \in R$, and so $\omega(\Gamma) = \Gamma$. Thus $\omega(\Gamma)$ is not necessarily part of $\bar{\Gamma}\backslash\Gamma$.

(2.21) Exercise. Determine the ω-sets for the orbits of the examples of flows in Chapter 1. For instance, show that any orbit of an irrational flow on T^2 has T^2 as ω-set.

Notice that in the definition (2.20) one could equally well have taken the intersection over any subset of $\mathbf{R}$ that is unbounded above, for example $N = \{1, 2, 3, \ldots\}$. Notice also that a point y is in $\omega(x)$ if and only if there is a real sequence (t_n) such that $t_n \to \infty$ and $t_n.x \to Y$ as $n \to \infty$.

Similarly, we define the α-set $\alpha(x)$ of a point $x \in X$ by

$$\alpha(x) = \bigcap_{t \in \mathbf{R}} \overline{\phi_x(J_t)}$$

where $J_t =]-\infty, t]$, and the α-*set* $\alpha(\Gamma)$ of an orbit Γ by $\alpha(\Gamma) = \alpha(x)$ for any $x \in \Gamma$. The α-set is the subset that $t.x$ approaches as $t \to -\infty$. To avoid continual repetition, we confine attention in the results that follow to ω-sets, since the corresponding results for α-sets are always exactly analogous in statement and proof. In fact, we can give a precise formulation for this correspondence in terms of the reverse flow ϕ^- of ϕ described in Example 1.26, as follows:

(2.22) Proposition. *Let Γ by an orbit of ϕ and let $\alpha^-(\Gamma)$ and $\omega^-(\Gamma)$ denote the α-set and ω-set of Γ as an orbit of ϕ^-. Then $\alpha^-(\Gamma) = \omega(\Gamma)$ and $\omega^-(\Gamma) = \alpha(\Gamma)$.*

We now derive a few simple properties of ω-sets (of a flow ϕ on a topological space X unless otherwise stated). First we show that any topological equivalence maps ω-sets onto ω-sets.

(2.23) Theorem. *Let $h : X \to Y$ be a topological equivalence from ϕ to ψ. Then, for each orbit Γ of ϕ, h maps $\omega(\Gamma)$ onto $\omega(h(\Gamma))$, the ω-set of the orbit $h(\Gamma)$ of ψ.*

Proof. Let $\lambda : \mathbf{R} \to \mathbf{R}$ be an increasing homeomorphism such that $h\phi_x = \psi_{h(x)}\lambda$ where $x \in \Gamma$. Then

$$h(\omega(\Gamma)) = h\left(\bigcap_{t \in \mathbf{R}} \overline{\phi_x(I_t)} \right) = \bigcap_{t \in \mathbf{R}} h\overline{\phi_x(I_t)} \qquad \text{(since } h \text{ is injective)}$$

$$= \bigcap_{t \in \mathbf{R}} \overline{h\phi_x(I_t)} \qquad \text{(since } h \text{ is a homeomorphism)}$$

$$= \bigcap_{t \in \mathbf{R}} \overline{\psi_{h(x)}\lambda(I_t)} = \bigcap_{t \in \mathbf{R}} \overline{\psi_{h(x)}(I_{\lambda(t)})} \qquad \text{(since } \lambda \text{ is increasing)}$$

$$= \omega(h(\Gamma)). \qquad \qquad \Box$$

(2.24) Proposition. *Let Γ be an orbit of ϕ. Then $\omega(\Gamma)$ is a closed subset of X, and $\omega(\Gamma) \subset \bar{\Gamma}$.* $\qquad \Box$

(2.25) Exercise. Show that if Γ is an orbit of a flow on X, then $\bar{\Gamma} = \Gamma \cup \alpha(\Gamma) \cup \omega(\Gamma)$. Deduce that if $\alpha(\Gamma) = \omega(\Gamma) = \varnothing$, then Γ is a closed subset of X.

(2.26) Proposition. *Let Γ and Δ be orbits of ϕ such that $\Gamma \subset \omega(\Delta)$. Then $\omega(\Gamma) \subset \omega(\Delta)$.*

Proof. By Proposition 2.24, $\omega(\Gamma) \subset \bar{\Gamma}$, and so $\omega(\Gamma) \subset \overline{\omega(\Delta)}$. But $\overline{\omega(\Delta)} = \omega(\Delta)$, by Proposition 2.24. $\qquad \Box$

Any union of orbits of a dynamical system is called an *invariant set* of the system (a subset is *invariant under a map* if the map takes it onto itself; a union of orbits is invariant under the maps ϕ^t for all $t \in \mathbf{R}$).

(2.27) Proposition. *Any ω-set of ϕ is an invariant set of ϕ.*

Proof. We have to show that, for all $s \in \mathbf{R}$ and for all $p \in \omega(\Gamma)$, $s.p \in \omega(\Gamma)$, or, equivalently, that $\phi^s(\omega(\Gamma)) = \omega(\Gamma)$. Now, for any $x \in \Gamma$,

$$\phi^s(\omega(\Gamma)) = \phi^s\left(\bigcap_{t \in \mathbf{R}} \overline{\phi_x(I_t)} \right) = \bigcap_{t \in \mathbf{R}} \overline{\phi^s\phi_x(I_t)}$$

$$= \bigcap_{t \in \mathbf{R}} \overline{\phi_x(I_{s+t})} = \omega(\Gamma). \qquad \qquad \Box$$

Thus, for instance, if $\omega(\Gamma)$ is a single point q, then q is a fixed point of ϕ.

As we have seen in the examples of Chapter 1, an orbit may have an empty ω-set. This phenomenon seems to be associated with the orbit "going to infinity" so it is reasonable to suppose that if we introduce some compactness condition we can ensure non-empty ω-sets. What can we say about connectedness of ω-sets? At first sight it seems plausible that $\omega(\Gamma)$ inherits connectedness from Γ. However, with a little imagination one can visualize a flow on $\mathbf{R}^2$ having orbits with non-connected ω-sets. If we wish to ensure connected ω-sets, the simplest answer is, once again, a compactness condition. In considering these, and later questions, it is convenient to quote two purely topological lemmas, whose statements and proofs are given in the appendix to this chapter.

(2.28) Proposition. *Let K be a compact subset of X, such that, for all $n \in \mathbf{N}$, $\overline{\phi_x(I_n)} \cap K \neq \varnothing$. Then $\omega(x) \cap K \neq \varnothing$.*

Proof. For all $n \in \mathbf{N}$, $F_n = \overline{\phi_x(I_n)} \cap K$ is a closed subset of the compact subset K and hence is compact. By Lemma 2.44 of the appendix, $\bigcap F_n$ is non-empty. $\qquad\square$

(2.29) Proposition. *Let K be a compact subset of X, such that, for some $r \in N$, $\phi_x(I_r) \subset K$. Then $\omega(x)$ is a non-empty compact subset of K. If in addition X is Hausdorff, then $\omega(x)$ is connected.*

Proof. By definition $\omega(x) \subset \overline{\phi_x(I_r)} \subset K$. Also $\omega(x)$ is closed in X, by Proposition 2.24, hence closed in K and hence compact. By Proposition 2.28, $\omega(x)$ is non-empty. Finally, if X is Hausdorff, the connectedness of $\omega(x)$ follows from Lemma 2.44 of the appendix with $F_n = \overline{\phi_x(I_{n+r})}$. $\qquad\square$

(2.30) Corollary. *If X is compact Hausdorff then, for any orbit Γ, $\omega(\Gamma)$ is non-empty, compact and connected.* $\qquad\square$

(2.31) Proposition. *Let X be compact. Then for any neighbourhood U of $\omega(x)$, there exists $n \in \mathbf{N}$ such that $\phi_x(I_n) \subset U$.*

Proof. We may suppose that U is open, whence $X \backslash U$ is closed and compact. The result now follows immediately from Proposition 2.28 with $K = X \backslash U$. $\qquad\square$

This is one of several results that we shall improve upon in the appendix when we discuss compactification (see Exercise 2.61).

(2.32) Note. Suppose that U and V are open subsets of X and Y respectively and that $h : U \rightarrow V$ is a topological equivalence from $\phi|U$ to $\psi|V$. We should like to know that h takes ω-sets of orbit components of ϕ onto ω-sets of orbit components of ψ. There are various awkward points to contend with. For example, an orbit in U may have part of its ω-set outside U. Again, an orbit wholly in U may be mapped by h to a component of an orbit that leaves V. However, suppose that X and Y are Hausdorff, that K is a compact (and hence closed) subset of U, and let Γ be an orbit of ϕ such that, for some $x \in \Gamma$ and $t \in \mathbf{R}$, $\phi_x(I_t) \subset K$. Thus, by Proposition 2.29, $\omega(\Gamma)$ is a non-empty subset of K. One may prove that if, by abuse of notation, we denote by $h(\Gamma)$ the orbit of ψ containing $h\phi_x(I_t)$, then the orbit component of $h(\Gamma)$ containing $h\phi_x(I_t)$ contains $\psi_{h(x)}(I_{t'})$ for some $t' \in \mathbf{R}$, and $\omega(h(\Gamma)) = h(\omega(\Gamma))$.

We commented earlier that if Γ is a fixed point or periodic orbit then $\omega(\Gamma) = \Gamma$. We now investigate this property more thoroughly.

(2.33) Lemma. *Let X be Hausdorff and let Γ be a compact orbit of ϕ. Then $\omega(\Gamma) = \Gamma$.*

Proof. The set Γ is closed in X. Hence, by Proposition 2.29, $\omega(\Gamma)$ is a non-empty subset of Γ. Since it is an invariant set, it must be the whole of Γ. $\square$

(2.34) Proposition. *Let X be compact and Hausdorff. Then an orbit Γ of ϕ is closed in X if and only if $\omega(\Gamma) = \Gamma$.*

Proof. Necessity is a special case of Lemma 2.33. If $\omega(\Gamma) = \Gamma$ then, by Proposition 2.24, Γ is closed. $\square$

(2.35) Theorem. *Let X be Hausdorff. Then an orbit Γ of ϕ is compact if and only if it is a fixed point or a periodic orbit.*

Proof. Sufficiency is immediate. Suppose, then, that Γ is compact and that Γ is neither a fixed point nor a periodic orbit. Then, for any $x \in \Gamma$, ϕ_x is injective. For each $n \in \mathbf{N}$, $C_n = \phi_x([-n, n])$ is a compact subset of X, and hence closed. Moreover, $\omega(\Gamma) = \Gamma$ by Lemma 2.33, and so, for all $p \in \Gamma$, each neighbourhood of p contains points in $\phi_x(I_{n+1}) \subset \Gamma \backslash C_n$. We may therefore apply Lemma 2.45 of the appendix with $Y = A = \Gamma$, and deduce that Γ is not locally compact, which is a contradiction. $\square$

(2.36) Corollary. *Let X be compact and Hausdorff. Then the following three conditions on an orbit Γ of ϕ are equivalent:*
   (i) *Γ is a closed subset of X,*
   (ii) *Γ is a fixed point or a periodic orbit,*
   (iii) *$\omega(\Gamma) = \Gamma$.* $\square$

As we commented in Chapter 1, this result partially justifies the use of the term "closed orbit" as a synonym for "periodic orbit". See Exercise 2.63 of the appendix for a generalization of Corollary 2.36 to non-compact X.

(2.37) Exercise. A *minimal* set of a dynamical system is a non-empty closed invariant set that does not contain any closed invariant proper subset. Use Zorn's Lemma (see Lang [2]) to prove that if, for any orbit Γ of ϕ, $\bar{\Gamma}$ is compact then it contains a minimal set. Give an example to show that Γ may have a non-empty ω-set that is not minimal.

VI. LIMIT SETS OF HOMEOMORPHISMS

The theory of α- and ω-limit sets may also be developed in the context of discrete dynamical systems. If f is a homeomorphism of a topological space, then the ω-set $\omega(x)$ of $x \in X$ with respect to f is defined by

$$\omega(x) = \bigcap_{n \in \mathbf{N}} \overline{\{f^r(x) : r \geq n\}}.$$

The α-set $\alpha(x)$ of x is the ω-set of x with respect to f^{-1}. All results of the previous section have analogues, with the one obvious exception that ω-sets of homeomorphisms need not be (and seldom are) connected.

VII. NON-WANDERING SETS

The fundamental equivalence relations in the theory of dynamical systems are indisputably topological equivalence (for flows) and topological conjugacy (for homeomorphisms). However, when the going gets rough in the classification problem, one tends to cast about for a new (but still natural) equivalence relation with respect to which classification may be easier. We shall describe some attempts in this direction in Chapter 7. One of these, generally accepted to be the most important, is concerned with a certain invariant set known as the *non-wandering set*, and now is a suitable time to explain this concept. The definition is due to David Birkhoff [1] and the logic behind it is as follows. If one compares phase portraits of dynamical systems, for example the two in Figure 2.38, it seems that certain parts are qualitatively more important than others. If one were asked to pick out the significant features of the left hand picture, one would inevitably begin by mentioning the fixed points and closed orbit. Generally speaking, qualitative features in a phase portrait of a dynamical system ϕ can usually be traced

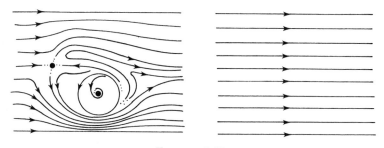

FIGURE 2.38

back to sets of points that exhibit some form of recurrence. The strongest
form of recurrence is periodicity, where a point resumes its original position
arbitrarily often, but there are weaker forms that are also important. One
uses the technical term *recurrent point* for a point that belongs to its own
ω-limit set. For example, all points of the torus T^2 are recurrent with respect
to an irrational flow (Example 1.25), although none are periodic. By
definition, a point is recurrent if and only if, for all neighbourhoods U of x,
$t.x \in U$ for arbitrarily large $t \in G$. We define x to be *non-wandering* if, for all
neighbourhoods U of x, $(t.U) \cap U$ is non-empty for arbitrarily large $t \in G$.
Thus we have a form of recurrence that is weaker than technical recurrence.
To see that it is strictly weaker, observe that any point of the non-recurrent
orbit Γ in Figure 2.39 is a non-wandering point. Rather more importantly

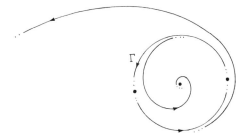

FIGURE 2.39

(because it is in a stabler configuration) consider any point $x \neq 0$ on the stable
manifold of the fixed point 0 of the hyperbolic toral automorphism (Example 1.30). It is clearly not recurrent, since $\omega(x) = \{0\}$, but it is non-wandering,
since any neighbourhood of it contains periodic points.

The term "non-wandering point" is an unhappy one, since not only may
the point wander away from its original position but, as we have seen, it may
never come back again. K. Sigmund has put forward the attractive alternative *nostalgic point*, for although the point itself may wander away, its
thoughts (represented by U) keep coming back. Non-wandering points are
also called Ω-*points*, and the set of all non-wandering points of ϕ, the
non-wandering set of ϕ, is denoted $\Omega(\phi)$ (or $\Omega(f)$ if ϕ is discrete and $f = \phi^1$).

(2.40) Exercise. Prove that for any homeomorphism f of X, $\Omega(f^{-1}) = \Omega(f)$.
Note the distinction between ω-sets and the Ω-set. All limit points are
Ω-points, but the converse is false, as the following exercise shows:

(2.41) Exercise. Prove that if $y \in \omega(x)$ for some $x \in X$, then $y \in \Omega(\phi)$. Sketch
the phase portrait of a flow on $\mathbf{R}^2$ such that (i) $\Omega(\phi) = \mathbf{R}^2$, but (ii) for some
$x \in \mathbf{R}^2$, x is neither an α-limit point nor an ω-limit point of any orbit of ϕ.

The following result sums up some elementary properties of Ω-sets:

(2.42) Theorem. *For any dynamical system ϕ on X, $\Omega(\phi)$ is a closed invariant subset of X, and is non-empty if X is compact. Topological conjugacies and equivalences preserve Ω-sets.*

Proof. The complement of $\Omega(\phi)$ is open in X, for if x has an open neighbourhood U such that $(t \cdot U) \cap U$ is empty for all sufficiently large t, then so has every point of U. Thus $\Omega(\phi)$ is closed in X. Moreover, for all $s \in G$, x has such a neighbourhood U if and only if $s \cdot x$ has a neighbourhood (namely $s \cdot U$) with a similar property. Thus $X \backslash \Omega(\phi)$, and hence also $\Omega(\phi)$, is an invariant set. If X is compact, then, by Proposition 2.29, any orbit of ϕ has a non-empty ω-set, and by Exercise 2.41, this is part of $\Omega(\phi)$.

Finally, suppose that $h : X \to Y$ is a topological conjugacy or equivalence from ϕ to a dynamical system ψ on Y. Let $p \in \Omega(\phi)$. Let V be a neighbourhood of $q = h(p)$ in Y and let $t_0 \in G$. We have to prove that, for some $t \geq t_0$, and some $y \in V$, $t \cdot y \in V$. This is trivially true if $V \cap (t_0 \cdot V)$ is non-empty, so suppose that it is empty. Let $h^{-1}(t_0 \cdot q) = t_1 \cdot p$. By continuity of ϕ, there is a neighbourhood U of p in $h^{-1}(V)$ such that $t_1 \cdot U \subset h^{-1}(t_0 \cdot V)$. Since $p \in \Omega(\phi)$, there exists $x \in U$ and $t_2 \in G$ with $t_2 > t_1$, such that $t_2 \cdot x \in U$. Let $h(t_1 \cdot x) = t_0 \cdot y$. Then $y \in V$. Also, since h is orientation preserving, $h(t_2 \cdot x) = t \cdot y$ for some $t > t_0$, and, since $t_2 \cdot x \in U$, $t \cdot y \in V$ (see Figure 2.42). $\qquad\square$

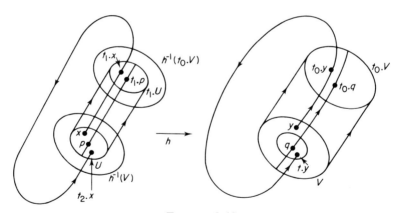

FIGURE 2.42

The new equivalence relations mentioned at the beginning of the section, are called Ω-*equivalence* (for flows) and Ω-*conjugacy* (for homeomorphisms). They are just the old ones, topological equivalence and conjugacy, restricted to Ω-sets. Thus if $\phi | \Omega(\phi)$ denotes the restriction of the flow ϕ to $\Omega(\phi)$, defined by $(\phi | \Omega(\phi))(t, x) = \phi(t, x)$ for all $(t, x) \in \mathbf{R} \times \Omega(\phi)$, then ϕ is

Ω-*equivalent* to ψ if and only if $\phi \mid \Omega(\phi)$ is topologically equivalent to $\psi \mid \Omega(\psi)$. Similarly homeomorphisms f and g are Ω-*conjugate* if and only if their restrictions $f|\Omega(f)$ and $g|\Omega(g)$ are topologically conjugate. By Theorem 2.42 topological equivalence (resp. conjugacy) is stronger than Ω-equivalence (resp. conjugacy).

(2.43) Exercise. Prove that topological equivalence is strictly stronger than Ω-equivalence. That is to say, give examples of flows that are Ω-equivalent but not topologically equivalent.

Appendix 2

In this appendix we prove the two topological results quoted in the main body of the chapter, we further discuss some points arising from orientation of orbits, and we extend results about limit sets from compact to locally compact spaces by the process of compactification.

I. TWO TOPOLOGICAL LEMMAS

Let X be a topological space. A sequence $(F_n)_{n \in \mathbf{N}}$ of subsets of X is *decreasing* if, for all m, $n \in \mathbf{N}$, $m \geq n$ implies $F_m \subset F_n$. For the definition of *increasing*, reverse the inclusion. Thus, in the above context of ω-sets, $\overline{(\phi_x(I_n))}_{n \in \mathbf{N}}$ is a decreasing sequence.

(2.44) Lemma. *Let (F_n) be a decreasing sequence of closed compact, non-empty subsets of X. Then $F = \bigcap_{n \in \mathbf{N}} F_n$ is non-empty. If further, X is Hausdorff and each F_n is connected, then F is connected.*

Proof. Without loss of generality we may assume that $X = F_1$, and hence that X is compact. Suppose that $F = \varnothing$. Then $\{X \backslash F_n\}_{n \in \mathbf{N}}$ is an open cover of X. Since X is compact, this cover has a finite subcover. But $X \backslash F_m \supset X \backslash F_n$ for all $m \geq n$. Hence $X = X \backslash F_{n_0}$ for some $n_0 \in N$, and so $F_{n_0} = \varnothing$, which is a contradiction.

Now let X be Hausdorff, and let each F_n be connected. Suppose F is not connected. Then $F = G \cup H$, where G and H are non-empty disjoint sets that are closed in the closed set F, and hence closed in X. Now X, being compact and Hausdorff, is also normal. Thus, there exist disjoint open subsets U and V of X containing G and H respectively. For all $n \in \mathbf{N}$, $F_n \cap U$ and $F_n \cap V$ are non-empty, and hence $F_n \not\subset U \cup V$, because F_n is connected. Let $K_n = F_n \cap (X \backslash (U \cup V))$. Applying the first part to (K_n), we deduce that $K = \bigcap_{n \in \mathbf{N}} K_n$ is non-empty. But $K \subset X \backslash (U \cap V)$ and also $K \subset F \subset U \cap V$. Therefore $K = \varnothing$, which is a contradiction. $\square$

Our second lemma is a version of the Baire category theorem. A topological space is a *Baire space* if the intersection of any countable sequence of dense open subsets of the space is a dense subset. Baire's theorem asserts that locally compact Haudsorff spaces (and complete metric spaces) are Baire spaces. We give a direct proof of an equivalent but slightly more technical statement which is particularly convenient for our applications.

(2.45) Lemma. *Let the topological space Y be the union of a finite or countably infinite sequence $(C_n)_{n>0}$ of closed subsets. Let A be a non-empty subset of Y such that, for all $a \in A$ and for all $n > 0$, $a \in \overline{A \backslash C_n}$. Then Y is not locally compact Hausdorff.*

Proof. Suppose that Y is locally compact Hausdorff, and hence regular. We choose inductively a sequence (a_n) of points of A and a decreasing sequence (F_n) of compact closed neighbourhoods F_n of a_n, such that, for all $n > 0$, $C_n \cap F_n$ is empty. Then, by Lemma 2.44, $\bigcap F_n$ is non-empty, and yet, by construction, it contains no point of $Y = \bigcup C_n$, which is a contradiction.

We may as well start the induction by taking $C_0 = \varnothing$, $a_0 = $ any point of A and $F_0 = $ any compact closed neighbourhood of a_0. Suppose that a_{n-1} and F_{n-1} have been constructed. Then int F_{n-1} contains a point a_n of $A \backslash C_n$. Choose F_n as any closed neighbourhood of a_n in (int $F_{n-1}) \cap (Y \backslash C_n)$. □

II. ORIENTED ORBITS IN HAUSDORFF SPACES

Let ϕ and ψ be flows on topological spaces X and Y respectively, and let $h : X \to Y$ be a homeomorphism mapping orbits of ϕ onto orbits of ψ. Recall that h preserves the orientation of an orbit Γ of ϕ if there is, for some $x \in \Gamma$, an increasing homeomorphism $\alpha : \mathbf{R} \to \mathbf{R}$, such that, for all $t \in \mathbf{R}$, $h(t.x) = \alpha(t).h(x)$. Similarly, h reverses the orientation of Γ if there is a decreasing homeomorphism $\beta : \mathbf{R} \to \mathbf{R}$ such that $h(t.x) = \beta(t).h(x)$. According to these definitions, h may both preserve and reverse the orientation of Γ.

(2.46) Exercise. Prove that h both preserves and reverses the orientation of Γ if and only if Γ is a fixed point.

A more worrying possibility is that h may neither *preserve* nor *reverse* the orientation of Γ.

(2.47) Example. Let ϕ be the flow given by $\phi(t, x) = x + t$ on the space $\mathbf{R}^\flat$ of real numbers with indiscrete topology. Then the topological equivalence $h : \mathbf{R}^\flat \to \mathbf{R}^\flat$ from ϕ itself defined by

$$h(x) = x \quad \text{for} \quad x \neq \pm 1, \qquad h(\pm 1) = \mp 1$$

neither preserves nor reverses the orientation of the unique orbit of ϕ.

The situation revealed by this example is, of course, pathological, and we shall show that the phenomenon cannot occur in Hausdorff spaces. The proof is rather tricky; a much easier result of the same type is:

(2.48) Exercise. Prove that if Γ is a closed orbit of ϕ then h either preserves or reserves the orientation of Γ. (*Hint*: See the first part of Exercise 2.4.)

We begin by proving a slightly off-beat property of real intervals (or, more generally, of connected, locally connected, locally compact, Hausdorff spaces).

(2.49) Lemma. *Let I be a real interval. If I is the union of a sequence (C_n) of two or more disjoint non-empty closed subsets, then the sequence is uncountably infinite.*

Proof. Suppose that $(C_n)_{n>0}$ is a finite or countably infinite sequence with the given property. Let A_n be the frontier of C_n in I. Then A_n is a non-empty (because I is connected) subset of C_n. Let $A = \bigcup A_n$, and let $a \in A_n$ for some n. Then, for all $m \neq n$, $a \in A \backslash A_m$. We prove that $a \in \overline{A \backslash A_n}$. Let V be any neighbourhood of a in I. Then V meets C_n and also C_m for some $m \neq n$. We may assume that V is connected, in which case, since V meets both C_m and its complement in I, V meets the frontier A_m of C_m. Since $A_m \subset A \backslash A_n$, we conclude that $a \in \overline{A \backslash A_n}$. Lemma 2.45 now gives a contradiction. $\square$

(2.50) Theorem. *If X is a Hausdorff space and the homeomorphism $h : X \to Y$ maps an orbit Γ of ϕ onto an orbit of ψ, then h either preserves or reverses the orientation of Γ.*

Proof. The theorem is trivial if Γ is a fixed point, and is just the result of Exercise 2.48 if Γ is a closed orbit. It remains to prove the result when ϕ_x is a continuous injection. In this case, by Theorem 2.35, $\psi_{h(x)}$ is also a continuous injection, and therefore it induces a continuous bijection from $\mathbf{R}$ onto $h(\Gamma)$. Let $\mu : h(\Gamma) \to \mathbf{R}$ be the inverse of this bijection. The map μ is not necessarily continuous; nevertheless, we assert that $\lambda = \mu h \phi_x : \mathbf{R} \to \mathbf{R}$ is a homeomorphism. This latter statement implies, of course, that λ is increasing or that λ is decreasing, and the conclusion of the theorem follows immediately.

To prove our assertion about λ, it is enough to show that λ is continuous at each $t \in \mathbf{R}$. Let J be a compact interval such that $t \in \text{int } J$. Then $h \phi_x (J)$ is a compact subset of the Hausdorff space Y, and is therefore closed in Y. Since $\psi_{h(x)}$ is continuous, $\lambda (J) = K$, say, is closed in $\mathbf{R}$. In fact, we shall prove that K is compact. This implies that $\psi_{h(x)}$ maps K homeomorphically onto $h \phi_x (J)$, hence that λ maps J homeomorphically onto K, and in particular that λ is continuous at t.

Suppose first that k contains an unbounded interval, say $[a, \infty[$. Then, by Proposition 2.29, $\omega(h(\Gamma))$ is non-empty and contained in $h\phi_x(J)$. Hence, using Proposition 2.27, $h\phi_x(J) = h(\Gamma) = h\phi_x(\mathbf{R})$. But $h\phi_x$ is injective and

$J \neq \mathbf{R}$, so we have a contradiction. Similarly, K can contain no unbounded interval of the form $]-\infty, b[$. Thus there exists an increasing sequence $(a_n)_{n \in \mathbf{Z}}$ of points of $\mathbf{R} \backslash K$ such that $a_n \to \infty$ as $n \to \infty$ and $a_n \to -\infty$ as $n \to -\infty$. Thus K is the union of the sequence $(K \cap [a_{n-1}, a_n])_{n \in \mathbf{Z}}$ of disjoint compact sets. These are mapped homeomorphically by λ^{-1} into compact sets whose union is J. It follows from Lemma 2.49 that K is a single compact set contained in $[a_{n-1}, a_n]$ for some n. $\square$

Note that a similar theorem holds when $h : U \to V$ is a homeomorphism between open subsets U of X and V of Y, and both Γ and $h(\Gamma)$ are orbit components.

The difficulty in proving the above theorems stems from the fact that the continuous bijection $\phi_x : \mathbf{R} \to X$ is not necessarily an embedding. To see that it is not, consider any irrational flow on the torus (Example 1.25). One has, in fact, the following simple criterion for ϕ_x to be an embedding.

(2.51) Theorem. *Let X be Hausdorff, and let ϕ have stabilizer $\{0\}$ at x. Then $\phi_x : R \to X$ is an embedding if and only if its image Γ is locally compact.*

Proof. We may as well take $X = \Gamma$. Necessity is trivial. For sufficiency, we have to prove that the inverse of the continuous bijection $\phi_x : \mathbf{R} \to \Gamma$ is continuous. Suppose not. Let F be a closed subset of $\mathbf{R}$ such that $\phi_x(F)$ is not closed. Let p be a point of $\overline{\phi_x(F)} \backslash \phi_x(F)$. For all $n > 0$, ϕ_x maps $F \cap [-n, n]$ homeomorphically onto a closed subset. Hence p is in the closure of either $\phi_x(F \cap [n, \infty[)$ or $\phi_x(F \cap]-\infty, -n])$. We deduce that p is in either $\alpha(\Gamma)$ or $\omega(\Gamma)$, say the latter. Thus $\omega(\Gamma) = \Gamma$. This is impossible, by the proof of Theorem 2.35. $\square$

(2.52) Exercise. Make the suggested generalization of Lemma 2.49.

(2.53) Exercise. Let G be a *second countable* (i.e. having a countable basis for its topology), locally compact topological group acting on a Hausdorff space X. Prove that, for any $x \in X$, the map $G/G_x \to G.x$ taking gG_x to $g.x$ is a homeomorphism if and only if $G.x$ is locally compact. (*Hint:* either generalize the proof of Theorem 2.51 or prove directly that the map is open, using a version of Baire's theorem, as, for example, in § 3 of Chapter 2 of Helgason [1]).

(2.54) Exercise. Let G and H be second countable, locally compact, locally connected topological groups acting *freely* (i.e. with trivial stabilizers) and *transitively* (i.e. with only one orbit) on a Hausdorff space X. Prove that, for all $x \in X$, the map $\psi_x^{-1}\phi_x : G \to H$ is a homeomorphism (where ϕ and ψ are the actions). The ω-set argument in the proof of Theorem 2.50 does not seem to generalize easily, but it can be replaced, as in Irwin [1].

III. COMPACTIFICATION

Let X be a non-compact topological space. There is a standard procedure for associating with X a compact topological space, called the *one-point compactification* X^* of X. Let ∞ denote some point not in X. We define the set X^* to be $X \cup \{\infty\}$. To turn X^* into a topological space, we define a subset U of X^* to be open in X^* if and only if either U is an open subset of X or $X^* \backslash U$ is a closed compact subset of X. It is an easy exercise to verify that X^* is indeed compact, and that it is Hausdorff if and only if X is locally compact and Hausdorff (see Proposition 8.2 of Chapter 3 of Hu [1]).

(2.55) Example. Let $X = \mathbf{R}^n$. Then X^* is homeomorphic to S^n, the unit sphere in $\mathbf{R}^{n+1}$. A particular homeomorphism h may be constructed by mapping $\mathbf{R}^n$, identified with the hyperplane $x_{n+1} = -1$ of $\mathbf{R}^{n+1}$, onto $S^n \backslash \{e_{n+1}\}$ by stereographic projection from e_{n+1}, and ∞ to e_{n+1}, where $e_{n+1} = (0, \ldots, 0, 1)$. See Figure 2.55.

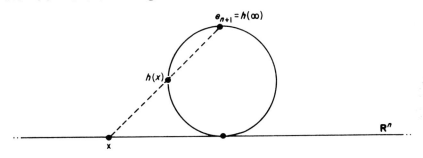

FIGURE 2.55

Now let ϕ be a dynamical system on X. We define the *one-point compactification* of ϕ to be the map $\phi^* : G \times X^* \to X^*$ defined by $\phi^*(g, x) = \phi(g, x)$ if $x \neq \infty$, and $\phi(g, \infty) = \infty$. For instance the phase portraits of ϕ^* for the flows ϕ of Examples 1.8 and 1.9 are illustrated in Figure 2.56 (identifying $\mathbf{R}^*$ with S^1, as above).

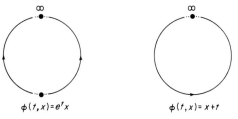

$$\phi(t, x) = e^t x \qquad \qquad \phi(t, x) = x + t$$

FIGURE 2.56

(2.57) Exercise. Sketch the phase portrait of ϕ^* near ∞ for the flows ϕ of Examples 1.14–1.17.

We now prove that ϕ^* is a dynamical system on X^*. This is trivial if $G = \mathbf{Z}$, so we consider the flow case.

(2.58) Lemma. *Let* $\Phi : \mathbf{R} \times X \to \mathbf{R} \times X$ *be defined by* $\Phi(t, x) = (t, \phi(t, x))$. *Then* Φ *is a homeomorphism.*

Proof. The inverse Φ^{-1} of Φ is given by $\Phi^{-1}(t, y) = (t, \phi(-t, y))$. Moreover if π_1 and π_2 denote projection of $\mathbf{R} \times X$ onto its factors, then $\pi_1 \Phi = \pi_1$ and $\pi_2 \Phi = \phi$. Since both π_1 and ϕ are continuous, it follows that Φ is continuous. Similarly, Φ^{-1} is continuous.   □

(2.59) Theorem. *The one-point compactification ϕ^* of a dynamical system on X is a dynamical system on X^*.*

Proof. It is immediate that ϕ^* satisfies conditions 1.1. We have to prove that ϕ^* is continuous. More precisely, we prove that, for all $(t, x) \in \mathbf{R} \times X^*$ and for any open neighbourhood V of $\phi^*(t, x)$ in X^*, there exists an open neighbourhood U of (t, x) in $\mathbf{R} \times X^*$ such that $\phi^*(U) \subset V$. There are two cases to consider.

(i) Suppose $x \in X$. Then we may assume that V is open in X. Thus $U = (\phi^*)^{-1}(V) = \phi^{-1}(V)$ is open in $\mathbf{R} \times X$ and hence in $\mathbf{R} \times X^*$.

(ii) Suppose that $x = \infty$. The set V is an open neighbourhood of ∞ in X^*, and so $F = X^* \backslash V$ is a compact closed subset of X. Let

$$K = \phi^{-1}(F) \cap ([t-1, t+1] \times X)$$
$$= \Phi^{-1}([t-1, t+1] \times F),$$

where Φ is the homeomorphism of Lemma 2.58. Hence K is a compact closed subset of $\mathbf{R} \times X$, and $\pi_2(K)$ is a compact subset of X. Moreover, since $\pi_1(K)$ is compact, $\pi_2 | \pi_1(K) \times X$ is proper, and so $\pi_2(K)$ is closed. Let $W = X^* \backslash \pi_2(K)$. Then W is an open neighbourhood of ∞ in X^*. Finally, we take $U = \,]t-1, t+1[\times W$, and check that $\phi^*(U) \subset V$.   □

One-point compactifications of spaces and dynamical systems are useful in extending results on compact spaces to locally compact spaces. One applies the known theory to ϕ^* and deduces properties of ϕ. It is immediate that any orbit Γ or ϕ is also an orbit of ϕ^*. It may happen, however, that $\omega^*(\Gamma)$ (the ω-set of Γ regarded as an orbit of ϕ^*) contains the point ∞. For instance, this is always so when $\omega(\Gamma) = \varnothing$.

(2.60) Exercise. Prove that $\omega(\Gamma) \subset \omega^*(\Gamma) \subset \omega(\Gamma) \cup \{\infty\}$. Deduce that if $\omega(\Gamma) = \varnothing$ then $\omega^*(\Gamma) = \{\infty\}$.

(2.61) Exercise. Let X be locally compact Hausdorff and let $\omega(\Gamma)$ be compact and non-empty. Prove that $\omega^*(\Gamma) = \omega(\Gamma)$. Deduce that for any $x \in \Gamma$ and any open neighbourhood U of $\omega(\Gamma)$ there exists $t \in \mathbf{R}$ such that $\phi_x(I_t) \subset U$.

(2.62) Exercise. Prove that if X is locally compact Hausdorff and if $\omega(\Gamma)$ has a compact connected component W, then $\omega(\Gamma) = W$.

(2.63) Exercise. Prove that if X is locally compact Hausdorff then the following four statements are equivalent:

   (i) $\omega(\Gamma) = \Gamma$,
   (ii) $\omega^*(\Gamma) = \Gamma$,
   (iii) Γ is compact,
   (iv) Γ is a fixed point or a periodic orbit.

CHAPTER 3

Integration of Vector Fields

At the end of the introduction we said a few words about the motion of a liquid whose velocity at any point of space is independent of time. This example was used to motivate the axioms for a flow. Conversely, given a flow ϕ on a space X, we can recover a notion of *velocity at a point* provided that X has the structure of a differentiable manifold and ϕ has a certain degree of smoothness. The velocity of ϕ is a *vector field* on X (a fact that is often obscured in Euclidean space, which is commonly identified with its own tangent bundle). In this chapter we discuss existence and uniqueness of a flow whose velocity is a prescribed vector field. The local problem is, using a chart, equivalent to the problem of existence and uniqueness of solutions (integral curves) of a system of ordinary differential equations.

In terms of the moving liquid model, an integral curve is the path of a given particle of fluid. Picard's theorem deals with the local existence of such integral curves. The idea behind its proof is as follows. One chooses at random a curve γ_0 (a continuous map from time to position in space). Taking the velocity of the liquid at $\gamma_0(t)$ for each t gives a vector valued function v of time. Now consider a journey with the same starting point as γ_0 but with velocity $v(t)$ at time t. Let γ_1 be the path of this journey (see Figure 3.1). Applying this process again with γ_0 replaced by γ_1 we obtain a new curve γ_2, and so on. The curves form an infinite sequence (γ_i).

At each stage we modify the previous curve to fit in with the velocity of the fluid along it, so it is not surprising that, for large i, γ_i approaches an integral curve. What we have described is rather like holding onto one end of a light thread while the thread gradually drifts into streamline position, but it is a discrete rather than a continuous process.

Convergence of such a sequence of paths may be thought of in terms of convergence of the sequence $(\gamma_i(t))$ of positions, for all times t. A more fruitful approach, however, is to consider convergence in an infinite

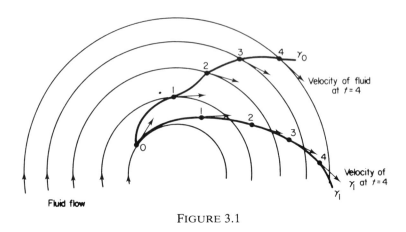

FIGURE 3.1

dimensional Banach space, each of whose points is a continuous map from time to position. We shall make use of map spaces at many crucial stages in this and in later chapters. This is not sophistication for its own sake; it leads to considerable simplifications of proofs. For example, most of the complications in the classical proof that the integral of a smooth vector field is a smooth function of initial position vanish once we have at our disposal a few elementary results on map spaces. We describe the requisite theory in Appendices B and C. The latter contains a version of Banach's contraction mapping theorem. This extremely useful tool is applied to avoid long-drawn-out successive approximation arguments, for example in Picard's theorem and in other parts of the integration theory below.

I. VECTOR FIELDS

Let X be a differentiable manifold, and let I be a real interval. A *vector field* on X is a map $v : X \to TX$ associating with each point x of X a vector $v(x)$ in the tangent space $T_x X$ to $\dot{X}$ at x. We think of $T_x X$ as the space of all possible velocities at x of a particle moving along paths on the manifold X. Thus we have in mind a picture like Figure 3.2, where X is the submanifold S^2 of $\mathbf{R}^3$ and $T_x X$ is embedded in $\mathbf{R}^3$ as a plane geometrically tangent to S^2 at x. However we really need an intrinsic definition of $T_x X$, independent of any particular embedding of the manifold X in Euclidean space. This is given in Appendix A, together with further notes on vector fields. Notice that a particle moving along I can have any real velocity at $t \in I$, and so $T_t I$ is a copy of the real line. In fact one may identify $T_t I$ with $\{t\} \times \mathbf{R}$ in $I \times \mathbf{R}$, and hence

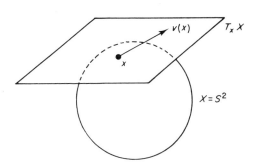

FIGURE 3.2

identify TI with $I \times \mathbf{R}$. Similarly we show in Example A.31 that if U is an open subset of Euclidean space $\mathbf{R}^n$ then TU may be identified with $U \times \mathbf{R}^n$. In this case, if v is a C^r vector field on U then the first component of $v(x)$ with respect to the product is always x, signifying that $v(x)$ is in the tangent space at x. Thus $v(x) = (x, f(x))$ for some C^r map $f: U \to \mathbf{R}^n$. Conversely, any C^r map $f: U \to \mathbf{R}^n$ corresponds to a C^r vector field (id, f) on U. The map f that determines and is determined by v in this way is called the *principal part* of v. It is common, and usually perfectly harmless, to blur the distinction between vector fields and principal parts. We shall resist the temptation to do so, however, at least for the rest of this chapter. Note that all the above remarks hold equally well when $\mathbf{R}^n$ is replaced by any Banach space $\mathbf{E}$.

(3.3) Example. (*Gradient vector fields*). Let $f: X \to \mathbf{R}$ be a C^{r+1} function $(r > 0)$, and suppose that X has a Riemannian structure (see Appendix A). Then we may associate with the linear tangent map $Tf: TX \to T\mathbf{R}$ a C^r vector field ∇f (or grad f) on X, called the *gradient vector field* of f. Example A.57 of Appendix A describes in detail the construction of ∇f. If $X = \mathbf{R}^n$, $f(x)$ has principal part $(\partial f/\partial x_1, \ldots, \partial f/\partial x_n)$. Thus, for example, if $f(x_1, x_2) = x_1^2 x_2$, then $\nabla f(x_1, x_2)$ has principal part $(2x_1 x_2, x_1^2)$. If $X = S^2$, embedded as

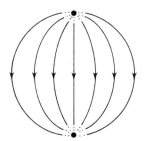 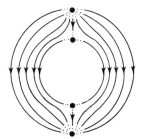

FIGURE 3.3

the unit sphere in $\mathbf{R}^3$, and f is the *negative height function* $f(x_1, x_2, x_3) = -x_3$, then $\nabla f(x)$ is tangent to the path taken by a raindrop on S^2, that is to say the *line of steepest descent* through x, which is a meridian of longitude. Figure 3.3 illustrates this, and also ∇f for f the negative height function on the embedding of T^2 as a tyre inner tube.

(3.4) Example. Two rather trivial vector fields needed later are the *zero vector field* on X, $x \mapsto 0_x$, where 0_x is the zero vector in T_xX (0_x is identified with the point x itself in Figure 3.2), and the *unit vector field* on I, $t \mapsto 1_t$, where 1_t is the vector $(t, 1)$ in $T_tI = \{t\} \times \mathbf{R}$. Now consider the product manifold $I \times X$. At any point (t, x), the velocity of a moving particle has components in the I and X directions, so we may identify $T(I \times X)$ with $TI \times TX$ (see Exercise A.33). We denote by u the *unit vector field* on $I \times X$ in the positive t direction. That is to say, $u(t, x) = (1_t, 0_x)$ (see Figure 3.4).

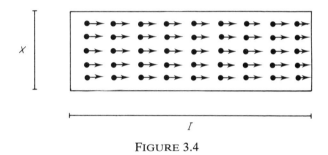

FIGURE 3.4

II. VELOCITY VECTOR FIELDS AND INTEGRAL FLOWS

Let $\phi: I \times X \to X$ be any map such that $\phi^t: X \to X$ is a homeomorphism for all $t \in I$. Given $x \in X$ and $t \in I$, let $y = (\phi^t)^{-1}(x)$. Then $\phi_y(t) = \phi(t, y) = x$. If we regard $\phi_y: I \to X$ as determining the position of a particle moving on X, then the particle reaches the point $x \in X$ at time t. Now suppose that the function ϕ_y is differentiable at t. Then the velocity of the particle when it reaches x is the vector $\phi_y'(t)$ of T_xX (see Example A.37). We call this vector the *velocity of ϕ at the point x at time t*. Thus, for example, if ϕ is defined for $t > 0$ and $x \in \mathbf{R}$ by $\phi(t, x) = \sinh(tx)$ then the above y is $(1/t)\sinh^{-1}x$, and differentiating ϕ with respect to t at (t, y) gives the velocity of ϕ at x at time t as the vector $(1/t)\sinh^{-1}x \cosh(\sinh^{-1}x)$ in the tangent space to $\mathbf{R}$ at x. Note that if ϕ is C^1 then its velocity at x at time t is $(T\phi)u(t, y)$, where u is as above.

We shall show below (Theorem 3.9) that of ϕ is a flow on X then its velocity at x is the same at all times t. Thus, in this case, ϕ gives rise to a vector field v, called the *velocity vector field* of f, where $v(x)$ is the velocity of ϕ at x at any time t. Since $\phi_x(0) = x$, the simplest formula for v is $v(x) = \phi_x'(0)$. For example, if $X = \mathbf{R}$ and ϕ is the flow $t \cdot x = (x^{1/3} + t)^3$, we differentiate with respect to t at $t = 0$, and obtain $v(x) = (x, 3x^{2/3})$. If ϕ is C^1 then $v(x) = (T\phi)u(0, x)$, so the diagram

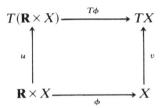

commutes. If ϕ is $C^r(r > 1)$, then v is C^{r-1}.

(3.5) Exercise. Find the velocity vector fields of the flows in Examples 1.14–1.17 and Exercise 1.18.

Conversely, if v is a given vector field on X, we call any flow ϕ on X an *integral flow* of v if v is the velocity vector field of ϕ, and say that v is *integrable* if such a flow exists.

(3.6) Example. Let $X = \mathbf{R}$ and let $v: X \to TX = \mathbf{R}^2$ be the vector field $v(x) = (x, -x)$. Then $\phi: \mathbf{R} \times X \to X$ given by $\phi(t, x) = x\,e^{-t}$ is an integral flow of ϕ.

(3.7) Exercise. Show that the zero vector field on any manifold X has the trivial flow as its unique integral flow.

(3.8) Exercise. Find an integral flow of the unit vector field u on $\mathbf{R} \times X$ defined above.

(3.9) Theorem. *Let ϕ be a flow on X such that, for all $x \in X$, the map $\phi_x : \mathbf{R} \to X$ is differentiable. Then the velocity of ϕ at any point is independent of time. Thus ϕ has a well defined velocity vector field.*

Proof. We prove that the velocity at x at time t equals the velocity at x at time 0. Let $y = \phi^{-t}(x)$. Then, by the basic property of flows $\phi_y(u) = \phi_x(-t + u)$ for all $u \in \mathbf{R}$. Differentiating with respect to u at $u = t$, using the chain rule for the right-hand side, we obtain $\phi_y'(t) = \phi_x'(0)$, as required. ☐

We wish to find conditions on a vector field v on X that ensure its integrability. Our main global theorem (Theorem 3.43) states that if X is compact and v is $C^r(r \geqslant 1)$ then v has a unique integral flow ϕ, which is itself

C'. The technique used in the proof of this result can be visualized as gluing ϕ together from fragments constructed locally at each point (t, x) of $\mathbf{R} \times X$. These fragments of map coincide on the intersections of their domains of definition by a uniqueness theorem (Theorem 3.34). We need some local definitions to get started on this programme.

An *integral curve* of v is a C^1 map $\gamma : I \to X$, where I is any real interval, satisfying $\gamma'(t) = v\gamma(t)$ for all $t \in I$. If $0 \in \text{int } I$, we call γ an integral curve *at* $\gamma(0)$. Thus if ϕ is a flow on X and v is the velocity vector field of ϕ, then the orbit of ϕ through $x \in X$ is the image of an integral curve of v at x. A *local integral* of v is a map $\phi : I \times U \to X$, where I is an interval neighbourhood of 0 and U is a non-empty open subset of X, such that, for all $x \in U$, $\phi_x : I \to U$ is an integral curve of v at x (see Figure 3.10). For all $x \in U$, we call ϕ a local integral *at x*, and say that v is *integrable at x* if such a local integral exists. If it does so, and is C^1, then the diagram

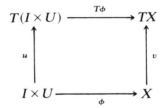

commutes. If further $I = \mathbf{R}$ and $U = X$ then the local integral is, as we shall show later (Corollary 3.38) a flow on X, and is hence an integral flow of v. The term *local integral* should not be confused with the term *local first integral*. See the appendix to this chapter for a discussion of *first integrals*.

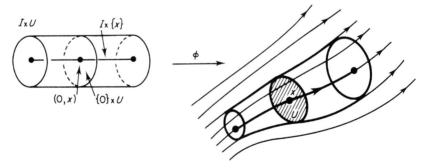

FIGURE 3.10

In investigating the existence of local integrals, we may restrict our attention to vector fields on (open subsets of) Banach spaces. For suppose that we wish to show that a vector field v on a manifold X is integrable at a point x. We take an admissible chart $\xi: V \to V' \subset \mathbf{E}$ at x, where X is modelled on $\mathbf{E}$. The C^∞ diffeomorphism ξ induces, from $v|V$, a vector field $\xi_*(v)$ on V', which is integrable at $\xi(x)$ if and only if v is integrable at x. This is a consequence of the following result:

(3.11) Theorem. *Let v be a vector field on X, and let $h: X \to Y$ be a C^1 diffeomorphism. If γ is an integral curve of v then $h\gamma$ is an integral curve of the vector field $h_*(v) = (Th)vh^{-1}$ induced on Y from v by h. If $\phi: I \times U \to X$ is a local integral of v at x, then $h\phi(id \times (h|U)^{-1})$ is a local integral of $h_*(v)$ at $h(x)$.*

Proof. The diagram

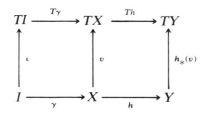

commutes, where ι is the positive unit vector field on I, and $T(h\gamma) = ThT\gamma$. Thus $h\gamma$ is an integral curve of $h_*(v)$. The proof of the second statement is similar. $\qquad\square$

III. ORDINARY DIFFERENTIAL EQUATIONS

Let V be an open subset of a Banach space $\mathbf{E}$, and let v be a C^r vector field on $V (r \geqslant 0)$. Suppose that $\phi: I \times U \to V$ is a local integral of v. Let us make explicit the relation so established between ϕ and the principal part f of v. According to the definition of local integral, for all $(t, x) \in I \times U$,

(3.12) $$\gamma'(t) = v\phi(t, x) = (\phi(t, x), f\phi(t, x)).$$

Hence $D_1\phi(t, x)(1) = f\phi(t, x)$. We make the usual abuse of notation that identifies a linear map from $\mathbf{R}$ to $\mathbf{E}$ with its value at 1, and write

(3.13) $$D_1\phi = f\phi.$$

Conversely, any map $\phi: I \times U \to V$ satisfying (3.13) also satisfies (3.12). We are led to make definitions of local integrals and integral curves of any map $f: V \to \mathbf{E}$ in the obvious manner. That is to say, $\phi: I \times U \to V$ is a *local integral* of f at $x \in U$ if ϕ^0 is the inclusion and $D_1\phi = f\phi$ on $I \times U$. An *integral curve* of f is a C^1 map $\gamma: I \to V$ such that $\gamma' = f\gamma$ on I. We say that γ is an integral curve *at* x if $0 \in \text{int } I$ and $\gamma(0) = x$. Trivially:

(3.14) Proposition. *A map is a local integral (resp. integral curve) of a vector field v on V if and only if it is a local integral (resp. integral curve) of the principal part of v.*   □

We observe that the relation $\gamma' = f\gamma$ implies that any integral curve of a C^0 vector field is C^1. More generally, we obtain, by induction, the following result:

(3.15) Proposition. *Any integral curve of a C^r vector field is C^{r+1}.*   □

In classical notation, with $\mathbf{E} = \mathbf{R}^n$, the relations $D_1\phi = f\phi$ and $\gamma' = f\gamma$ are both written in the form $dx/dt = f(x)$, which is called an *autonomous system of first order ordinary differential equations*. Here the term "autonomous" refers to the fact that the variable t is not explicitly present on the right-hand side. It is a "system of equations" since there are n scalar equations of the form $dx_i/dt = f_i(x_1, \ldots, x_n)$. The function γ is called a solution of the equations, and ϕ is described as a *solution regarded as a function of initial conditions*, or *general solution*.

(3.16) Example. Let v be the vector field on $\mathbf{R}^2$ with principal part $f: \mathbf{R}^2 \to \mathbf{R}^2$ given by $f(x, y) = (-y, x)$. The corresponding pair of scalar differential equations is

$$\frac{dx}{dt} = -y, \qquad \frac{dy}{dt} = x.$$

The solution $x = \cos t$, $y = \sin t$ of these equations is, in our parlance, the integral curve $\gamma: \mathbf{R} \to \mathbf{R}^2$ given by $\gamma(t) = (\cos t, \sin t)$. The general solution $x = A \cos t - B \sin t$, $y = A \sin t + B \cos t$, where A and B are "arbitrary constants", is in our parlance, the local integral (and, indeed, integral flow) $\phi: \mathbf{R} \times \mathbf{R}^2 \to \mathbf{R}$ given by

$$\phi(t, (x, y)) = (x \cos t - y \sin t, x \sin t + y \cos t).$$

(3.17) *Note. Non-autonomous systems.* In the present context, we lose no generality by restricting our attention to autonomous differential equations. Any non-autonomous system $dx/dt = g(t, x)$, where g maps an open subset V of $\mathbf{R} \times \mathbf{E}$ to $\mathbf{E}$, gives rise to an autonomous system $dy/dt = f(y)$, where $f: V \to \mathbf{R} \times \mathbf{E}$ is defined by $f(u, x) = (1, g(u, x))$. In other words, we are replacing the given differential equation by $du/dt = 1$, $dx/dt = g(u, x)$. Any solution of $dy/dt = f(y)$ yields a solution of $dx/dt = g(t, x)$. Specifically, suppose that we want a solution δ of $dx/dt = g(t, x)$ satisfying the boundary condition $\delta(t_0) = p$. Suppose that we have found an integral curve $\gamma: I \to V$ of f at (t_0, p). Then we assert that $\delta(t) = \gamma_2(t - t_0)$ defines a suitable solution δ, where the index denotes the component in the product $\mathbf{R} \times \mathbf{E}$. For observe that $\delta(t_0) = \gamma_2(0) = p$ and

$$\delta'(t) = \gamma_2'(t - t_0) = f_2\gamma(t - t_0) = g(\gamma_1(t - t_0), \delta(t)).$$

But, since $\gamma_1'(t) = f_1\gamma(t) = 1$ and $\gamma_1(0) = t_0$, $\gamma_1(t) = t + t_0$, and so $\delta'(t) = g(t, \delta(t))$.

We should also point out that an mth order ordinary differential equation in "normal form"

$$\frac{d^m y}{dt^m} = h\left(t, y, \frac{dy}{dt}, \ldots, \frac{d^{m-1}y}{dt^{m-1}}\right)$$

may be reduced by the substitutions

$$x_1 = y, x_2 = \frac{dy}{dt}, \ldots, x_m = \frac{d^{m-1}y}{dt^{m-1}}$$

to the form $dx/dt = g(t, x)$, where $x = (x_1, \ldots, x_m) \in \mathbf{E}^m$, and

$$g(t, x) = (x_2, x_3, \ldots, x_m, h(t, x_1, \ldots, x_m)).$$

Moreover, the implicit mapping theorem (Exercise C.14) can often be used locally to reduce more general mth order equations to the above normal form.

(3.18) Example. The simple pendulum equation $\theta'' = -g \sin \theta$ of the introduction was reduced to $\theta' = \omega$, $\omega' = -g \sin \theta$ by the substitution $\omega = \theta'$.

(3.19) Example. In conservative mechanics, *Lagrange's equations* of motion are usually written

$$\frac{d}{dt}\left(\frac{\partial L}{\partial q_i}\right) - \frac{\partial L}{\partial q_i} = 0,$$

where q_i $(i = 1, 2, \ldots, n)$ are "generalized coordinates", the *Lagrangian L* is $T - V$, T is the *kinetic energy* and V is the *potential energy*. If we choose q_i such that the positive definite quadratic form

$$T = \sum_{i=1}^{n} \tfrac{1}{2}m_i(q_i')^2,$$

where the m_i are "generalized masses", then the equations of motion take the form

$$m_i q_i'' = -\frac{\partial V}{\partial q_i}.$$

The substitution $m_i q_i' = p_i$, the "generalized momentum", converts the equations of motion to $p_i' = -\partial V/\partial q_i$, and the substitution and equations of motion may be written together as *Hamilton's equations*

$$q_i' = \frac{\partial H}{\partial p_i}, \qquad p_i' = -\frac{\partial H}{\partial q_i},$$

where the *Hamiltonian* $H = T + V$ is the total energy as a function of p_i and q_i.

(3.20) Example. *Van der Pol's equation.* This is, amongst other things, an important mathematical model for electronic oscillators. It has the form

$$x'' - \alpha x'(1 - x^2) + x = 0$$

where α is a positive constant. This is equivalent to the pair of first order equations

$$x' = y, \qquad y' = -x + \alpha y(1 - x^2)$$

where α is a positive constant. The phase portrait of this vector field on $\mathbf{R}^2$ is topologically equivalent to the reverse flow of the flow in Example 1.27. It has a unique closed orbit, which is the ω-set of all orbits except the fixed point at the origin (see Figure 3.20; a proof of this assertion may be found,

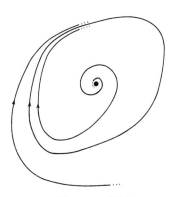

FIGURE 3.20

for example, in Hirsch and Smale [1] and in Simmons [1]). The system is said to be *auto-oscillatory*, since all solutions (except one) tend to become periodic as time goes by.

IV. LOCAL INTEGRALS

Let p be a point of a Banach space $\mathbf{E}$. Suppose that we have a vector field v with principal part f defined on some neighbourhood B' of p. We wish to show that for some neighbourhood B of p there are unique integral curves at each point of B. We must first make some comments. To stand any chance of proving uniqueness of an integral curve at the point x, we must fix the interval on which the curve is to be defined (since restricting to a smaller interval gives, strictly speaking, a different integral curve). If we wish to fit the integral curves together to form a local integral ϕ at p, we may as well define them all on the same interval I. We cannot have I arbitrarily large, since the orbit of p may not be wholly contained in B'. Similarly, however small I, we usually need B strictly inside B', since points x near the frontier of B' will leave B' at some time in I.

Actually, we adopt a slightly different approach, for ease of presentation. We start with given sets B' and B, balls with centre p in $\mathbf{E}$. We insist that the radius d of B is strictly less than the radius d' of B'. Note, however, that the argument works equally well with $B = B' = \mathbf{E}$, so we include this as an exceptional possibility $d = d' = \infty$. We prove that there is some interval I on which the above situation holds (see Figure 3.21). For this we need to make

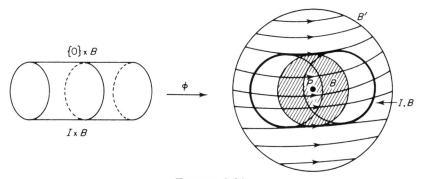

FIGURE 3.21

some assumption about the smoothness of f. Continuity of f ensures existence of integral curves, but not their uniqueness. For uniqueness we need something stronger, and the relevant property is what is known as a *Lipschitz condition* (see Appendix C). This is about half way between continuity and differentiability; it implies continuity and it is certainly satisfied by any C^1 map on B'.

Any Lipschitz map is bounded on a ball of finite radius. From now on we shall be making full use of the map space notation and theory of Appendix B and the contraction mapping theorem of Appendix C. In particular, $|f|_0$ denotes sup $\{|f(x)|: x \in B'\}$. Our version of Picard's theorem is:

(3.22) Theorem. *Let $f: B' \to E$ be Lipschitz with constant κ, and let I be the interval $[-a, a]$, where $a < (d' - d)/|f|_0$ if $d' < \infty$. Then for each $x \in B$, there exists a unique integral curve $\phi_x: I \to B'$ of f at x. Let $\phi: I \times B \to B'$ send (t, x) to $\phi_x(t)$. Then for all $t \in I$, $\phi': B \to B'$ is uniformly (in t) Lipschitz, and is C^r iff f is C^r.*

Proof. We give a proof under the additional restriction $a < 1/\kappa$, which may be removed by an easy technical modification (see Exercise 3.24 below). We define a map $\chi: B \times C^0(I, B') \to C^0(I, B')$ by, for all $t \in I$,

$$(3.23) \qquad \chi(x, \gamma)(t) = x + \int_0^t f\gamma(u) \, du,$$

where $C^0(I, B')$ is as defined in Appendix B. We must check, of course, that $\chi(x, \gamma)$ is indeed in $C^0(I, B')$. It is continuous by elementary analysis, and, by the inequalities

$$\left| x + \int_0^t f\gamma(u) \, du - p \right| \le |x - p| + |t| \, |f|_0 \le d + a|f|_0 \le d',$$

its image is in B'.

Now, for all $x \in B$ and $\gamma, \tilde{\gamma} \in C^0(I, B')$,

$$|\chi^x(\gamma) - \chi^x(\tilde{\gamma})|_0 = \sup \left\{ \left| \int_0^t (f\gamma(u) - f\tilde{\gamma}(u)) \, du \right| : t \in I \right\}$$

$$\le \sup \left\{ \left| \int_0^t |f\gamma(u) - f\tilde{\gamma}(u)| \, du \right| : t \in I \right\}$$

$$\le \kappa a |\gamma - \tilde{\gamma}|_0.$$

By the contraction mapping theorem (Theorem C.5) there exists a unique map $g: B \to C^0(I, B')$ such that, writing $g(x) = \phi_x$ for all $x \in B$,

$$\phi_x(t) = x + \int_0^t f\phi_x(u) \, du$$

for all $t \in I$, or (equivalently) such that ϕ_x is an integral curve of f at x. The uniqueness of g gives the uniqueness of the integral curve at x, since another integral curve at x would give a possible alternative for the value $g(x)$.

Since χ is uniformly Lipschitz on the first factor, with constant 1, it follows by Theorem C.7 that g is Lipschitz. Thus ϕ', which is g composed with the continuous linear evaluation map $ev^t: C^0(I, B') \to B'$, is Lipschitz, where

$ev^t(\gamma) = \gamma(t)$. Moreover, since $|ev^t| = 1$, the Lipschitz constant of g is also a Lipschitz constant for each map ϕ^t.

Suppose, finally, that f is C^r. We observe that χ may be expressed as the sum of two maps. The first, taking (x, γ) to the constant map with value x, is continuous linear (and hence C^∞). The second is the composite of projection onto the second factor, the map

$$\gamma \mapsto f\gamma : C^0(I, B') \to C^0(I, E)$$

and the map

$$\mu \mapsto \left(t \mapsto \int_0^t \mu(u)\, du\right) : C^0(I, E) \to C^0(I, E).$$

The outside maps of the composite are continuous linear, while the middle one is C^r, by Theorem B.10. Thus, by Theorem C.7, g, and hence ϕ^t, is C^r.   □

(3.24) Exercise. The assumption $a < 1/\kappa$ in the proof of Theorem 3.22 was made to ensure that the maps χ^x were contractions. They lose this property for larger a, but regain it if they are conjugated with a suitable automorphism of the space $C^0(I, E)$. In fact, the map $\gamma \mapsto (t \mapsto \text{sech } \alpha t \cdot \gamma(t))$ has the desired effect for $\alpha > \kappa$. The exercise, then, is to eliminate the assumption $a < 1/\kappa$, by considering instead of χ, the map $\psi : B \times C \to C$ defined, for all $t \in I$, by

$$\psi(x, \delta)(t) = \text{sech } \alpha t \left(x + \int_0^t f(\cosh \alpha u \cdot \delta(u))\, du\right),$$

where C is the closed subset $\{\delta \in C^0(I, E) : \delta(t) \cosh \alpha t \in B' \text{ for all } t \in I\}$, the image of $C^0(I, B')$ under the above mentioned automorphism.

(3.25) Exercise. Derive a local integrability theorem for non-autonomous ordinary differential equations (see Note 3.17).

(3.26) Exercise. Show that the map $f : \mathbf{R} \to \mathbf{R}$ sending x to $x^{2/3}$ is not Lipschitz on any neighbourhood of 0, but that nevertheless f has a continuous local integral at 0.

The way in which the local integral ϕ depends separately on time t and initial position x emerges very clearly and easily from Proposition 3.15 and Theorem 3.22. One is naturally led to ask how ϕ depends on t and x together. Here the answers are just as simple. As we shall now prove, the map ϕ is locally Lipschitz, and it is as smooth as the function f.

(3.27) Theorem. *If $d' < \infty$ then the map ϕ of Theorem 3.22 is Lipschitz. In any case, it is locally Lipschitz.*

Proof. For all (t, x) and $(t', x') \in I \times B$,

(3.28) $|\phi(t', x') - \phi(t, x)| \leq \lambda |x' - x| + |\phi_x(t') - \phi_x(t)|,$

where λ is, for all $t \in I$, a Lipschitz constant for ϕ^t (which is uniformly Lipschitz by Theorem 3.22). When d' is finite, f is bounded (being Lipschitz) and thus so is $D\phi_x (= f\phi_x)$. Hence ϕ_x is uniformly (in x) Lipschitz, and we deduce from (3.28) that ϕ is Lipschitz. If d' is not finite, then (3.28) nevertheless gives that ϕ is continuous. Thus $f\phi$ is locally bounded, ϕ_x is locally uniformly Lipschitz and hence ϕ is locally Lipschitz.   □

(3.29) Exercise. Find an example of a Lipschitz map $f : \mathbf{E} \to \mathbf{E}$ for which there is no corresponding Lipschitz $\phi : I \times \mathbf{E} \to \mathbf{E}$, for any I.

(3.30) Exercise. Suppose that $\gamma : [0, \tau] \to \mathbf{R}$ is continuous, and that, for all $t \in [0, \tau]$,

$$\gamma(t) - \kappa \int_0^t \gamma(u) \, du \leq A + Bt + C e^{\alpha t},$$

where $\kappa > 0$, $\alpha \neq \kappa$ and where A, B and C are all constants. Prove that

$$\gamma(t) \leq A e^{\kappa t} + B\kappa^{-1} (e^{\kappa t} - 1) + C(\alpha - \kappa)^{-1}(\alpha e^{\alpha t} - \kappa e^{\kappa t}).$$

(This deduction is called "integrating the inequality".) Show that if $\phi : I \times U \to V$ is a local integral of a map $f : V \to \mathbf{E}$ with Lipschitz constant κ, then, for all $t \in I$, ϕ^t satisfies the inequality

$$\left| |\phi^t(x) - \phi^t(x')| - \kappa \left| \int_0^t |\phi^u(x) - \phi^u(x')| du \right| \right| \leq |x - x'|.$$

Deduce that ϕ^t is Lipschitz with constant $e^{\kappa |t|}$.

We now turn to the situation when f is smooth. The only substantial point is to show that the partial derivatives of the local integral ϕ with respect to x are continuous in t and x together. We deal with this first.

(3.31) Lemma. *Let f be C^r and let ϕ be the map of Theorem 3.22. Then the partial derivatives $(D_2)^s \phi : I \times B \to L_s(\mathbf{E}, \mathbf{E})$ exist and are continuous for $1 \leq s \leq r$.*

Proof. Since we may "slow down" f by multiplying it by a small positive constant α, we may assume that the Lipschitz constant κ of f is less than 1. The local integral for f is recovered by "speeding up" the local integral of αf. More precisely, we multiply the time parameter t by $1/\alpha$ and then apply the local integral of αf. This procedure does not affect the smoothness class of the integral. We leave the details as an exercise. We also continue to assume that $a < 1/\kappa$, having already indicated in Exercise 3.24 how to deal with this point.

We observe that the relation (3.23) defines a map

$$\chi : B \times C^1(I, B') \to C^1(I, B')$$

which is still a uniform contraction on the second factor. The extra calculation needed is

$$|D(\chi^x(\gamma) - \chi^x(\tilde{\gamma}))|_0 = |f\gamma - f\tilde{\gamma}|_0 \leq \kappa |\gamma - \tilde{\gamma}|_0,$$

and, by assumption, $\kappa < 1$. Moreover, χ_γ is still uniformly Lipschitz, and χ is itself C^r. To prove this latter assertion, consider the inclusion ι of $C^1(I, B')$ in $C^0(I, \mathbf{E})$, followed by $f_* : \text{im } \iota \to C^0(I, \mathbf{E})$. Since f_* is composition with f, $f_* \iota$ is C^r, by Theorem B.10. Also integrating back into $C^1(I, \mathbf{E})$ is a continuous linear operation, so $\chi : B \times C^1(I, B') \to C^1(I, B')$ is C^r as claimed. Thus the fixed point map $g : B \to C^1(I, B')$ corresponding to our new χ is C^r. As in Theorem 3.22, we define ϕ by $\phi(t, x) = g(x)(t)$. By the uniqueness of integral curves in Theorem 3.22, the present map ϕ is the map ϕ of Theorem 3.22.

Now for all $t \in I$, ϕ^t is the composite of $g : B \to C^1(I, B')$ and the continuous linear map $ev^t : C^1(I, B') \to B'$. Thus, for $1 \leq s \leq r$, $(D_2)^s \phi$ exists, and is given by

$$(D_2)^s \phi(t, x) = ev^t \circ D^s g(x)$$

(recall that $D^s g(x)$ is an s-linear map from $\mathbf{E}$ to $C^1(I, \mathbf{E})$ so we may talk of its composite with $ev^t : C^1(I, \mathbf{E}) \to \mathbf{E}$). Thus $(D_2)^s \phi = \text{comp}(ev^\cdot \times D^s g)$, where $ev^\cdot : I \to L(C^1(I, \mathbf{E}), \mathbf{E})$ is the continuous map of Corollary B.17 and comp is the continuous bilinear composition map from $L(C^1(I, \mathbf{E}), \mathbf{E}) \times L_s(\mathbf{E}, C^1(I, \mathbf{E}))$ to $L_s(\mathbf{E}, \mathbf{E})$. Thus $(D_2)^s \phi$ is continuous. □

Our main theorem on smoothness of local integrals now follows easily.

(3.32) Theorem. *If the map f of Theorem 3.22 is $C^r(r \geq 1)$, then the local integral ϕ is C^r. Moreover, $D_1(D_2)^r \phi$ exists and equals*

$$(D_2)^r(f\phi) : I \times \mathbf{E} \to L_s(\mathbf{E}, \mathbf{E}).$$

Proof. Let f be C^r. We prove by induction on s the statement that ϕ is C^s, for $0 \leq s \leq r$. By Theorem 3.27, ϕ is locally Lipschitz, and hence C^0. Now assume that ϕ is C^j, for some j with $0 \leq j < s$. We prove that ϕ is C^{j+1}, by showing that its jth partial derivatives are C^1. There are two cases.

(i) With the exception of $(D_2)^j \phi$, all jth partial derivatives can be expressed, by the symmetry of $D^j \phi$, as $(j-1)$st partial derivatives of $D_1 \phi$. Since $D_1 \phi = f\phi$, which is C^j, these partial derivatives are C^1.

(ii) By Lemma 3.31 we know that $(D_2)^j\phi$ is continuous, and also that $(D_2)^{j+1}\phi$ exists and is continuous. Since $(D_2)^j D_1\phi = (D_2)^j(f\phi)$ is continuous, $D_1(D_2)^j\phi$ exists and equals $(D_2)^j D_1\phi$. (To see this, observe that

$$(D_2)^j D_1\phi(t, x) = D_1 \int_0^t (D_2)^j D_1\phi(u, x)\, du$$

$$= D_1(D_2)^j \int_0^t D_1\phi(u, x)\, du = D_1(D_2)^j\phi(t, x)\,.\,)$$

Hence, by induction, ϕ is C^s, and hence C^r. Finally, since $(D_2)^r D_1\phi = (D_2)^r(f\phi)$ is continuous, $D_1(D_2)^r\phi$ exists and equals $(D_2)^r(f\phi)$. $\square$

Since any C^1 vector field is locally Lipschitz (Proposition C.2), we obtain the following local result for manifolds:

(3.33) Corollary. *Any C^r vector field $(r \geq 1)$ on a manifold has a C^r local integral at each point of the manifold.* $\square$

A very natural question arises at this stage. How does the local integral ϕ depend on the vector field v? If we make small perturbations of f and its derivatives, are the corresponding alterations in ϕ and its derivatives small? We deal with this problem in the appendix to this chapter.

V. GLOBAL INTEGRALS

Our main tool for extending local integrability results to global ones is the uniqueness of integral curves proved in Theorem 3.22. We first extend this into a global uniqueness theorem.

(3.34) Theorem. *Let v be a locally Lipschitz vector field on a manifold X, and let $\alpha: J \to X$ and $\beta: J \to X$ be integral curves of v. If, for some $t_0 \in J$, $\alpha(t_0) = \beta(t_0)$, then $\alpha = \beta$.*

Proof. Let $\alpha(t_0) = \beta(t_0)$, where $t_0 \in J$. We prove that $\alpha(t) = \beta(t)$ for all $t \in J$ with $t \geq t_0$; the proof for $t \leq t_0$ follows by reversing v (that is, replacing v by $-v$). Let $\tau = \sup\{t \in J: \alpha = \beta \text{ on } [t_0, t]\}$, and suppose that $\tau \in \text{int } J$. By the continuity of α and β, $\alpha(\tau) = \beta(\tau) = p$, say. Now identify a neighbourhood U of p in X with an open subset U' of the model space $\mathbf{E}$ of X, by a chart $\xi: U \to U'$. Choose balls B and B' in U with centre p and obtain a corresponding interval $I = [-a, a]$ as in Theorem 3.22. We may take a small enough for $\tau + a$ to be in J. If $\tilde{\alpha}: I \to X$ and $\tilde{\beta}: I \to X$ take t to $\alpha(t + \tau)$ and $\beta(t + \tau)$ respectively then $\tilde{\alpha} = \tilde{\beta}$ by Theorem 3.22, since both are integral curves at p with domain I. Thus $\alpha = \beta$ on $[t_0, \tau + a]$, contrary to the definition of τ. Thus τ is the right end-point of J. If $\tau \in J$ then, as above, $\alpha(\tau) = \beta(\tau)$. $\quad\square$

(3.35) Exercise. Let $f: \mathbf{R} \to \mathbf{R}$ take x to $x^{2/3}$. Find two different integral curves of f at 0 with the same domain.

(3.36) Exercise. Prove that all integral curves of a locally Lipschitz vector field v at a point p are constant functions if and only if p is a singular point of v (p is a *singular point*, or *zero*, of v if $v(p) = 0_p$).

Earlier in the chapter we commented that the velocity of a flow is independent of time, and so may be regarded as a vector field. Conversely, a local integral of a vector field is locally a flow in the following sense:

(3.37) Theorem. *Let $\phi: I \times U \to X$ be a local integral of a locally Lipschitz vector field on a manifold X. Then for all $x \in U$ and all $s, t \in I$ such that $s + t \in I$ and $\phi(t, x) \in U$, $\phi(s, \phi(t, x)) = \phi(s + t, x)$.*

Proof. Apply Theorem 3.34 to integral curves $\alpha: [0, s] \to X$ and $\beta: [0, s] \to X$ defined by $\alpha(u) = \phi(u, \phi(t, x))$ and $\beta(u) = \phi(u + t, x)$. Since α and β agree at 0, they agree at s. $\quad\square$

(3.38) Corollary. *If $\phi: \mathbf{R} \times X \to X$ is a local integral of a locally Lipschitz vector field on a manifold X, then it is a flow on X.* $\quad\square$

(3.39) Theorem. *Any Lipschitz vector field v on a Banach space $\mathbf{E}$ has a unique integral flow ϕ. If v is C^r then ϕ is C^r.*

Proof. For all $a > 0$, v has a unique local integral $\phi: [-a, a] \times \mathbf{E} \to \mathbf{E}$, by Theorem 3.22. Any two such agree on the intersection of their domains. Hence the collection of all such local integrals defines a local integral $\phi: \mathbf{R} \times \mathbf{E} \to \mathbf{E}$, which is a flow by Corollary 3.38. If v is C^r then ϕ is C^r by Theorem 3.32 and uniqueness. $\quad\square$

(3.40) Exercise. (i) Prove that the vector field on $\mathbf{R}$ with principal part $f(x) = x^2$ does not have an integral flow. (*Hint:* The vector field is so large that the orbits escape to infinity in finite time.)

(ii) Prove that any C^1 vector field v on a manifold X gives rise to a partial flow $\phi: D \to X$ on X, such that, for all $x \in X$, $\phi_x: D_x \to X$ is the maximal integral curve of v at x.

We have been at some pains to prove that the local integrals that we have obtained in our existence theorems are as smooth as the vector fields we integrate. This behaviour is completely general, as we now show.

(3.41) Theorem. *Any local integral of a C^r vector field $(r \geqslant 1)$ on a manifold X is C^r. Similarly any integral of a locally Lipschitz vector field is locally Lipschitz.*

Proof. Let v be a C^r vector field on X, and let $\phi : I \times U \to X$ be a local integral of v. Let $a > 0$ be the right end-point of I (possibly $a = \infty$). Fix $x_0 \in U$, and let $T = \{t > 0 : \phi$ is C^r at (u, x_0) for all $u \in [0, t]\}$. By Corollary 3.33, T is non-empty. Let $\tau = \sup T$. It is enough to show that $\tau = a$, and that $a \in T$ if $a \in I$, for then a similar argument for negative t completes a proof that ϕ is C^r at all points of $I \times \{x_0\}$.

Suppose that $\tau \in \operatorname{int} I$. By Corollary 3.33 there exists a C^r local integral $\psi :]-b, b[\times V \to X$ of v at $\phi(\tau, x_0)$, for some $b > 0$. By continuity of ϕ_{x_0} at τ, there is some $t_0 \in T$ with $t_0 > \tau - b$ and $\phi(t_0, x_0) \in V$. By continuity of ϕ^{t_0} at x_0 there is some neighbourhood W of x_0 in U with $\phi^{t_0}(W) \subset V$. Let $c = \min\{b, a - t_0\}$, so that $t_0 + c > \tau$. Then, for all $x \in W$ and all u with $0 \leqslant u < c$, $\phi(t_0 + u, x) = \psi(u, \phi(t_0, x))$. To check this, we have only to fix x and apply Theorem 3.34 to integral curves α and β whose values at $u \in [0, c[$ are respectively $\phi(t_0 + u, x)$ and $\psi(u, \phi(t_0, x))$. Since ψ is C^r and ϕ^{t_0} is C^r at x_0, it follows that ϕ is C^r at $(t_0 + u, x_0)$ for all $u \in]0, c[$, which contradicts the definition of τ. Thus $\tau = a$.

If $a \in I$, we obtain ψ, t_0 and W as above, and observe that, for all $x \in W$ and all u with $0 \leqslant u \leqslant a - t_0$, $\phi(t_0 + u, x) = \psi(u, \phi(t_0, x))$. This shows that ϕ is C^r at (a, x_0), as required.

An identical approach works for the locally Lipschitz case. $\square$

To complete our integration programme, we prove that every smooth vector field on a compact manifold has an integral flow. We make use of a lemma which is of some interest in its own right.

(3.42) Lemma. *Let v be a locally Lipschitz vector field on a manifold X. Suppose, that for some $a > 0$ and for all $x \in X$, v has an integral curve at x with domain $[-a, a]$. Then v has a unique integral flow ϕ, which is locally Lipschitz. Furthermore, if v is $C^r (r \geqslant 1)$ then ϕ is C^r.*

Proof. We have to show that for all $x \in X$ there is an integral curve $\phi_x : \mathbf{R} \to X$ of v at x. Suppose that for some x this is not the case. Reversing v if necessary, we have that

$$\tau = \sup\{t \in \mathbf{R} : v \text{ has an integral curve at } x \text{ with domain } [0, t]\}$$

is finite. By hypothesis $\tau > 0$. Choose t_0 with $\tau - a < t_0 < \tau$. Then there is an integral curve $\gamma : [0, t_0] \to X$ at x, and, by hypothesis, an integral curve $\delta : [-a, a] \to X$ at $\gamma(t_0)$. Now $\tilde{\gamma}$ defined by $\tilde{\gamma}(t) = \delta(t - t_0)$ is an integral curve on $[t_0 - a, t_0 + a]$. Using Theorem 3.34, we may extend the domain of γ to $[0, t_0 + a]$ by defining $\gamma = \tilde{\gamma}$ on $[t_0 - a, t_0 + a]$. This contradicts the definition of τ. We have, then, a local integral $\phi : \mathbf{R} \times X \to X$, and its various properties come from Theorem 3.34, Corollary 3.38 and Theorem 3.41. $\square$

(3.43) Theorem. *Any locally Lipschitz vector field v on a compact manifold*

X has a unique integral flow φ, which is locally Lipschitz. If v is $C^r (r \geq 1)$ then φ is C^r.

Proof. For all $x \in X$, there is, by Theorem 3.22, a local integral at x. Let this be $\phi_x: I_x \times U_x \to X$, where $I_x = [-a_x, a_x]$, with $a_x > 0$, and U_x is an open neighbourhood of x in X. Choose a finite subcovering $U_{x_1}, \ldots, U_{x_n}$ of the open covering $\{U_x : x \in X\}$ of X, and let $a = \min\{a_{x_1}, \ldots, a_{x_n}\}$. Then any point x of X is in some U_{x_i}, and the corresponding ϕ_{x_i} provides an integral curve at x with domain $[-a_{x_i}, a_{x_i}]$, and hence, by restriction, one with domain $[-a, a]$. The result now follows from Lemma 3.42.   □

(3.44) Exercise. One does not normally expect a vector field on a non-compact manifold to have an integral flow, since integral curves may not be defined on the whole of **R** (see Exercise 3.40). However, we can remedy this situation without affecting the orbit structure as follows. Let X be a (paracompact) manifold admitting C^r partitions of unity $(r \geq 1)$. Let v be a C^r vector field on X. Prove that for some positive C^r function $\alpha: X \to \mathbf{R}$, the vector field $x \to \alpha(x)v(x)$ has an integral flow $\phi: \mathbf{R} \times X \to X$ with the same orbits as v. (Here the *orbit* of v *through* x is defined to be the image of the integral curve of v at x with maximal domain.) A solution to this problem may be found in Renz [1].

By virtue of the correspondence between velocity vector fields and integral flows, the qualitative theories of smooth vector fields and smooth flows are one and the same subject. For example we define two smooth vector fields to be *flow equivalent* if their integral flows are flow equivalent and *topologically equivalent* if, after the modification in Exercise 3.44 if necessary, they have topologically equivalent integral flows. It is sometimes easier to describe the theory in terms of vector fields, sometimes in terms of flows. We are free to use whichever terminology we find the more convenient in any given situation.

Appendix 3

I. INTEGRALS OF PERTURBED VECTOR FIELDS

We have already found spaces of maps to be a great technical convenience, and we now put them to a new use. We discuss the map that takes vector fields, regarded as points in a map space, to their local integrals, similarly interpreted. The properties of this map are clearly of considerable interest. For example, its continuity at a point (= vector field) corresponds to small perturbations in the vector field producing small perturbations in the integral. In this context, the precise nature of a "small perturbation" depends, of course, on the topology that we put on the map spaces. We do not attempt a comprehensive treatment of the subject, but give instead a couple of typical theorems that can be proved along these lines.

Suppose that notations are as in Theorem 3.22. For the purposes of the following theorem $N'(r \geqslant 0)$ denotes the subset of $C'(B', \mathbf{E})$ consisting of Lipschitz maps h with $|h|_0 < ((d'-d)/a) - |f|_0$ when $d' < \infty$. Thus if f is C' and $h \in N'$, $f + h$ has, by Theorem 3.32 a C' integral $\psi : I \times B \to B'$, say.

(3.45) Theorem. *Suppose that f is $C'(r \geqslant 0)$, with Df C'^{-1}-bounded if $r \geqslant 1$. Then $g^h - g^0$ is C'-bounded, where $g^h : B \to C^0(I, B')$ denotes the map taking x to ψ_x. If, further, f is uniformly C' then the map $\alpha : N' \to C'(B, C^0(I, \mathbf{E}))$ taking h to $g^h - g^0$ is continuous at 0. In this case, for all $t \in I$, the map $\beta : N' \to C'(B, \mathbf{E})$ taking h to $\psi^t - \phi^t$ is continuous at 0.*

Proof. We may assume that the Lipschitz constant of $f + h$ is strictly less than $1/a$ (otherwise we apply the technique of Exercise 3.24). We define a map

$$\chi : N' \times B \times C^0(I, B') \to C^0(I, \mathbf{E})$$

by

$$\chi(h, x, \gamma)(t) = x + \int_0^t (f+h)\gamma(u)\, du,$$

for $t \in I$. For each $h \in N^r$, we have, as in the proof of Theorem 3.22, a fixed point map $g^h : B \to C^0(I, B')$. Now $\chi^h - \chi^0$ is bounded by $a|h|_0$, and $D\chi^h$ is C^{r-1}-bounded, by Corollary B.11. Thus, by Theorem C.10, $g^h - g^0$ is C^r-bounded. We may express $\chi^h - \chi^0$ as a composite

$$B \times C^0(I, B') \longrightarrow C^0(I, B') \overset{h_*}{\longrightarrow} C^0(I, \mathbf{E}) \longrightarrow C^0(I, \mathbf{E}),$$

where the first map is the product projection, and the third is the continuous linear integration map sending γ to $(t \mapsto \int_0^t \gamma(u)\, du)$. By Corollary B.12, the map for N^r to $C^r(C^0(I, B'), C^0(I, \mathbf{E}))$ taking h to h_* is continuous linear. Hence by Lemmas B.4 and B.13, so is the map from N^r to

$$C^r(B \times C^0(I, B'), C^0(I, \mathbf{E}))$$

taking h to $\chi^h - \chi^0$. Suppose now that f is uniformly C^r. Then χ^0 is uniformly C^r, by Theorem B.10. Thus we may apply Theorem C.10 to deduce that α is continuous at 0. So also is $\beta = (ev^t)_*\alpha$, where

$$ev^t : C^0(I, B') \to B'$$

is the continuous linear evaluation map.

The situation with purely Lipschitz perturbations is not so clear. However, we shall only make use of the following results, which gives continuity at f of the operation of taking local integrals in the case when f is linear.

(3.46) Theorem. *Let f be a continuous linear endomorphism of $\mathbf{E}$ and let $h : V \to \mathbf{E}$ be Lipschitz, where V is open in $\mathbf{E}$. Suppose that $\phi : I \times U \to V$ and $\psi : I \times U \to V$ are local integrals of f and $f + h$ respectively. Then, for all $t \in I$, $\psi^t - \phi^t$ is Lipschitz, with a constant that tends to zero with the Lipschitz constant of h.*

Proof. We may suppose that $t > 0$ (otherwise reverse the vector fields). Let θ^t denote $\psi^t - \phi^t$. For all $x, x' \in U$,

$$|\theta^t(x) - \theta^t(x')| = \left| \int_0^t ((f+h)\psi^u(x) - f\phi^u(x) - (f+h)\psi^u(x') + f\phi^u(x'))\, du \right|$$

$$\leqslant \kappa \int_0^t |\theta^u(x) - \theta^u(x')|\, du + \lambda \int_0^t |\psi^u(x) - \psi^u(x')|\, du,$$

where κ and λ are positive Lipschitz constants for f and h respectively. Let $\mu = \kappa + \lambda$. Then μ is a Lipschitz constant for $f + h$, and thus, by Exercise 3.30, ψ^u is Lipschitz with constant $e^{\mu u}$. Thus

$$|\theta^t(x) - \theta^t(x')| - \kappa \int_0^t |\theta^u(x) - \theta^u(x')|\, du \leqslant \mu^{-1}(e^{\mu t} - 1)\lambda |x - x'|.$$

Integrating this inequality (see Exercise 3.30 again), we obtain

$$|\theta'(x) - \theta'(x')| \leqslant \mu^{-1} \left(\frac{\mu e^{\mu t} - \kappa e^{\kappa t}}{\mu - \kappa} - e^{\kappa t} \right) \lambda |x - x'| = (e^{\mu t} - e^{\kappa t})|x - x'|,$$

and so θ' is Lipschitz with constant $e^{\mu t} - e^{\kappa t}$. The proof is now complete, since $\mu \to \kappa$ as $\lambda \to 0$. $\square$

II. FIRST INTEGRALS

Let v be a vector field on a differentiable manifold X and let $g: X \to R$ be a C^r function ($r \geqslant 1$). The *derivative of* g in *the direction of* v is the C^{r-1} function $L_v g$ defined by

$$Tg(v(x)) = (g(x), L_v g(x)).$$

If γ is an integral curve of v at x, then $L_v g(x)$ is, by the chain rule, the derivative with respect to t at $t = 0$ of $f\gamma(t)$. In fact, it is the rate of change of f at x at any time t as we move along an integral curve through x with the prescribed velocity v. If X has a given Riemannian structure $\langle \ , \ \rangle$ then the gradient vector field ∇g of g is defined as in Example 3.3, and in terms of this $L_v g(x) = \langle v(x), \nabla g(x) \rangle$. Thus, if X is Euclidean space $\mathbf{R}^n$,

$$L_v g(x) = v_1(x) \frac{\partial g}{\partial x_1} + \cdots + v_n(x) \frac{\partial g}{\partial x_n}.$$

This explains the notation

$$v_1(x) \frac{\partial}{\partial x_1} + \cdots + v_n(x) \frac{\partial g}{\partial x_n}$$

sometimes used for a vector field on $\mathbf{R}^n$. The notation is justifiable since we can recover v from L_v by applying L_v to the coordinate functions $x \mapsto x_i$. In fact, the notion of a vector field is quite commonly defined in terms of the directional derivative property (see, for example, Warner [1]).

A *first integral* of v is a C^1 function $g: X \to \mathbf{R}$ such that $L_v g = 0$, the zero function. Thus a first integral is constant along any integral curve of v. Of course all constant functions on X are trivially first integrals of any vector field X. For many vector fields these are the only first integrals. When a non-constant first integral g does exist, it is of considerable interest. The reason for this is that, by Sard's theorem the level sets of g are usually submanifolds of X of codimension 1 (see Theorem 7.3 and Example A.7, below, and Hirsch [1]). Since the integral curve at any point lies in the level set through that point, $v(x)$ is tangential to the submanifold at every point x. Thus for each non-critical value of g, we obtain a vector field on a manifold of dimension 1 less than X (assuming X to be finite dimensional). Since this ought to be easier to integrate than the original vector field v on X, we have made a reasonable first attempt at integrating v, hence the name *first integral* given to g.

(3.47) Exercise. Prove that the vector field on **R** with principal part $f(x) = x$ has no non-constant first integral.

(3.48) Exercise. Find a non-constant first integral for the vector field on $\mathbf{R}^2$ with principal part (i) $f(x, y) = (x, -y)$, (ii) $f(x, y) = (-y, x)$.

(3.49) Exercise. Prove that $g: \mathbf{R}^2 \to \mathbf{R}$ defined by $g(x, y) = x^2 + y^2$ is a first integral for the vector fields on $\mathbf{R}^2$ with principal part
   (i) $f_1(x, y) = (-y(x^2 + y^2), x(x^2 + y^2))$,
   (ii) $f_2(x, y) = (-xy^2, x^2 y)$,
   (iii) $f_3(x, y) = (-x^2 y^3, x^3 y^2)$,
   (iv) $f_4(x, y) = (x^3 y^4, -x^4 y^3)$.
Sketch the phase portraits of the vector fields. Are any of the vector fields topologically equivalent?

(3.50) Exercise. Let $U = \{(x, y) \in \mathbf{R}^2 : x \neq 0\}$. Prove that the function $f: U \to \mathbf{R}$ defined by $f(x, y) = (x^2 + y^2)/x$ is a first integral for the vector field on $\mathbf{R}^2$ with principal part $v(x, y) = (2xy, y^2 - x^2)$. Sketch the phase portrait of the vector field.

Prove that the function $g = \mathbf{R}^3 \to \mathbf{R}$ defined by $g(x, y, z) = x^2 + y^2 + z^2$ is a first integral of the vector field on $\mathbf{R}^3$ with principal part

$$w(x, y, z) = 2xyz, (y^2 - x^2)z, -(x^2 + y^2)y).$$

By comparing v and w, write down another non-constant first integral $h: U \times \mathbf{R} \to \mathbf{R}$ of w. Sketch the phase portrait of w on the hemisphere $x^2 + y^2 + z^2 = 1$, $x \geq 0$, as viewed from a point $(a, 0, 0)$ for large positive a.

(3.51) Example. The Hamiltonian $H = T + V$ is a first integral of the Hamiltonian vector field of conservative mechanics (see Example 3.19). The *principle of conservation of energy* is the observation that H is constant along integral curves of the vector field.

Although a vector field may have no non-constant global first integrals, it always has non-constant *local first integrals* in the neighbourhood of any non-singular point. This is an easy consequence of the rectification theorem (Theorem 5.8 below).

CHAPTER 4

Linear systems

In a naive approach to the theory of dynamical systems on Euclidean space $\mathbf{R}^n$, we might hope to obtain a complete classification of all systems, starting with the very simplest types and gradually building up an understanding of more complicated ones. The simplest conceivable diffeomorphisms of R^n are *translations*, maps of the form $x \mapsto x + p$ for some constant $p \in R^n$. It is an easy exercise to show that two such maps are *linearly conjugate* (topologically conjugate by a linear automorphism of $\mathbf{R}^n$) providing the constant vectors involved are non-zero. Similarly, any two non-zero constant vector fields on $\mathbf{R}^n$ are linearly flow equivalent. The next simplest systems are linear ones, discrete dynamical systems generated by linear automorphisms of $\mathbf{R}^n$ and vector fields on $\mathbf{R}^n$ with linear principal part. Rather surprisingly, the classification problem for such systems is far from trivial. It has been completely solved for vector fields (Kuiper [2]) but not as yet for diffeomorphisms, although Kuiper and Robbin [1] have made a great deal of progress in this latter case.

The difficulties that one encounters even at the linear stage underline the complexity and richness of the theory of dynamical systems, and this is part of its attraction. They also indicate the need for a less ambitious approach to the whole classification problem. Instead of aiming at a complete solution, we attempt to classify a "suitably large" class of dynamical systems. In the case of linear systems on $\mathbf{R}^n$, it is easy to give a precise definition of what we mean by "suitably large", as follows.

The set $L(\mathbf{R}^n)$ of linear endomorphisms of $\mathbf{R}^n$ is a Banach space with the *induced norm* given by

$$|T| = \sup \{|T(x)|: x \in \mathbf{R}^n, |x| = 1\}.$$

A "suitably large" subset $\mathcal{S}$ of linear vector fields on $\mathbf{R}^n$ is one that is both open and dense in $L(\mathbf{R}^n)$ (from now on we shall not usually bother to distinguish between a vector field on $\mathbf{R}^n$ and its principal part). The density of such a set $\mathcal{S}$ implies that we can approximate any element of $L(\mathbf{R}^n)$ arbitrarily closely by an element of $\mathcal{S}$; the fact that $\mathcal{S}$ is open implies that any element of $\mathcal{S}$ remains in $\mathcal{S}$ after any sufficiently small perturbation. In general, of these two desirable properties we tend to regard density as essential and openness as negotiable.

The set $GL(\mathbf{R}^n)$ of linear automorphisms of $\mathbf{R}^n$ inherits the subspace topology from $L(\mathbf{R}^n)$. "Suitably large" in the context of discrete linear systems means open and dense in $GL(\mathbf{R}^n)$.

How, then, do we select a candidate for the set $\mathcal{S}$? There are very good reasons for our choice, as follows. Suppose that X is a topological space and that $\sim$ is an equivalence relation on X. We say that a point $x \in X$ is *stable* with respect to $\sim$ if x is an interior point of its $\sim$ class. The *stable set* Σ of $\sim$ is the set of all stable points in X. Of course Σ is automatically an open subset of X. Moreover, if the topology of X has a countable basis, then Σ contains points of only countably many equivalence classes. Thus there is some chance that Σ will be classifiable. The snag is that Σ may fail to be dense in X. In the present context, $X = GL(\mathbf{R}^n)$ (or $L(\mathbf{R}^n)$), $\sim$ is topological conjugacy (or topological equivalence), and Σ is called the set of *hyperbolic* linear automorphisms (or the *hyperbolic* linear vector fields). In each case the stable set Σ is dense and is easily classifiable, as we shall see.

There is another vital reason for considering these particular subsets: in the next chapter we shall find that hyperbolic linear automorphisms (and hyperbolic linear vector fields) are stable not only under small linear perturbations but also, more generally, under small smooth perturbations. Now the basic idea of differential calculus is one of approximating a function locally by a linear function. Similarly, when discussing the local nature of a dynamical system we consider its "linear approximation", which is a certain linear system. The important question is whether this linear system is a good approximation in the sense that it is (locally) qualitatively the same as the original system. This question is obviously bound up with the stability (in the wider sense) of the linear system. In fact, we shall obtain a satisfactory linear approximation theory precisely when the linear system is stable.

We begin with a review of linear systems on $\mathbf{R}^n$. The bulk of the chapter, however, deals with stable linear systems on a general Banach space. For this we need a certain amount of spectral theory, which we include in an appendix to the chapter. The material of $\mathbf{R}^n$ provides a background of concrete examples against which the more general results on Banach spaces may be tested and placed in perspective.

I. LINEAR FLOWS ON $\mathbf{R}^n$

Let T be any linear endomorphism of $\mathbf{R}^n$. Then we may think of T as a vector field on $\mathbf{R}^n$, in which case we call it a *linear vector field*. The corresponding ordinary differential equation is

(4.1) $x' = T(x)$,

where $x \in \mathbf{R}^n$. One may immediately write down the integral flow. Since $L(\mathbf{R}^n)$ is a Banach space, it makes sense to talk of the infinite series

$$\exp(tT) = id + tT + \frac{1}{2} t^2 T^2 + \cdots + \frac{1}{n!} t^n T^n + \cdots$$

where *id* is the identity map on $\mathbf{R}^n$ and $t \in \mathbf{R}$. This series converges for all t and T, and the integral flow ϕ of T is given by

(4.2) $\phi(t, x) = \exp(tT)(x)$,

(term-by-term differentiation makes this plausible; another proof is indicated later in the chapter). We call a flow ϕ *linear* if ϕ^t is a linear automorphism varying smoothly with t. Thus ϕ is linear if and only if its velocity vector field is linear.

(4.3) **Example.** If $T = a(id)$ for some $a \in R$ then

$$\exp(tT) = id + at(id) + \cdots + \frac{1}{n!}(at)^n id^n + \cdots$$

$$= (1 + at + \cdots + \frac{1}{n!}(at)^n + \cdots)id$$

$$= e^{at}id.$$

(4.4) **Example.** If T has matrix

$$\begin{bmatrix} 0 & 1 \\ 0 & 0 \end{bmatrix} \text{ then } T^2 \text{ has matrix } \begin{bmatrix} 0 & 0 \\ 0 & 0 \end{bmatrix},$$

so the series for $\exp(tT)$ terminates as $id + tT$.

We may define a notion of *linear equivalence* for linear flows, as follows. Let ϕ and ψ be linear flows with velocities S and $T \in L(\mathbf{R}^n)$ respectively. Then ψ is *linearly equivalent* to ϕ, written $\psi \sim_L \phi$, if for some $\alpha \in \mathbf{R}$, $\alpha > 0$, and some linear automorphism $h \in GL(\mathbf{R}^n)$ of $\mathbf{R}^n$, the diagram

(4.5)

$$
\begin{array}{ccc}
\mathbf{R} \times \mathbf{R}^n & \xrightarrow{\ \ \phi\ \ } & \mathbf{R}^n \\
\Big\downarrow{\scriptstyle \alpha \times h} & & \Big\downarrow{\scriptstyle h} \\
\mathbf{R} \times \mathbf{R}^n & \xrightarrow[\ \ \psi\ \ ]{} & \mathbf{R}^n
\end{array}
$$

commutes. That is, $h\phi(t, x) = \psi(\alpha t, h(x))$ for all $(t, x) \in \mathbf{R} \times \mathbf{R}^n$. Equivalently $\psi \sim_L \phi$ if and only if $S = \alpha(h^{-1}Th)$. The problem of classifying linear flows up to linear equivalence is therefore the same as the problem of classifying $L(\mathbf{R}^n)$ up to similarity (linear conjugacy). Now the latter is solved by the theory of real Jordan canonical form (see Hirsch and Smale [1] or Gantmacher [1]) which we recall briefly.

Every $T \in L(\mathbf{R}^n)$ is similar to a direct sum $T_1 \oplus \cdots \oplus T_q$ of linear endomorphisms $T_j \in L(V_j)$, $j = 1, \ldots, q$, where $V_1 \oplus \cdots \oplus V_q$ is some direct sum decomposition of $\mathbf{R}^n$, and, for some basis of V_j, the matrix of T_j takes the form

$$
M_j = \begin{bmatrix}
\lambda_j & 1 & 0 & \cdot & & \cdot & \cdot & 0 \\
0 & \lambda_j & 1 & 0 & \cdot & \cdot & \cdot & 0 \\
\cdot & \cdot & \cdot & \cdot & \cdot & \cdot & \cdot & \cdot \\
0 & \cdot & \cdot & \cdot & \cdot & 0 & \lambda_j & 1 \\
0 & \cdot & \cdot & \cdot & \cdot & \cdot & 0 & \lambda_j
\end{bmatrix}
$$

where $\lambda_j \in \mathbf{C}$. For each M_j there are two possibilities:

(i) $\lambda_j \in \mathbf{R}$, in which case the entries of M_j are real numbers, and T_j is associated with dim V_j repeated eigenvalues λ_j of T.

(ii) $\lambda_j = a + ib$, where $a, b \in \mathbf{R}$, $b > 0$, in which case the entries of M_j are real 2×2 submatrices

$$
\lambda_j = \begin{bmatrix} a & -b \\ b & a \end{bmatrix}, \qquad 1 = \begin{bmatrix} 1 & 0 \\ 0 & 1 \end{bmatrix} \quad \text{and} \quad 0 = \begin{bmatrix} 0 & 0 \\ 0 & 0 \end{bmatrix},
$$

and T_j is associated with $\frac{1}{2}$ dim V_j pairs of repeated complex conjugate eigenvalues λ_j and $\bar{\lambda}_j$ of T.

Notice that there may be more than one T_j having the same eigenvalue λ_j. The set of numbers dim V_j and the corresponding eigenvalues λ_j (or λ_j, $\bar{\lambda}_j$) determine the similarity class of T.

It follows that the integral flow ϕ of T can be decomposed as

$$
\phi_1 \oplus \cdots \oplus \phi_q, \qquad \text{where} \qquad \phi_j(t, x_j) = \exp{(tT_j)}(x_j),
$$

and $x = (x_1, \ldots, x_q) \in V_1 \oplus \cdots \oplus V_q = \mathbf{R}^n$. With respect to the above basis of V_j, $(\phi_j)^t \colon V_j \to V_j$ has matrix

$$N_j = e^{at}\begin{bmatrix} e^{ibt} & t\,e^{ibt} & \cdot & \cdot & \cdot & \cdot & \dfrac{t^{m-1}\,e^{ibt}}{(m-1)!} \\ 0 & e^{ibt} & t\,e^{ibt} & \cdot & \cdot & \cdot & \cdot \\ \cdot & \cdot & \cdot & \cdot & \cdot & \cdot & \cdot \\ 0 & \cdot & \cdot & \cdot & 0 & e^{ibt} & t\,e^{ibt} \\ 0 & \cdot & \cdot & \cdot & \cdot & 0 & e^{ibt} \end{bmatrix},$$

where $\lambda_j = a + ib$, $a, b \in \mathbf{R}$, and, if $b > 0$, $(t^r/r!)\,e^{ibt}$ must be interpreted as the 2×2 block

$$\frac{t^r}{r!}\begin{bmatrix} \cos bt & -\sin bt \\ \sin bt & \cos bt \end{bmatrix}.$$

Thus, for example if $\dim V_j = 4$ and $\lambda_j = i$, then

$$M_j = \begin{bmatrix} 0 & -1 & 1 & 0 \\ 1 & 0 & 0 & 1 \\ 0 & 0 & 0 & -1 \\ 0 & 0 & 1 & 0 \end{bmatrix}$$

and

$$N_j = \begin{bmatrix} \cos t & -\sin t & t\cos t & -t\sin t \\ \sin t & \cos t & t\sin t & t\cos t \\ 0 & 0 & \cos t & -\sin t \\ 0 & 0 & \sin t & \cos t \end{bmatrix}.$$

Let us examine the case $a < 0$, $b = 0$, $\dim V_j = 2$. The phase portrait in the plane V_j is illustrated in Figure 4.6. Every orbit has the point 0 as its unique ω-limit point. This leads us to suspect that any two negative values of a give

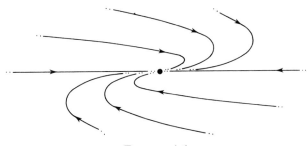

FIGURE 4.6

topologically equivalent flows. How can we prove this? Suppose it were true that for some value of a the distance $|t . x|$ decreased with t. Then we could define a map $h\colon V_j \to V_j$ as the identity on 0 and on the unit circle S^1 and, for all $x \in S^1$ taking $t . x$ to $e^{-t}x$ (see Figure 4.7). It is not hard to show that h is a

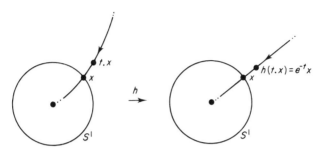

FIGURE 4.7

homeomorphism, and hence that it is a flow equivalence from the given vector field to the vector field $-id$ on V_j. Now $|t . x|$ decreases if and only if the scalar product of x and $T_j(x)$ is negative definite. This quadratic form is $a(x_1^2 + x_2^2) + x_1 x_2$ which is certainly *not* negative definite for small negative a. However we now note that, for any $\varepsilon > 0$, a linear conjugacy with

$$\begin{bmatrix} 1 & 0 \\ 0 & \varepsilon \end{bmatrix} \text{ changes } \begin{bmatrix} a & 1 \\ 0 & a \end{bmatrix} \text{ to } \begin{bmatrix} a & \varepsilon \\ 0 & a \end{bmatrix},$$

and, for given a, we can always take ε small enough for the form

$$a(x_1^2 + x_2^2) + \varepsilon x_1 x_2$$

to be negative definite. Thus

$$\begin{bmatrix} a & \varepsilon \\ 0 & a \end{bmatrix} \text{ is flow equivalent to } \begin{bmatrix} -1 & 0 \\ 0 & -1 \end{bmatrix}, \text{ and hence so is } \begin{bmatrix} a & 1 \\ 0 & a \end{bmatrix},$$

since linear conjugacy trivially implies flow equivalence. Thus, as we suspected, negative values of a give topologically equivalent vector fields.

We may go considerably further than this, using similar techniques. We may take the direct sum V_- of all subspaces corresponding to eigenvalues ξ_j with negative real part, and show that the vector field on this subspace is flow equivalent to the vector field $-id$. Thus any two linear *contractions* on a subspace of R^n are flow equivalent. Similarly for positive eigenvectors we get a vector field on a subspace V_+ which is flow equivalent to the vector field *id*. We do not go into details at this stage, since we prove a more general version later in the chapter.

There remains the linear subspace V_0 obtained by combining those summands V_j for which the real part of λ_j is zero. If $V_0 = \{0\}$, then $\mathbf{R}^n = V_+ \oplus V_-$ and the vector field T is flow equivalent to the vector field on $\mathbf{R}^n$ with matrix $I_r \oplus -I_s = \text{diag}\,(1, \ldots, 1, -1, \ldots, -1)$, where $r = \dim V_+$ and $s = \dim V_-$. Such a linear vector field T is called a *hyperbolic linear vector field* and its integral flow ϕ is called a *hyperbolic linear flow*. Hyperbolic linear vector fields are sometimes referred to as *elementary* linear vector fields, which explains our notation $EL(\mathbf{R}^n)$ for the set of all hyperbolic linear vector fields on $\mathbf{R}^n$.

We do not give a detailed analysis of the flow on V_0, since it would involve a fair amount of work and the hyperbolic case is of more fundamental importance. Instead we state the main conclusion reached by Kuiper. The fact is that if ϕ and ψ are linear flows given by linear endomorphisms S and T of $\mathbf{R}^n$ whose eigenvalues all have zero real part then ϕ is topologically equivalent to ψ if and only if ϕ is linearly equivalent to ψ. More generally, two linear flows ϕ and ψ on $\mathbf{R}^n$ with decompositions $U_+ \oplus U_- \oplus U_0$ and $V_+ \oplus V_- \oplus V_0$ are topologically equivalent if and only if $\phi|U_0$ is linearly equivalent to $\psi|V_0$, $\dim U_+ = \dim V_+$ and $\dim U_- = \dim V_-$. For full details see Kuiper [2].

(4.8) Example. *Linear flows on $\mathbf{R}$.* Any linear endomorphism of the real line $\mathbf{R}$ is of the form $x \mapsto ax$ for some real number a, which is the eigenvalue of the map. Thus the map is a hyperbolic vector field if and only if $a \neq 0$. The integral flow is $t \cdot x = x e^{at}$, and there are exactly three topological equivalence classes, corresponding to the cases $a < 0$, $a = 0$ and $a > 0$ (see Figure 4.8).

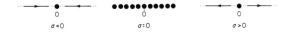

$$a<0 \qquad\qquad a=0 \qquad\qquad a>0$$

FIGURE 4.8

(4.9) Example. *Linear flows on $\mathbf{R}^2$.* The real Jordan form for a real 2×2 matrix is one of the following three types:

$$\text{(i)}\ \begin{bmatrix} \lambda & 0 \\ 0 & \mu \end{bmatrix}, \qquad \text{(ii)}\ \begin{bmatrix} \lambda & 1 \\ 0 & \lambda \end{bmatrix}, \qquad \text{(iii)}\ \begin{bmatrix} a & -b \\ b & a \end{bmatrix},$$

where λ, μ, a, b are real numbers and $b > 0$. The corresponding eigenvalues are λ, μ and $a \pm ib$. It follows that a linear vector field on $\mathbf{R}^2$ is hyperbolic if and only if $\lambda\mu \neq 0$ (case (i)), $\lambda \neq 0$ (case (ii)) or $a \neq 0$ (case (iii)). We have three topological equivalence classes of hyperbolic flows, and Figure 4.9

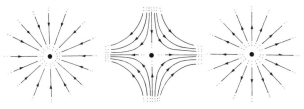

FIGURE 4.9

illustrates representatives

$$\begin{bmatrix} -1 & 0 \\ 0 & -1 \end{bmatrix}, \quad \begin{bmatrix} 1 & 0 \\ 0 & -1 \end{bmatrix} \text{ and } \begin{bmatrix} 1 & 0 \\ 0 & 1 \end{bmatrix}$$

of these classes. (N.b. Many classical texts on ordinary differential equations employ a finer classification than ours, giving rise to terminology such as *proper node, improper node* and *focus*.) There are also five topological equivalence classes of non-hyperbolic linear flows. Case (i) with $\mu = 0$ gives three classes corresponding to $\lambda < 0$, $\lambda = 0$ and $\lambda > 0$. If $\lambda = 0$ we have the trivial flow; the other two cases are illustrated in Figure 4.10. These three are, of course, product flows. The remaining two, which are irreducible, are case (ii) with $\lambda = 0$ and case (iii) with $a = 0$ (see Figure 4.11).

$\lambda < 0 \quad , \quad \mu = 0$ $\lambda > 0 \quad , \quad \mu = 0$

FIGURE 4.10

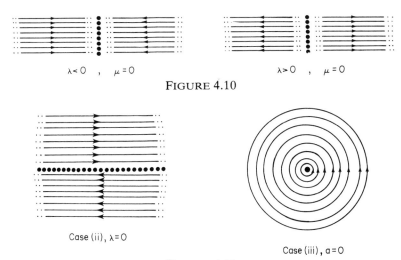

Case (ii), $\lambda = 0$

Case (iii), $a = 0$

FIGURE 4.11

(4.12) Exercise. Show that there are seventeen topological equivalence classes of linear flows on $\mathbf{R}^3$. Show also that, for all $n > 3$, there is a continuum of topological equivalence classes of linear flows on $\mathbf{R}^n$.

(4.13) Exercise. Classify up to topological equivalence affine vector fields on $\mathbf{R}^n$ for $n = 1, 2, 3$.

II. LINEAR AUTOMORPHISMS OF $\mathbf{R}^n$

Discrete linear dynamical systems on $\mathbf{R}^n$ are determined by linear automorphisms of $\mathbf{R}^n$. There are two obvious equivalence relations to compare, topological conjugacy and *linear conjugacy*, which is similarity in the ordinary sense of linear algebra. Recall that if S and $T \in GL(\mathbf{R}^n)$, S is similar to T if for some $P \in GL(\mathbf{R}^n)$, $T = PSP^{-1}$. The relation between linear and topological conjugacy is more difficult to pin down than that between linear and topological equivalence for flows. Kuiper and Robbin [1] have made a detailed study of the problem. We content ourselves here with the observation that there is an open, dense subset $HL(\mathbf{R}^n)$ of $GL(\mathbf{R}^n)$ analogous to $EL(\mathbf{R}^n)$ in the flow case. Elements of $HL(\mathbf{R}^n)$ are called *hyperbolic linear automorphisms* of $\mathbf{R}^n$. For any $T \in GL(\mathbf{R}^n)$, $T \in HL(\mathbf{R}^n)$ if and only if none of its eigenvalues lies on the unit circle S^1 in $\mathbf{C}$. Note that T is a hyperbolic vector field if and only if exp T is a hyperbolic automorphism. It is unfortunate that the term "hyperbolic" must carry two meanings, but the context should always make clear which is to be taken. If $T \in HL(\mathbf{R}^n)$ then we find that $\mathbf{R}^n$ decomposes into two direct summands corresponding to eigenvalues of T inside S^1 and eigenvalues outside S^1. The classification of $HL(\mathbf{R}^n)$ up to topological conjugacy is bound up with the dimensions of these invariant subspaces, and so resembles the classification of $EL(\mathbf{R}^n)$ up to topological equivalence. In fact, elements S and T of $EL(\mathbf{R}^n)$ are topologically equivalent if and only if exp S and exp T are topologically conjugate.

(4.14) Example. *Automorphisms of* $\mathbf{R}$. As we have seen, any element T of $GL(\mathbf{R})$ is of the form $T(x) = ax$, where a is a non-zero real number. There are exactly six topological conjugacy classes, given by (i) $a < -1$, (ii) $a = -1$, (iii) $-1 < a < 0$, (iv) $0 < a < 1$, (v) $a = 1$, (vi) $a > 1$. Of these, all but (ii) and (v) are hyperbolic.

(4.15) Example. *Automorphisms of* $\mathbf{R}^2$. Let $T \in GL(\mathbf{R}^2)$. Referring to Example 4.9, we can suppose that the matrix of T is in Jordan form (i), (ii) or (iii). Now T is hyperbolic if and only if $\lambda^2 \neq 1 \neq \mu^2$, in cases (i) and (ii), or $a^2 + b^2 \neq 1$, in case (iii). There are eight topological conjugacy classes of hyperbolic maps. All are product of automorphisms of $\mathbf{R}$, all occur in case (i) and six occur only there. Case (i) yields also eleven classes of non-hyperbolic maps. These are the products of the six classes of $GL(\mathbf{R})$ with the identity map *id* on $\mathbf{R}$, and of five of them (*id* being excluded) with the antipodal map $-id$ on $\mathbf{R}$. We get two further classes from case (ii) with $\lambda = \pm 1$. The non-hyperbolic automorphisms of case (iii) give a continuum of different topological conjugacy classes, as we now show.

When $a^2 + b^2 = 1$ in case (iii), T is a rotation through an angle θ, say, where $0 < \theta < \pi$. Rotation through θ is topologically conjugate to rotation through $-\theta$, by conjugating with a reflection. The question is: Can rotations T_1 and T_2 through distinct angles θ_1 and θ_2 be topologically conjugate if $0 < \theta_i < \pi$, $i = 1, 2$? We first observe that if $\theta_1 = 2\pi p/q$, with $q > 0$ and p coprime integers, then every point $x \neq 0$ is a periodic point of T_1 of period q. Thus T_2 cannot be topologically conjugate to T_1 unless $\theta_2 = 2\pi p'/q$, where p' and q are coprime. Suppose, then, that θ_2 has this form, and let $h : \mathbf{R}^2 \to \mathbf{R}^2$ be a topological conjugacy from T_1 to T_2. Let S_r denote the circle with centre 0 and radius r. Choose r large enough for S_r to contain the compact set $h(S_1)$ in its interior. By a highly non-trivial theorem of topology, there is a homeomorphism, f say, from the closed region D between $h(S_1)$ and S_r to the closed annulus A between S_1 and S_2. We may assume that f takes S_r to S_2 and preserves the standard orientation. Let g be the homeomorphism of A induced by f from $T_2|D$ (i.e. $g = f(T_2|D)f^{-1}$). Then g is pointwise periodic of period q, since $T_2|D$ is. Let $C = [1, 2] \times \mathbf{R}$, and let $\pi : C \to A$ be the covering defined by $\pi(t, u) = t e^{2\pi i u}$. We may lift g to a homeomorphism $G : C \to C$ such that $\pi G = g\pi$, by a similar technique to that used in Exercise 2.4. By the periodicity of g, $G^q(t, u) = (t, u + n(t, u))$, where $n(t, u)$ is an integer that varies continuously with (t, u). Now $g|S_2$ has rotation number $[p'/q]$, since it is topologically conjugate to $T_2|S_r$ by $f|S_r$. Thus $G^q(2, u) = (2, u + mq + p')$ for some fixed integer m and all $u \in \mathbf{R}$. One easily deduces that since $n(t, u)$ is a continuously varying integer, $n(t, u) = mq + p'$ for all (t, u). In particular the rotation number of $g|S_1$ is $[p'/q]$. But $g|S_1$ is topologically conjugate to $T_1|S_1$ by $gh|S_1$. Hence $[p/q] = [p'/q]$.

We have now proved that if $0 < \theta_i < \pi$ ($i = 1, 2$), $\theta_1/2\pi$ is rational, and T_1 and T_2 are topologically conjugate then $\theta_1 = \theta_2$. This leaves the case when $\theta_1/2\pi$ and $\theta_2/2\pi$ are irrational, and, fortunately, this is much easier than the rational case. If h is a conjugacy between T_1 and T_2 and if $x \in S^1$ then h maps the closure of the T_1-orbit of x onto the closure of the T_2-orbit of $h(x)$. But the first of these sets is S_1 and the second is S_r for some $r > 0$. Thus h restricts to a conjugacy between $T_1|S_1$ and $T_2|S_r$. Therefore, $\theta_1/2\pi = \theta_2/2\pi$, since these are the rotation numbers of the two circle homeomorphisms.

As one might expect from the above details, it is rotations, and, in particular, periodic rotations that cause the most trouble when one attempts to deal with automorphisms of $\mathbf{R}^n$ for large n. In fact, Kuiper and Robbin in their paper [1] reduced the classification of automorphisms of $\mathbf{R}^n$ to the so called *periodic rotation conjecture*, that any two periodic orthogonal maps of

$\mathbf{R}^n$ are topologically conjugate if and only if they are linearly conjugate. It is this conjecture that is unsolved at the time of writing, although some results are known in the positive direction.

III. THE SPECTRUM OF A LINEAR ENDOMORPHISM

We now embark on a study of linear dynamical systems on a Banach space $\mathbf{E}$ (real or complex). As we have seen above, when $\mathbf{E} = \mathbf{R}^n$ the eigenvalues of the system (vector field or automorphism) play a vital role in the theory. In infinite dimensions, the analogue of the set of eigenvalues of a linear endomorphism T is called the *spectrum* $\sigma(T)$ of T. In the complex case, it is the set of all complex numbers λ for which $T - \lambda(id)$ is *not* an automorphism. It is a compact subset of $\mathbf{C}$, and is only empty in the trivial case $\mathbf{E} = \{0\}$. We shall need a certain amount of spectral theory for our investigation of linear systems. For those unacquainted with this theory, most of the necessary material is gathered in the appendix to this chapter. We give here a brief summary.

Recall that the set $L(\mathbf{E})$ of (continuous) linear endomorphisms of $\mathbf{E}$ is a Banach space, with norm defined by

$$|T| = \sup\{|T(x)|: x \in \mathbf{E}, |x| \leqslant 1\}.$$

The spectral radius $r(T)$ of T is the radius of the smallest circle in the Argand diagram with centre 0 containing the spectrum $\sigma(T)$. That is to say

$$r(T) = \sup\{|\lambda|: \lambda \in \sigma(T)\}.$$

It measures the eventual size of T under repeated iteration, in the sense that it equals $\lim_{n \to \infty} |T^n|^{1/n}$. One may always pick a norm of $\mathbf{E}$ equivalent to the given one, such that, with respect to the corresponding new norm of $L(\mathbf{E})$, T has norm precisely $r(T)$. This is of especial interest to us when $r(T) < 1$. In this case we call T a *linear contraction*, and with respect to the new norm T is a contraction in the sense of metric space theory (Appendix C). If T is an automorphism, and T^{-1} is a contraction, we call T an *expansion*. Notice that T is an automorphism if and only if $0 \notin \sigma(T)$, and that

$$r(T^{-1}) = (\inf\{|\lambda|: \lambda \in \sigma(T)\})^{-1},$$

since, in fact, $\sigma(T^{-1}) = \{\lambda^{-1}: \lambda \in \sigma(T)\}$.

Now let D be a contour in the Argand diagram, symmetrical about the real axis if $\mathbf{E}$ is a real Banach space, such that $\sigma(T) \cap D$ is empty. Suppose that $0 \notin \sigma(T)$ and that D separates $\sigma(T)$ into two subsets $\sigma_s(T)$ and $\sigma_u(T)$. Then, by a corollary of Dunford's spectral mapping theorem, $\mathbf{E}$ splits uniquely as a direct sum $\mathbf{E}_s \oplus \mathbf{E}_u$ of T-invariant closed subspaces, such that, if $T_s: \mathbf{E}_s \to \mathbf{E}_s$ and $T_u: \mathbf{E}_u \to \mathbf{E}_u$ are the restrictions of T, then $\sigma(T_s) = \sigma_s(T)$ and $\sigma(T_u) = \sigma_u(T)$. Moreover, the splitting of $\mathbf{E}$ depends analytically on $T \in L(\mathbf{E})$. More precisely, the subspace $\mathbf{E}_s$, for example, is obtained as the image of a projection $S \in L(\mathbf{E})$ (*projection* means $S^2 = S$), and S varies in $L(\mathbf{E})$ as an analytic function of T in $\{T \in L(\mathbf{E}): \sigma(T) \cap D \text{ empty}\}$.

IV. HYPERBOLIC LINEAR AUTOMORPHISMS

Let $T \in GL(\mathbf{E})$, the open subset of $L(\mathbf{E})$ consisting of all linear automorphisms of $\mathbf{E}$. We say that T is a *hyperbolic linear* automorphism if $\sigma(T) \cap S^1$ is empty, where S^1 is the unit circle in the Argand diagram $\mathbf{C}$. The set of all hyperbolic linear automorphisms of $\mathbf{E}$ is denoted by $HL(\mathbf{E})$.

(4.16) Example. The map $T: R^2 \to R^2$ given by

$$T(x, y) = (\tfrac{1}{2}x, 2y)$$

is in $HL(\mathbf{R}^2)$. Its spectrum is the set $\{\tfrac{1}{2}, 2\}$ of eigenvalues of T and neither of these points of $\mathbf{C}$ is on S^1. The effect of T is suggested in Figure 4.16, where

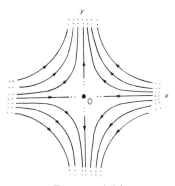

FIGURE 4.16

the hyperbolae and their asymptotes are invariant submanifolds of T. The x-axis is stable in the sense that positive iterates of T take its points into bounded sequences (in fact they converge to the origin 0). The y-axis is unstable (meaning, in this context, stable with respect to T^{-1}). These axes together generate $\mathbf{R}^2$, of course. A direct sum decomposition into stable and unstable summands (one of which may, however, be $\{0\}$) is typical of hyperbolic linear automorphisms.

(4.17) Exercise. Prove that if $T \in GL(\mathbf{R}^n)$ and if $a \in \mathbf{R}$ is sufficiently near but not equal to 1, then aT is hyperbolic. Deduce that $HL(\mathbf{R}^n)$ is dense in $GL(\mathbf{R}^n)$.

(4.18) Exercise. Let $T \in HL(\mathbf{E})$. Prove that $0 \in \mathbf{E}$ is the unique fixed point of T.

Let $T \in HL(\mathbf{E})$ and let $\sigma_s(T)$ and $\sigma_u(T)$ be the parts of $\sigma(T)$ inside and outside the unit circle in the Argand diagram. By the spectral decomposition theorem (Corollaries 4.56 and 4.58 of the appendix) $\mathbf{E}$ splits as a direct sum $\mathbf{E}_s \oplus \mathbf{E}_u$ of T-invariant subspaces, where the spectra of the restrictions $T_s \in GL(\mathbf{E}_s)$ and $T_u \in GL(\mathbf{E}_u)$ of T are respectively $\sigma_s(T)$ and $\sigma_u(T)$. Thus T_s is a contraction and T_u is an expansion. Let b be any number less than 1 but greater than the larger of the spectral radii of T_s and T_u^{-1}. By Theorem 4.47 of the appendix we have:

(4.19) Theorem. *If $T \in HL(\mathbf{E})$ then there is an equivalent norm $\|\ \|$ such that*

(i) *for all $x = x_s + x_u \in E$, $\|x\| = \max\{\|x_s\|, \|x_u\|\}$,*

(ii) $\max\{\|T_s\|, \|T_u^{-1}\|\} = a \leqslant b.$ □

Thus, with respect to the new norm $\|\ \|$, T_s is a metric contraction and T_u is a metric expansion. We call a the *skewness* of T with respect to $\|\ \|$. The *stable summand* $\mathbf{E}_s$ *of* $\mathbf{E}$ *with respect to* T is equivalently characterized as $\{x \in \mathbf{E}: T^n(x) \text{ is bounded as } n \to \infty\}$ and $\{x \in \mathbf{E}: T^n(x) \to 0 \text{ as } n \to \infty\}$. We also call $\mathbf{E}_s$ the *stable manifold* of 0 with respect to T and T_s the *stable summand of* T. A similar characterization holds for the *unstable summand* $\mathbf{E}_u$, with T^{-n} replacing T^n.

The main aim of this section is to prove that hyperbolic linear automorphisms of $\mathbf{E}$ are stable in $L(\mathbf{E})$ with respect to topological conjugacy. A first step in this direction is:

(4.20) Theorem. $HL(\mathbf{E})$ *is open in* $GL(\mathbf{E})$, *and hence in* $L(\mathbf{E})$.

Proof. Let $T \in HL(\mathbf{E})$. Choose a norm $\|\ \|$ as in Theorem 4.19. Let T have skewness a. We assert that, for all $T' \in GL(\mathbf{E})$ with $\|T' - T\| < 1 - a$, $T' \in HL(\mathbf{E})$. Let $\lambda \in S^1$. Then $\lambda \notin \sigma(T)$, so $T - \lambda(id)$ is an automorphism. In fact we may explicitly write down the components of $(T - \lambda(id))^{-1}$ in the $\mathbf{E}_s$ and $\mathbf{E}_u$ directions; they are $-\sum_{r=0}^{\infty} \lambda^{-r-1}(T_s)^r$ and $\sum_{r=0}^{\infty} \lambda^r(T_u)^{-r-1}$. From these expressions it is easy to see that $\|(T - \lambda(id))^{-1}\| \leqslant (1 - a)^{-1}$. Thus, if T' satisfies the given inequality, then the series

$$\sum_{r=0} (T - \lambda(id))^{-1}((T - T')(T - \lambda(id))^{-1})^r$$

converges and is an inverse of $T' - \lambda(id)$. Thus $\lambda \notin \sigma(T')$, and hence T' is hyperbolic. □

Our next step is to study the direct sum decomposition of $\mathbf{E}$ associated with a hyperbolic linear automorphism. We say that $T \in HL(\mathbf{E})$ is *isomorphic to* $T' \in HL(\mathbf{E}')$ if there exists a (topological) linear isomorphism from $\mathbf{E}$ to $\mathbf{E}'$ taking $\mathbf{E}_s(T)$ onto $\mathbf{E}'_s(T')$ and $\mathbf{E}_u(T)$ onto $\mathbf{E}_u(T')$. Equivalently, T is isomorphic to T' if and only if there are linear isomorphisms of $\mathbf{E}_s(T)$ onto $\mathbf{E}'_s(T')$ and of $\mathbf{E}_u(T)$ onto $\mathbf{E}'_u(T')$. Thus, for example, there are exactly $n+1$ isomorphism classes in $HL(\mathbf{R}^n)$. We prove:

(4.21) Theorem. *Any $T \in HL(\mathbf{E})$ is stable with respect to isomorphism.*

Proof. By the spectral theory of the appendix to the chapter, there is a continuous (in fact, analytic) map $f : HL(\mathbf{E}) \to L(\mathbf{E})$ such that for all $T \in HL(\mathbf{E})$, $f(T)$ is a projection with image $\mathbf{E}_s(T)$. Let $T \in HL(\mathbf{E})$. Then $f(T)f(T') \to f(T)^2 (= f(T))$ as $T' \to T$ in $HL(\mathbf{E})$. Since the restriction of $f(T)^2 (= f(T))$ to $\mathbf{E}_s(T)$ is the identity, we deduce that if T' is sufficiently near T then $f(T)f(T')$ restricts to a (topological linear) automorphism of $\mathbf{E}_s(T)$. Thus $f(T')$ maps $\mathbf{E}_s(T)$ injectively into $\mathbf{E}_s(T')$. Similarly, since $f(T')f(T) - f(T')^2 \to 0$ as $T' \to T$ and since $f(T')^2$ is the identity on $\mathbf{E}_s(T')$, $f(T')f(T)$ is an automorphism of $\mathbf{E}_s(T')$ for T' sufficiently close to T. Thus $f(T')$ maps $\mathbf{E}_s(T)$ surjectively to $\mathbf{E}_s(T')$. Thus $\mathbf{E}_s(T)$ is isomorphic to $\mathbf{E}_s(T')$. A similar argument applies to the unstable summands. Thus T' is isomorphic to T. $\square$

(4.22) Exercise. Let $T \in GL(\mathbf{R}^n) \backslash HL(\mathbf{R}^n)$. By comparing aT with bT, for $0 < a < 1 < b$, show that any neighbourhood of T in $GL(\mathbf{R}^n)$ contains elements of more than one isomorphism class.

The foregoing theory enables us to make a connection between the stability of a hyperbolic automorphism and the stability of its restriction to its stable and unstable summands. So let us now suppose, for the time being, that T is a linear contraction of $\mathbf{E}$, and that a norm has been chosen so that T is a metric contraction. Then T maps the unit sphere S_1 in $\mathbf{E}$ into the interior of the unit ball B_1. The closed annulus A between S_1 and $T(S_1)$ is called a *fundamental domain* of T. Its images under positive and negative powers of T cover $\mathbf{E} \backslash \{0\}$ (see Figure 4.23). Moreover, for $r > s$, $T^r(A)$ intersects $T^s(A)$ only when $s = r + 1$, in which case the intersection is $T^{r+1}(S_1)$.

It is easy to show that any topological conjugacy $h : \mathbf{E} \to \mathbf{E}'$ between T and some homeomorphism f of a Banach space $\mathbf{E}'$ is completely determined by its values on A. For, to find $h(x)$ for any $x \in E \backslash \{0\}$, we map x into A by some power T^n, then map by h and finally map by f^{-n}. Since h is a conjugacy, we have now reached $h(x)$, since $h = f^{-n}hT^n$. (Because h is a conjugacy, $fh = hT$ and so $f^r h = hT^r$ for all integers r.)

Conversely, let T and T' be contracting linear homeomorphisms of $\mathbf{E}$ and $\mathbf{E}'$ respectively, and let A and A' be corresponding fundamental domains.

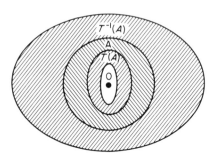

FIGURE 4.23

Suppose that $h: A \to A'$ is a homeomorphism such that

(4.23)   for all $x \in S_1$, $hT(x) = T'h(x)$.

Then we may extend h to a conjugacy from T to T' by putting $h(0) = 0$ and, for all $x \in \mathbf{E} \backslash \{0\}$, $h(x) = (T')^{-n}hT^n(x)$, where $T^n(x) \in A$. By (4.23), if $T^n(x) \in S_1$ then $(T')^{-n}hT^n(x) = (T')^{-n-1}hT^{n+1}(x)$, and so h is well-defined. In like fashion we can construct an inverse $h^{-1}: \mathbf{E} \to \mathbf{E}$ from $h^{-1}: A' \to A$. Continuity of h (and of h^{-1}) is in doubt only at 0. To deal with this, observe that h maps $\{T^n(B_1): n \in \mathbf{Z}\}$, which is a basis of neighbourhoods of 0 in $\mathbf{E}$, onto a similar basis $\{(T')^n(B_1): n \in \mathbf{Z}\}$ in $\mathbf{E}'$. The result that we have now proved may be formulated, rather loosely, as follows:

(4.24) Theorem. *Contracting linear homeomorphisms are topologically conjugate if and only if they are topologically conjugate on fundamental domains.*   □

In the following important case, we may explicitly construct a conjugacy between fundamental domains.

(4.25) Theorem. *Let T and T' belong to the same path component of $GL(\mathbf{E})$ and have spectral radii <1. Then T and T' are topologically conjugate.*

Proof. Let T and T' be metric contractions with respect to norm $|\ |$ and $|\ |'$ respectively on $\mathbf{E}$, and let A and A' be the corresponding fundamental domains (between S_1 and $T(S_1)$ and between S_1' and $T'(S_1')$). Pick ε with $0 < \varepsilon < \min\{1 - |T|, 1 - |T'|'\}$, and let $I = [1 - \varepsilon, 1]$. Since $T'T^{-1}$ and the identity map id belong to the same path component of $GL(\mathbf{E})$, there exists a continuous map $g: I \to GL(\mathbf{E})$ with $g(1 - \varepsilon) = T'T^{-1}$ and $g(1) = id$. Identifying $I \times S_1$ and $I \times S_1'$ with annuli in $\mathbf{E}$, by putting $(t, x) = tx$, we have a homeomorphism $h: I \times S_1 \to I \times S_1'$ defined by

$$h(tx) = \frac{tg(t)(x)}{|g(t)(x)|'}.$$

We extend to a homeomorphism $h: A \to A'$, by mapping the line segment $[(1-\varepsilon)x, x/|T^{-1}(x)|]$ linearly onto the line segment

$$\left[(1-\varepsilon)\frac{T'T^{-1}(x)}{|T'T^{-1}(x)|'}, \frac{T'T^{-1}(x)}{|T^{-1}(x)|'}\right]$$

for each $x \in S_1$. Notice that for all $y \in S_1$,

$$y = \frac{T^{-1}(x)}{|T^{-1}(x)|}$$

for some $x \in S_1$, and that

$$hT(y) = h\left(\frac{x}{|T^{-1}(x)|'}\right) = \frac{T'T^{-1}(x)}{|T^{-1}(x)|'} = T'h\left(\frac{T^{-1}(x)}{|T^{-1}(x)|}\right) = T'h(y).$$

Thus (4.23) holds, and we can apply Theorem 4.24 to complete the argument. We leave the details of checking that $h: A \to A'$ is a homeomorphism to the reader. ☐

This result enables us to classify hyperbolic linear automorphisms up to topological conjugacy, in finite dimensional spaces. We concern ourselves particularly with $\mathbf{R}^n$. In the first place, we have

(4.26) Corollary. *For each $n > 0$ there are exactly two topological conjugacy classes of contracting linear homeomorphisms of $\mathbf{R}^n$, consisting respectively of orientation preserving and orientation reversing maps.*

Proof. $GL(\mathbf{R}^n)$ has exactly two path components, consisting of orientation preserving and orientation reversing maps (distinguished by the sign of the determinant). These give distinct topological conjugacy classes, since any topological conjugate of an orientation preserving homeomorphism is orientation preserving. This is most easily proved by extending such homeomorphisms to the one-point compactification S^n of $\mathbf{R}^n$, and using the fact that if $f: \mathbf{R}^n \to \mathbf{R}^n$ is a homeomorphism, and $f: S^n \to S^n$ is its extension to S^n, then the induced isomorphism of homology, maps a generator 1 of $H_n(S^n) \cong \mathbf{Z}$ to 1 or -1 according as f is orientation preserving or reversing (see Greenberg [1] for details). ☐

(4.27) Corollary. *Two hyperbolic linear homeomorphisms of $\mathbf{R}^n$ are topologically conjugate if and only if*

(i) *they are isomorphic,*

(ii) *their stable components are either both orientation preserving or both orientation reversing, and*

(iii) *their unstable components are either both orientation preserving or both orientation reversing.*

Proof. Suppose that T, $T' \in HL(\mathbf{R}^n)$ satisfy conditions (i) to (iii). For simplicity of notation, we put $\mathbf{R}^n = \mathbf{E}$. Let $L \in GL(\mathbf{E})$ map $\mathbf{E}_s(T)$ onto $\mathbf{E}_s(T')$ and $\mathbf{E}_u(T)$ onto $\mathbf{E}_u(T')$. By Corollary 4.26, the restrictions of T and $L^{-1}T'L$ to $\mathbf{E}_s(T)$ are topologically conjugate, and so are their restrictions to $\mathbf{E}_u(T)$. This enables us to construct a topological conjugacy from T to $L^{-1}T'L$, and hence to T'.

Conversely, let h be a topological conjugacy from $T \in HL(\mathbf{E})$ to $T' \in HL(\mathbf{E})$. Then h preserves 0, the unique fixed point of T and T'. By the continuity of h and of h^{-1}, $T^n(x) \to 0$ as $n \to \infty$ if and only if $hT^n(x) = (T')^n h(x) \to 0$ as $n \to \infty$. Thus h maps $\mathbf{E}_s(T)$ onto $\mathbf{E}_s(T')$, and similarly $\mathbf{E}_u(T)$ onto $\mathbf{E}_u(T')$. Since dimension is a topological invariant (§ 1 of Chapter 4 of Hu [2]) T and T' are isomorphic. Conditions (ii) and (iii) now follow by Corollary 4.26.   □

It follows immediately from this theorem that there are exactly $4n$ topological conjugacy classes of hyperbolic linear automorphisms of $\mathbf{R}^n$ $(n \geq 1)$. (See Examples 4.14 and 4.15.)

(4.28) Exercise. Prove that, for any $n > 0$, $GL(\mathbf{R}^n)$ has, as asserted above, precisely two path components. (*Hint*: move $T \in GL(\mathbf{R}^n)$ into the subspace $GL(\mathbf{R}) \times GL(\mathbf{R}^{n-1})$.)

(4.29) Exercise. Classify hyperbolic linear automorphisms of $\mathbf{C}^n$ up to topological conjugacy.

For infinite dimensional Banach spaces, the situation is less clear. Homeomorphic Banach spaces need not be linearly homeomorphic, so that topological conjugacy does not immediately imply isomorphism. Furthermore, the connectivity properties of $GL(\mathbf{E})$ are not fully understood for an arbitrary Banach space $\mathbf{E}$. In several important cases (for example if $\mathbf{E}$ is a Hilbert space—see Kuiper [1]) $GL(\mathbf{E})$ is path connected, and so *any* two linear contractions are topologically conjugate. However our main stability theorem, which we now prove, is, happily, valid for all Banach spaces.

(4.30) Theorem. *For any Banach space* $\mathbf{E}$, *any* $T \in HL(\mathbf{E})$ *is stable in* $L(\mathbf{E})$ *with respect to topological conjugacy.*

Proof. Consider the ball B in $L(\mathbf{E})$ with centre T and radius $d > 0$. Let $T^1 \in B$ and, for each $t \in [0, 1]$, define T^t by $T^t = (1-t)T + tT^1$. As in the proof of Theorem 4.21, we choose d so small that for all T^1 and $t \in [0, 1]$, the restrictions $P^t : \mathbf{E}_s(T) \to \mathbf{E}_s(T^t)$ and $Q^t : \mathbf{E}_s(T^t) \to \mathbf{E}_s(T)$ of the projections $f(T^t)$ and $f(T)$ respectively are isomorphisms. We now compare T_s and $(P^1)^{-1}(T^1)_s P^1$, both of which are in $GL(\mathbf{E}_s(T))$ and have spectral radius <1. We show that these two maps are in the same path component of $GL(\mathbf{E}_s(T))$. In fact we assert that $t \mapsto (P^t)^{-1}(T^t)_s P^t$ is a path joining the one to the other. The only difficulty is whether this map of $[0, 1]$ into $GL(\mathbf{E}_s(T))$ is continuous. Since f is continuous the map from $[0, 1]$ to $L(\mathbf{E}_s(T))$ taking t to $Q^t(T^t)_s P^t$ (which is the restriction of $f(T)(T^t)f(T^t)$) is continuous. Similarly the map taking t to $Q^t P^t$ is continuous. Hence the map taking t to $(Q^t P^t)^{-1} Q^t(T^t)_s P^t$ is continuous, as stated. To complete the proof of the theorem, we apply Theorem 4.25, giving that T_s is topologically conjugate to $(P^1)^{-1}(T^1)_s P^1$, and hence to $(T^1)_s$. A similar argument holds for the unstable components. Thus T is topologically conjugate to T^1. $\qquad\square$

(4.31) Corollary. *Let* $\mathbf{E}$ *be finite dimensional. Then* $HL(\mathbf{E})$ *is the stable set of* $GL(\mathbf{E})$ *with respect to topological conjugacy. Furthermore* $HL(\mathbf{E})$ *is an open dense subset of* $GL(\mathbf{E})$.

Proof. By Theorem 4.30 any hyperbolic map is stable with respect to topological conjugacy. If $T \in GL(\mathbf{E})$ is not hyperbolic, every neighbourhood of T meets two different isomorphism classes of $HL(\mathbf{E})$ (by Exercise 4.22), and hence meets two different topological conjugacy classes (by Corollary 4.27). Thus T is unstable. This proves that $HL(\mathbf{E})$ is the stable set of $GL(\mathbf{E})$, and also that $HL(\mathbf{E})$ is dense in $GL(\mathbf{E})$. It is open in $GL(\mathbf{E})$ by Theorem 4.20. $\qquad\square$

In the above theory we have not bothered to analyse the size of perturbation that a given hyperbolic map admits without change to its topological conjugacy class. We shall, in the next chapter, return to Theorem 4.30 and give it a new proof, which is better designed to deal with this point.

V. HYPERBOLIC LINEAR VECTOR FIELDS

We now turn to a study of linear vector fields. We are interested in determining which ones are stable in $L(\mathbf{E})$ with respect to topological equivalence, and once again we find we can give a complete description for finite dimensional $\mathbf{E}$. Let $T \in L(\mathbf{E})$. We say that T is a *hyperbolic linear vector field* if its spectrum $\sigma(T)$ does not intersect the imaginary axis of $\mathbf{C}$. We denote the set of all hyperbolic linear vector fields on $\mathbf{E}$ by $EL(\mathbf{E})$. As we have commented earlier, the notion of hyperbolicity depends on whether we are regarding T as a vector field or a diffeomorphism. It will, we hope, always be clear from the context in which sense we are using the term. The connection between the two notions is expressed in the following propositions, whose proofs are straightforward exercises in spectral theory (see Theorem 4.55 of the appendix). The special case $\mathbf{E} = \mathbf{R}^n$ has already been discussed above. The linear automorphism $\exp(tT)$ may be defined either by its power series or by a contour integral, as in (4.50) of the appendix.

(4.32) Proposition. *A linear vector field T is hyperbolic if and only if, for some (and hence all) non-zero real t, exp (tT) is a hyperbolic automorphism.* □

(4.33) Proposition. *The linear vector field T has integral flow* $\phi: \mathbf{R} \times \mathbf{E} \to \mathbf{E}$ *given by* $\phi(t, x) = \exp(tT)(x)$. □

(4.34) Example. The linear vector on $\mathbf{R}^2$ defined by $T(x, y) = (-x, y)$ is hyperbolic. The integral flow is illustrated by Figure 4.16, branches of the hyperbolae now being orbits of T, as are the positive and negative semi-axes and the point 0.

Suppose that $T \in EL(\mathbf{E})$. Let $\sigma_s(T) = \{\lambda \in \sigma(T): \text{re } \lambda < 0\}$ and $\sigma_u(T) = \{\lambda \in \sigma(T): \text{re } \lambda > 0\}$. Let D be a contour in the Argand diagram consisting of the line segment $[-ir, ir]$ and the semi-circle

$$\left\{ r e^{i\theta}: \frac{\pi}{2} \leqslant \theta \leqslant \frac{3\pi}{2} \right\},$$

where r is large enough for D to enclose σ_s. Then the spectral decomposition theorem gives us, correspondingly, a T-invariant direct sum splitting $\mathbf{E} = \mathbf{E}_s \oplus \mathbf{E}_u$ such that $\sigma(T_s) = \sigma_s(T)$ and $\sigma(T_u) = \sigma_u(T)$, where $T_s \in L(\mathbf{E}_s)$ and $T_u \in L(\mathbf{E}_u)$ are the restrictions of T. Once again we call $\mathbf{E}_s$ and T_s *stable summands*, and $\mathbf{E}_u$ and T_u *unstable summands*. Similarly $\mathbf{E}_s$ and $\mathbf{E}_u$ are again called *stable* and *unstable manifolds* at 0.

(4.35) Exercise. Prove that, for all $t > 0$, the stable summand of the hyperbolic automorphism exp (tT) is exp (tT_s). Similarly for unstable summands.

The map $\exp: L(\mathbf{E}) \to L(\mathbf{E})$ is continuous (see the appendix), and thus, since $HL(\mathbf{E})$ is open in $L(\mathbf{E})$, Proposition 4.32 implies that $EL(\mathbf{E})$ is open in $L(\mathbf{E})$. We define the term *isomorphic* for hyperbolic vector fields exactly as for hyperbolic automorphisms. Since the stable and unstable summands of $\mathbf{E}$ are the same with respect to $T \in EL(\mathbf{E})$ as they are with respect to exp $T \in HL(\mathbf{E})$ (by Exercise 4.35), the fact that $T \in EL(\mathbf{E})$ is stable with respect to isomorphism follows immediately from the corresponding result for $HL(\mathbf{E})$. We sum up these observations in the following proposition.

(4.36) Theorem. *Let* $T \in EL(\mathbf{E})$. *Then* $T = T_s \oplus T_u$, *where* $\sigma(T_s) = \sigma_s(T)$ *and* $\sigma(T_u) = \sigma_u(T)$. *The set* $EL(\mathbf{E})$ *is open in* $L(\mathbf{E})$, *and its elements are stable with respect to isomorphism.* □

Let $T \in EL(\mathbf{E})$. Since, for any $t \neq 0$, exp $(tT) \in HL(\mathbf{E})$, we may choose (by Theorem 4.19) a norm on $\mathbf{E}$ with respect to which the stable and unstable summands of exp (tT) are respectively a metric contraction and a metric expansion. We need a stronger result than this, however. We may deal with the stable and unstable manifolds separately.

(4.37) Theorem. *Let $T \in EL(\mathbf{E})$ have spectrum $\sigma(T) = \sigma_s(T)$. Then there exists a norm $\| \ \|$ on $\mathbf{E}$ equivalent to the given one such that, for any non-zero $x \in E$, the map from $\mathbf{R}$ to $]0, \infty[$ taking t to $\|\exp(tT)(x)\|$ is strictly decreasing and surjective. There exists $b > 0$ such that, for all $t > 0$, $\|\exp(tT)\| \leqslant e^{-bt}$.*

Proof. Let d be the distance from $\sigma(T)$ to the imaginary axis. By the spectral mapping theorem the spectral radius $r(\exp(tT))$ of $\exp(tT)$ is e^{-dt} for all $t \in \mathbf{R}$. Choose b and c with $0 < b < c < d$, and a norm $| \ |$ on $\mathbf{E}$ equivalent to the given one, such that $|\exp T| \leqslant e^{-c}$ (this is possible by Theorem 4.19). We first show that $\int_0^\infty |e^{bt} \exp(tT)| \, dt$ converges. Let $|\exp(tT)|$ be bounded by $A > 0$ for $t \in [0, 1]$. Then, for all integers m and n with $0 \leqslant m < n$,

$$\int_m^n |e^{bt} \exp(tT)| \, dt \leqslant \sum_{r=m}^{n-1} e^{b(r+1)} e^{-rc} A$$

$$\leqslant A e^b \sum_{r=m}^{n-1} e^{-r(c-b)}$$

$$\leqslant A e^b e^{-m(c-b)} (1 - e^{b-c})^{-1},$$

which tends to zero as $m \to \infty$.

One easily checks that the relation

$$\|x\| = \int_0^\infty |e^{bt} \exp(tT)(x)| \, dt$$

defines a norm $\| \ \|$ on $\mathbf{E}$. Moreover, for all $x \in \mathbf{E}$,

$$\|x\| < |x| \int_0^\infty |e^{bt} \exp(tT)| \, dt.$$

By the interior mapping theorem, $\| \ \|$ is equivalent to $| \ |$.

Finally, for all $t \in \mathbf{R}$,

$$\|\exp(tT)(x)\| = \int_0^\infty |e^{bu} \exp((u+t)T)(x)| \, du$$

$$= e^{-bt} \int_t^\infty |e^{bu} \exp(uT)(x)| \, du,$$

which shows that $\|\exp(tT)(x)\|$ decreases from ∞ to 0 as t increases from $-\infty$ to ∞ (for fixed $x \neq 0$). Also, for $t > 0$,

$$\|\exp(tT)(x)\| \leqslant e^{-bt} \|x\|,$$

as required. $\qquad\qquad\qquad\qquad\qquad\qquad\qquad\qquad\qquad\qquad\qquad\qquad$ $\square$

Theorem 4.37 gives a very precise hold on the phase portrait of T. Let S_1 be the unit sphere in E with respect to the norm $\| \ \|$ of Theorem 4.37. Then, with the exception of the fixed point 0, every orbit of T crosses S_1 in precisely one point. Thus S_1 is a sort of cross section to the integral flow of T. We may sharpen this remark as follows:

(4.38) Corollary. *If* $\phi: \mathbf{R} \times \mathbf{E} \to \mathbf{E}$ *is the integral flow of* T, *then* ϕ *maps* $\mathbf{R} \times S_1$ *homeomorphically onto* $\mathbf{E} \backslash \{0\}$.

Proof. We know that ϕ is continuous, and Theorem 4.37 shows that it maps $\mathbf{R} \times S_1$ bijectively onto $\mathbf{E} \backslash \{0\}$. Thus it remains to prove that the inverse of the restricted map is continuous. Consider, in $\mathbf{E} \backslash \{0\}$, a sequence (x_n) converging to a point x. Then $x_n = \phi(t_n, y_n)$ for some $t_n \in \mathbf{R}$ and $y_n \in S_1$. The sequence (t_n) is bounded, since $\|x_n\| \leqslant e^{-bt_n}$ for $t_n > 0$ and $(\|x_n\|)$ converges to the non-zero number $\|x\|$. Let (t_{n_k}) be a convergent subsequence of (t_n), converging to t, say. Then, as $k \to \infty$, $y_{n_k} = \phi(-t_{n_k}, x_{n_k})$ converges to $y = \phi(-t, x)$, which is thus some point of S_1. If the sequence (t_n) has some other cluster point t', we similarly deduce that $\phi(t', x)$ is in S_1, contrary to Theorem 4.37. Thus (t_n) converges to t, and hence $y_n = \phi(-t_n, x_n)$ converges to y. This proves continuity of the inverse map at x.   □

Now suppose that $T, T' \in EL(\mathbf{E})$ both have spectra in re $z < 0$. We can pick equivalent norms on $\mathbf{E}$ such that the unit spheres with respect to them are cross sections of the integral flows of T and T' restricted to $\mathbf{E} \backslash \{0\}$. This situation is illustrated in Figure 4.39, in which S_1 and S_1' are the unit spheres.

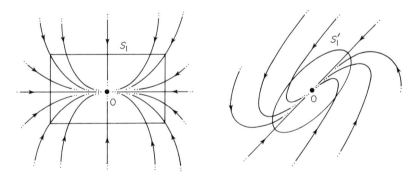

FIGURE 4.39

Since S_1 and S_1' are homeomorphic, there is an obvious way of constructing a flow equivalence from T to T'. Once the proposed flow equivalence is given on the unit spheres (mapping S_1 to S_1') its values elsewhere are determined up to the speeding up factor α in the definition of flow equivalence.

(4.40) Theorem. *Let* $T, T' \in HL(\mathbf{E})$ *have spectra in* re $z < 0$. *Then* T *and* T' *are flow equivalent.*

Proof. We show that there is a homeomorphism h of $\mathbf{E}$ such that, for all $(t, x) \in \mathbf{R} \times \mathbf{E}$,

(4.41) $$h\phi(t, x) = \psi(t, h(x))$$

where ϕ and ψ are the integral flows of T and T' respectively. Let $\| \; \|$ be as in the statement of Theorem 4.37, and let $\| \; \|'$ have the same property relative to T'. Let S_1 and S_1' be the unit spheres with respect to these norms. We define a homeomorphism $\bar{h}: S_1 \to S_1'$ by $\bar{h}(y) = y/\|y\|'$. Let $\bar{\phi}: \mathbf{R} \times S_1 \to \mathbf{E}\backslash\{0\}$ and $\bar{\psi}: \mathbf{R} \times S_1' \to \mathbf{E}\backslash\{0\}$ be the restrictions of ϕ and ψ. By Corollary 4.38, $\bar{\phi}$ and $\bar{\psi}$ are homeomorphisms. We define $h: \mathbf{E} \to \mathbf{E}$ by putting $h(0) = 0$ and letting h be the homeomorphism $\bar{\psi}(id \times \bar{h})\bar{\phi}^{-1}$ on $\mathbf{E}\backslash\{0\}$. Then h extends $\bar{h}$.

Also (4.41) holds trivially for $x = 0$. For $x \neq 0$, $x = \bar{\phi}(t', y)$ for some $t' \in \mathbf{R}$ and $y \in S_1$, and so, for all $t \in \mathbf{R}$,

$$\begin{aligned}
h\phi(t, x) &= h\phi(t + t', y) \\
&= \bar{\psi}(id \times \bar{h})(t + t', y) \\
&= \bar{\psi}(t + t', \bar{h}(y)) \\
&= \psi(t, \bar{\psi}(t', \bar{h}(y))) \\
&= \psi(t, \bar{\psi}(id \times \bar{h})\bar{\phi}^{-1}(\bar{\phi}(t', y))) \\
&= \psi(t, h(x)).
\end{aligned}$$

Finally we show that h is continuous at 0 (the proof for h^{-1} being similar). Let W be any neighbourhood of 0 in $\mathbf{E}$. By Theorem 4.37, there exists $t_0 > 0$ such that, for all $s \geq t_0$, $\psi^s(S_1') \subset W$. For all $x \in E$ with $\|x\| < \|\phi^{-t_0}\|^{-1}$, $\|\phi(-t_0, x)\| \leq 1$ and so, by Theorem 4.37, the number s such that $\phi(-s, x) \in S_1$ satisfies $s > t_0$. Thus, since $h(x) = \psi(s, h\phi(-s, x))$ and $h\phi(-s, x) \in S_1'$, $h(x) \in W$. $\square$

By applying this theorem to stable and unstable summands separately, we have:

(4.42) Corollary. *If T and $T' \in EL(\mathbf{E})$ are isomorphic, then they are flow equivalent (and thus topologically equivalent).* $\square$

By Theorem 4.36, we now deduce our main stability result:

(4.43) Corollary. *Any $T \in EL(\mathbf{E})$ is stable in $L(\mathbf{E})$ with respect to flow equivalence (and thus with respect to topological equivalence).* $\square$

(4.44) Exercise. Let $\mathbf{E}$ be finite dimensional. Prove
(i) $EL(\mathbf{E})$ is dense in $L(\mathbf{E})$,

  (ii) for all $T, T' \in EL(\mathbf{E})$ the following statements are equivalent:
      (a) T is flow equivalent to T',
      (b) T is topologically equivalent to T',
      (c) T is isomorphic to T',
  (iii) $EL(\mathbf{E})$ is the stable set of $L(\mathbf{E})$ with respect to
      (a) flow equivalence, and
      (b) topological equivalence.

Thus, for $\mathbf{E} = \mathbf{R}^n$ there are precisely $n + 1$ flow equivalence classes in $EL(\mathbf{E})$, classified by the dimension of the stable summand. (See Examples 4.8 and 4.9.)

Appendix 4

I. SPECTRAL THEORY

In this brief survey of spectral theory we omit a number of proofs that can be found in our main reference for the subject, which is Dunford and Schwartz [1]. We shall also assume without comment easy generalizations of certain parts of the theory of functions of one complex variable from **C**-valued to **E**-valued functions, where **E** is any complex Banach space. For these we refer to § 14 of Chapter 3 of Dunford and Schwartz [1].

Let **E** be a complex Banach space, with $\mathbf{E} \neq \{0\}$, and let $T \in L(\mathbf{E})$. The *resolvent* $\rho(T)$ of T is the set $\{\lambda \in \mathbf{C} : T - \lambda(id)$ is a homeomorphism$\}$. Clearly $\rho(T)$ is an invariant of the linear conjugacy class of T. The *spectrum* $\sigma(T)$ of T is the complement of $\rho(T)$ in **C**. For finite dimensional **E**, $\sigma(T)$ is the set of zeros of det $(T - \lambda(id))$, the eigenvalues of T. Spectral theory links properties of the map T with properties of its spectrum. For example, we shall find that the size of $\sigma(T)$ reflects the size of T itself (as an element of the Banach space $L(\mathbf{E})$), and that, when $\sigma(T)$ is disconnected, T splits into factors having the components of $\sigma(T)$ as spectra.

We define a map $R: \rho(T) \to L(\mathbf{E})$ by $R(\lambda) = (T - \lambda(id))^{-1}$. For fixed $\lambda \in \rho(T)$ and small $\eta \in \mathbf{C}$, the inverse of $T - (\lambda + \eta)id$ is the map

$$(T - \lambda(id))^{-1} \sum_{n=0}^{\infty} \eta^n (T - \lambda(id))^{-n}.$$

Thus $\rho(T)$ is open, and R is analytic. Notice the relation

(4.45) $\quad (T - \lambda(id))^{-1} - (T - \mu(id))^{-1} = (\lambda - \mu)(T - \lambda(id))^{-1}(T - \mu(id))^{-1}$

resulting from the trivial manipulation

$$L^{-1} - M^{-1} = M^{-1}ML^{-1} - M^{-1}LL^{-1} = M^{-1}(M - L)L^{-1}.$$

We may continue manipulating to obtain

$$(L^{-1} - M^{-1})(id + (M - L)L^{-1}) = L^{-1}(M - L)L^{-1}$$

and then, for fixed L and small $|M - L|$,

$$L^{-1} - M^{-1} = \sum_{r=0}^{\infty} (-1)^r L^{-1}((M - L)L^{-1})^{r+1}.$$

We use this in the particular case

(4.46) $(T - \lambda(id))^{-1} - (T' - \lambda(id))^{-1}$

$$= \sum_{r=0}^{\infty} (-1)^r (T - \lambda(id))^{-1}((T' - T)(T - \lambda(id))^{-1})^{r+1}$$

showing that $R(\lambda)$ depends analytically on T, for fixed λ.

We define a non-negative real number $r(T)$ by

$$r(T) = \lim_{n \to \infty} \sup |T^n|^{1/n}.$$

Notice that $r(T) \leq |T|$ (where $|\ |$ denotes both the given norm on **E** and the induced norm on $L(\mathbf{E})$). The series $-\sum_{n=0}^{\infty} \lambda^{-n-1} T^n$ converges to $R(\lambda)$ for $|\lambda| > r(T)$, and diverges for $|\lambda| < r(T)$. Since this is a Laurent series for R centred on 0, we deduce that $r(T)$ is precisely the *spectral radius*

$$\sup \{|\lambda| : \lambda \in \sigma(T)\}$$

of T. Since $\sigma(T)$ is bounded and closed it is compact. Moreover, since $\mathbf{E} \neq \{0\}$, $\sigma(T)$ is non-empty. For, otherwise, the domain of R is **C**, so the above Laurent series is a Taylor series, which is impossible, since the coefficient of λ^{-1} is *id*.

It is possible to simplify the formula for $r(T)$. We observe that, for all $n > 0$, $\lambda \in \sigma(T)$ implies $\lambda^n \in \sigma(T^n)$, because $T - \lambda(id)$ is a factor of $T^n - \lambda^n(id)$. In this case, $|\lambda^n| \leq |T^n|$, and so $|\lambda| \leq |T^n|^{1/n}$. Thus $r(T) \leq \lim_{n \to \infty} \inf |T^n|^{1/n}$, and it follows that $\lim_{n \to \infty} |T^n|^{1/n}$ exists and equals $r(T)$. We now give a very useful characterization of $r(T)$, due to Holmes [1].

(4.47) Theorem. *Let N be the set of all norms on $L(\mathbf{E})$ induced by norms on $\mathbf{E}$ equivalent to the given norm $|\ |$. Then*

$$r(T) = \inf \{\|T\| : \|\ \| \in N\}.$$

Proof. It follows from the definition of $r(T)$ that $r(T) \leq \|T\|$ for all $\|\ \| \in N$. Suppose that $r(T) < \kappa < |T|$. Then, for some integer $m > 0$ and all $n \geq m$, $|T^n| \leq \kappa^n$. We define a norm $\|\ \|$ on **E** by

$$\|x\| = \max \{|x|, |\kappa^{-1} T(x)|, \ldots, |\kappa^{-m+1} T^{m-1}(x)|\}.$$

It is easy to check that $\| \ \|$ is a norm. Moreover, $|x| \leq \|x\|$, so, by the interior mapping theorem, $|\ |$ is equivalent to $\| \ \|$. Now

$$\|T(x)\| = \max\{|T(x)|, |\kappa^{-1}T^2(x)|, \ldots, |\kappa^{-m+1}T^m(x)|\}$$

$$= \kappa \max\{|\kappa^{-1}T(x)|, \ldots, |\kappa^{-m+1}T^{m-1}(x)| \ |\kappa^{-m}T^m(x)|\}$$

$$\leq \kappa\|x\|,$$

since $|\kappa^{-m}T^m| \leq 1$. Thus $\|T\| \leq \kappa$. $\square$

In particular:

(4.48) Corollary. *The spectral radius $r(T)$ of T is strictly less than 1 if and only if there is an equivalent norm on $\mathbf{E}$ with respect to which T is a metric contraction.* $\square$

(4.49) Exercise. Prove that if T is a linear automorphism then $\sigma(T^{-1}) = \sigma(T)^{-1} (= \{\lambda^{-1}: \lambda \in \sigma(T)\})$. Deduce that $\sigma(T)$ lies outside the unit circle S^1 of $\mathbf{C}$ if and only if T is a metric expansion with respect to some equivalent norm on $\mathbf{E}$. Deduce that if $\mathbf{E}$ has a T-invariant direct sum decomposition $\mathbf{E} = \mathbf{E}_s \oplus \mathbf{E}_u$, and if T is a metric contraction on $\mathbf{E}_s$ and a metric expansion on $\mathbf{E}_u$, then T is hyperbolic (i.e. $\sigma(T) \cap S^1$ is empty).

If $p(z)$ is a complex polynomial, say $p(z) = \sum_{n=0}^{s} \alpha_n z^n$, where $\alpha_n \in \mathbf{C}$, then we can define a function $p: L(\mathbf{E}) \to L(\mathbf{E})$ (by an abuse of notation) by $p(T) = \sum_{n=0}^{s} \alpha_n T^n$, and we can handle power series in the same way. The technique of contour integration allows us to carry this process one step further. Let $f: U \to \mathbf{C}$, where U is open in $\mathbf{C}$, be an analytic function. Suppose that $\sigma(T) \subset K \subset U$, where the boundary ∂K of the compact set K is a union of contours (K need not be connected). We define $f(T) \in L(\mathbf{E})$ by

(4.50) $$f(T) = -\frac{1}{2\pi i} \int_{\partial K} f(\lambda) R(\lambda) \, d\lambda,$$

where ∂K is positively oriented with respect to K in the usual way. Note that $f(T)$ is well defined (by (4.45)), is independent of the choice of K, and that we obtain an analytic (by (4.46)) function f from the open subset $\{T \in L(\mathbf{E}): \sigma(T) \subset U\}$ into $L(\mathbf{E})$. The proofs of the following propositions appear in § 3 of Chapter 7 of Dunford and Schwartz [1]).

(4.51) Proposition. *For all $\alpha, \beta \in \mathbf{C}$ and all analytic functions $f, g: U \to \mathbf{C}$, $(\alpha f + \beta g)(T) = \alpha f(T) + \beta g(T)$.* $\square$

(4.52) Proposition. *For all analytic functions $f, g: U \to \mathbf{C}$, let $g.f: U \to \mathbf{C}$ be the map whose value at z is the product $g(z)f(z)$. Then $(g.f)(T)$ is the composite $g(T)f(T)$ of the maps $g(T)$ and $f(T)$.* $\square$

(4.53) Proposition. *If* $f(z) = \sum_{r=0}^{\infty} \alpha_r z^r$ *converges on a neighbourhood of* $\sigma(T)$, *then* $f(T) = \sum_{r=0}^{\infty} \alpha_r T^r$.   □

(4.54) Proposition. *If* $\sigma(T)$ *is contained in the domain of the composite gf of analytic maps f and g then* $(gf)(T) = g(f(T))$.   □

(4.55) Theorem. *(Dunford's spectral mapping theorem). For all analytic maps* $f: U \to \mathbf{C}$, $f(\sigma(T)) = \sigma(f(T))$.   □

(4.56) Corollary. *(Spectral decomposition theorem). Let* $T \in GL(\mathbf{E})$, *and let D be a contour in the Argand diagram* $\mathbf{C}$ *such that* $\sigma(T) \cap D$ *is empty. Let* $\sigma_s(T)$ *be the part of* $\sigma(T)$ *inside D and let* $\sigma_u(T)$ *be the part outside D. Then there is a direct sum decomposition* $\mathbf{E} = \mathbf{E}_s \oplus \mathbf{E}_u$ *into T-invariant closed subspaces such that, if* $T_s \in L(\mathbf{E}_s)$ *and* $T_u \in L(\mathbf{E}_u)$ *are the restrictions of T, then* $\sigma(T_s) = \sigma_s(T)$ *and* $\sigma(T_u) = \sigma_u(T)$.

Proof. We define the analytic complex function f on the complement of D in C by

$$f(\lambda) = \begin{cases} 1 & \text{if } \lambda \text{ is inside } D, \\ 0 & \text{if } \lambda \text{ is outside } D. \end{cases}$$

Let K_s (resp. K_u) be a compact subset inside (resp. outside) D and containing $\sigma_s(T)$ (resp. $\sigma_u(T)$), and let $K = K_s \cup K_u$. We suppose further that ∂K is a union of contours, so that $f(T)$ may be defined by the formula (4.50). By Proposition 4.52 since $f \cdot f = f, f(T)^2 = f(T)$. That is to say, $f(T)$ is a projection. Thus $\mathbf{E} = \mathbf{E}_s(T) \oplus \mathbf{E}_u(T)$, where $\mathbf{E}_s(T)$ is the kernel of $id - f(T)$ (or, equivalently, the image of $f(T)$), and $\mathbf{E}_u(T)$ is the kernel of $f(T)$. Since T commutes with $f(T)$ (by Proposition 4.52 *any* two functions of T commute), $\mathbf{E}_s(T)$ and $\mathbf{E}_u(T)$ are invariant under T. Moreover $Tf(T) = T_s \oplus 0 = h(T)$, where, by Proposition 4.52,

$$h(\lambda) = \begin{cases} \lambda & \text{if } \lambda \text{ is inside } D, \\ 0 & \text{if } \lambda \text{ is outside } D. \end{cases}$$

By Theorem 4.55,

$$\sigma(Tf(T)) = h(\sigma(T)) = \sigma_s(T) \cup \{0\}$$

(or $\sigma_s(T)$, if $\sigma_u(T)$ is empty). Since T_s is an automorphism $0 \notin \sigma(T_s)$, and hence $\sigma(T_s) = \sigma_s(T)$. Similarly $\sigma(T_u) = \sigma_u(T)$.   □

The above theory is for a complex Banach space. The corresponding theory for real spaces can be extracted easily from it, as follows.

Suppose that $\mathbf{E}$ is a real Banach space and that $T \in GL(\mathbf{E})$, the space of all (topological) $\mathbf{R}$-linear automorphisms of $\mathbf{E}$. We form the space $\tilde{\mathbf{E}} = \mathbf{E} \otimes_{\mathbf{R}} \mathbf{C}$ (any point of which can be written uniquely as $x \otimes 1 + y \otimes i$, where $x, y \in \mathbf{E}$).

The space $\tilde{E}$ has a natural complex vector space structure (in which scalar multiplication acts on the second factor C of $\tilde{E}$) and we may give it a norm $|x \otimes 1 + y \otimes i| = \max\{|x|, |y|\}$, which makes it into a Banach space. Then T induces a C-linear automorphism $\tilde{T} \in L(\tilde{E})$ by its action on the first factor E. We define the spectrum $\sigma(T)$ of T to be $\sigma(\tilde{T})$. The crucial fact that emerges is that, if the spectrum of $\tilde{T}$ decomposes as in Corollary 4.56 and if D is symmetric about the real axis, then the projection $f(\tilde{T})$ in the proof of the Corollary acts independently on the real and imaginary parts of $\tilde{E}$, and so decomposes the real space E. Once we have proved this, the real version of Corollary 4.56 follows immediately.

(4.57) Theorem. *Let the spectrum of $\tilde{T}$ decompose as in Corollary 4.56. Suppose that D is symmetric about the real axis. Then the projection $\tilde{P} = f(\tilde{T})$ is of the form $\tilde{P}(x \otimes 1 + y \otimes i) = P(x) \otimes 1 + P(y) \otimes i$, for all $x, y \in E$, where P is a projection of E.*

Proof. Any $R \in L(\tilde{E})$ may be written in matrix form as

$$R = \begin{bmatrix} A & B \\ C & D \end{bmatrix}$$

where the entries are in $L(E)$. Let $\bar{R}$ denote the matrix

$$\begin{bmatrix} A & -B \\ -C & D \end{bmatrix}.$$

Note that, for all $z \in \tilde{E}$, the complex conjugate $\overline{Rz} = \bar{R}\bar{z}$, and hence, for all $R_1, R_2 \in L(E)$, $\overline{R_1 R_2} = \bar{R}_1 \bar{R}_2$.

Now recall that $\tilde{P} = f(\tilde{T})$ is defined by (4.50), where K is as described in the proof of Corollary 4.56. Moreover we may choose K so that ∂K is symmetric about the real axis. Now, in the formula (4.50), $R(\lambda)$ denotes the resolvent $(\tilde{T} - \lambda(id))^{-1}$. By conjugating the relation $(\tilde{T} - \lambda(id))R(\lambda) = id$, we see that $\overline{R(\lambda)} = R(\bar{\lambda})$. Let ∂K_+ and ∂K_- be the portions of ∂K in the upper and lower half planes of the Argand diagram, with orientation induced by that of ∂V. Then by the symmetry, $\partial K_- = -\overline{\partial K_+}$. Thus

$$\tilde{P} = \frac{1}{2\pi i}\left(\int_{\partial K_+} R(\lambda)\, d\lambda + \int_{\partial K_-} R(\lambda)\, d\lambda \right)$$

$$= \frac{1}{2\pi i}\left(\int_{\partial K_+} R(\lambda)\, d\lambda - \int_{\partial K_+} \overline{R(\bar{\lambda})\, d\lambda} \right)$$

$$= \frac{1}{2\pi}\int_{\partial K_+} \left(\frac{1}{i}R(\lambda)\, d\lambda + \overline{\frac{1}{i}R(\lambda)\, d\lambda} \right).$$

Since the term in the bracket acts separately on the real and complex parts of $\tilde{\mathbf{E}}$, so does $\tilde{P}$. Thus $\tilde{P}$, being a projection, has the form

$$\tilde{P}(x \otimes 1 + y \otimes i) = P(x) \otimes 1 + Q(y) \otimes i.$$

But since $\tilde{P}$ is $\mathbf{C}$-linear $\tilde{P}(iz) = i\tilde{P}(z)$, and this implies that $P = Q$.   □

It follows immediately that the stable summand $\tilde{T}_s : \mathbf{E}_s \to \mathbf{E}_s$ of $\tilde{T}$ is the complexification of an $\mathbf{R}$-linear automorphism $T_s : \mathbf{E}_s \to \mathbf{E}_s$ restricting T. Similarly for the unstable summand. Summing up:

(4.58) Corollary. (*Real spectral decomposition theorem*) *Let* $\mathbf{E}$ *be a real Banach space, let* $T \in GL(\mathbf{E})$ *and let* D *be a contour in the Argand diagram, symmetric about the real axis, such that* $\sigma(T) \cap D$ *is empty. Let* $\sigma_s(T)$ *be the part of* $\sigma(T)$ *inside* D *and let* $\sigma_u(T)$ *be the part outside* D. *Then there is a direct sum decomposition* $\mathbf{E} = \mathbf{E}_s \oplus \mathbf{E}_u$ *into* T-*invariant closed subspaces such that if* $T_s \in L(\mathbf{E}_s)$ *and* $T_u \in L(\mathbf{E}_u)$ *are the restrictions of* T, *then* $\sigma(T_s) = \sigma_s(T)$ *and* $\sigma(T_u) = \sigma_u(T)$.   □

CHAPTER 5

Linearization

We can hardly expect to make much headway in the global theory of dynamical systems on a smooth manifold X if we are completely ignorant of how systems behave locally (that is to say, near a point). We should like to attempt a local classification, but, bearing in mind our difficulties with linear systems, we do not expect complete success. However, regular (as opposed to fixed) points present no problems. As we shall see in the following section, there is essentially only one type, represented in the case of diffeomorphisms by a translation of $\mathbf{E}$, the space on which X is modelled, and in the case of vector fields by any non-zero constant vector field on $\mathbf{E}$. Thus, in both cases, interesting local behaviour occurs only at fixed points.

As we have seen in the previous chapter, linear systems exhibit a wide variety of behaviour near the fixed point zero. Moreover, if we introduce higher order terms, we admit further possibilities. For example, the vector field $v(x) = x^2$ on $\mathbf{R}$ has a "one-way zero" (see Figure 5.1), but no linear

FIGURE 5.1

vector field has an orbit structure looking like this. In order to prevent our classification programme from getting out of hand, we pick on a class of fixed points that occur often enough to be thought of as "the sort that one usually comes across". In the case of linear systems we were able to give this rather vague notion a precise topological meaning in terms of the space of all linear systems on a given Banach space $\mathbf{E}$. When we generalize to smooth systems on a given manifold X we are still able to topologize the set of all systems, and to discuss open and dense subsets of the resulting space (see Chapter 7 and Appendix B below). Of course the space of all systems is infinite dimensional even when X is finite dimensional. Although the connection may seem tenuous at the moment, that is, in fact, one reason why we try to make our theory work for infinite dimensional **E** whenever possible.

Suppose that a given dynamical system ϕ has a fixed point p. If we are only interested in local structure, we may assume, after taking a chart at p, that $X = \mathbf{E}$ and $p = 0$. Suppose now that by altering ϕ slightly near 0 we always end up with another fixed point near 0, with a local phase portrait resembling that of ϕ at 0. In particular, then, the *linear approximation* of ϕ at 0 (if we can make sense of this term) must have a phase portrait near 0 resembling that of ϕ near 0, as also must all nearby linear systems. This rough and ready argument, together with the stability theorems of Chapter 4, suggests that we consider fixed points whose linear approximations are *hyperbolic systems* (automorphisms or vector fields as the case may be). Such points are said to be *hyperbolic fixed points*. We now give some precise definitions.

A fixed point p of a diffeomorphism f of X is *hyperbolic* if the tangent map $T_p f : T_p X \to T_p X$ is a hyperbolic linear automorphism. Corollary 4.27 has supplied us with a classification of hyperbolic linear automorphisms, and it turns out that this amounts to a local classification of their fixed points. We extend this to a classification of hyperbolic fixed points using the result due to Hartman that any hyperbolic fixed point p of f is topologically conjugate to the fixed point 0 of $T_p f$ (see Chapter 2 for the definition of local topological conjugacy). This theorem does not require X to be finite dimensional.

(5.2) Exercise. (i) Prove that p is a hyperbolic fixed point of f if and only if it is a hyperbolic fixed point of f^{-1}.

(ii) Find an example of a diffeomorphism f having a fixed point p that is not topologically conjugate to the fixed point 0 of $T_p f$.

As usual, an analogous situation exists with flows. A fixed point p of a local integral $\phi : I \times U \to X$ of a vector field v on X (or equivalently, a zero p of v) is said to be *hyperbolic* (or *elementary*) if for some (and hence as we shall see any), $t \neq 0$ in I, p is a hyperbolic fixed point of ϕ^t. Let us examine the implications of this definition. Any admissible chart $\xi : U \to U'$ at p induces a vector field $w = (T\xi)v\xi^{-1}$ on U'. We may assume $\xi(p) = 0$. Now p is hyperbolic if and only if 0 is a hyperbolic fixed point of $\psi^t = \xi\phi^t\xi^{-1}$, and by Theorem 3.11 ψ is a local integral of w at 0. Let f be the principal part of w. By Theorem 3.32, $T(t) = D\psi^t(0) \in L(\mathbf{E})$ satisfies the differential equation $DT = Df(0)T$. Moreover $D\psi^0(0) = id$. By Proposition 4.33, $D\psi^t(0) = \exp(tDf(0))$, and thus, by Proposition 4.32, p is hyperbolic if and only if $Df(0) \in EL(\mathbf{E})$. Summing up:

(5.3) Proposition. *The point p is a hyperbolic zero of a vector field v if and only if, for any chart ξ at p, the differential at $\xi(p)$ of (the principal part of) the induced vector field $(T\xi)v\xi^{-1}$ is a hyperbolic vector field.* □

An equivalent, but rather more sophisticated, approach is to consider $Tv: TX \to T(TX)$, which is a section of the vector bundle $T\pi_X: T(TX) \to TX$, where $\pi_X: TX \to X$ is the tangent bundle of X. Now $T\pi_X$ may be identified with the tangent bundle on TX, $\pi_{TX}: T(TX) \to TX$ by the canonical involution (see Example A.38 of Appendix A) and Tv then becomes a vector field on TX. With this identification $T_p v$ maps $T_p X$ into $T(T_p X)$. Thus $T_p v$ is a linear vector field on $T_p X$. We call this linear vector field the *Hessian* of v at p. Again, p is a hyperbolic zero of v if and only if the Hessian of v at p is a hyperbolic linear vector field.

(5.4) Exercise. Justify in detail the statements made in the preceding paragraph.

(5.5) Exercise. Let p be a critical point of a smooth function $g: R^n \to R$. Relate the Hessian of g at p in the classical sense to the Hessian of the gradient vector field $v = \nabla g$ (see Examples 3.3 and A.57) at p in the sense described above.

We prove in Corollary 5.26 below that if p is a hyperbolic fixed point of a vector field v then it is flow equivalent to the zero of the Hessian of v at p. Thus, using Exercise 4.44, we obtain a complete classification of hyperbolic fixed points of flows on real n-dimensional manifolds into $n + 1$ classes, distinguished from one another by the dimension of the stable manifold (of the Hessian, let us say at this stage). Once again, the importance of the classification lies in the comparative ubiquity of the type of point under consideration (see Chapter 7).

(5.6) Exercise. Give an example of a smooth vector field v having a zero p that is not topologically equivalent to the zero of the Hessian of v at p.

We complete the chapter by discussing the structural stability of periodic orbits. In the case of a diffeomorphism f there is very little to be said beyond remarking that a periodic point of f is a fixed point of some positive power f^n. On the other hand, closed orbits of flows have a more interesting theory, and once again we are able to classify the types that "usually occur" in finite dimensions.

I. REGULAR POINTS

Recall that a point p of X is a *regular point* of a dynamical system on X if it is not a fixed point of the system. Equivalently, it is a regular point of a vector field v if $v(p) \neq 0_p$. The theorems that follow show that regular points are uninteresting from a local point of view. See Chapter 2 for the definitions of the various types of local equivalence in use.

(5.7) Theorem. *If p is a regular point of a diffeomorphism $f: X \to X$ and g is translation of the model space $\mathbf{E}$ of X by a non-zero vector x_0 then $f|p$ is topologically conjugate to $g|0$.*

Proof. Let $\xi: U \to U'$ be a chart at p with $\xi(p) = 0$. We may assume that $f(U) \cap U$ is empty, and that the diameter of U' is less than $\frac{1}{2}|x_0|$. Then ξ together with $(g|U')\xi(f|U)^{-1}$ is a (smooth) conjugacy from f at p to g at 0. $\qquad\square$

Of course the above proof has little to do with the smoothness; it works in the context of homeomorphisms of Hausdorff topological spaces for any pair of regular points with homeomorphic neighbourhoods. Moreover, as we commented in Chapter 2, its simplicity could be taken as a criticism of our definition of local topological conjugacy. The analogous theorem for flows has a more secure status, and a correspondingly harder proof.

(5.8) Theorem. *(Rectification theorem) Let p be a regular point of a C^1 flow ϕ on X. Let x_0 be any non-zero vector of the model space $\mathbf{E}$ of X, and let ψ be the flow on $\mathbf{E}$ defined by $\psi(t, x) = x + tx_0$. Then $\phi|p$ is flow equivalent to $\psi|0$.*

Proof. By taking an admissible chart, we can assume that $p = 0 \in \mathbf{E}$ and that ϕ is the local integral of a vector field v on some open subset of $\mathbf{E}$. Moreover we can assume after a linear conjugacy that $\mathbf{E} = \mathbf{R} \times \mathbf{F}$ for some Banach space $\mathbf{F}$ and that $v(0) = x_0 = (1, 0) \in \mathbf{R} \times \mathbf{F}$. Let h be the C^1 map defined on some neighbourhood of 0 in $\mathbf{E} = \mathbf{R} \times \mathbf{F}$ by

$$h(u, y) = \phi(u, (0, y))$$

(see Figure 5.9).

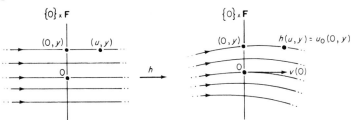

FIGURE 5.9

Then h preserves the action of $\mathbf{R}$, since for all sufficiently small s and $t \in \mathbf{R}$ and $y \in \mathbf{F}$,

$$h\psi(t, (s, y)) = h(s + t, y)$$
$$= \phi(s + t, (0, y))$$
$$= \phi(t, \phi(s, (0, y)))$$
$$= \phi(t, h(s, y)).$$

Since h is the identity on $\{0\} \times \mathbf{F}$, and since

$$D_1 h(0) = D_1 \phi(0, 0) = v(0) = (1, 0),$$

it is clear that $Dh(0): \mathbf{E} \to \mathbf{E}$ is the identity. Thus, by the inverse mapping theorem (see Exercise C.11 of Appendix C), h is a C^1 diffeomorphism between suitably restricted neighbourhoods of 0 in $\mathbf{E}$. Thus h is a (C^1) flow equivalence from ψ at 0 to ϕ at 0. □

(5.10) Remark. We emphasize that in both the above theorems the equivalence obtained is as smooth as the system under consideration.

II. HARTMAN'S THEOREM

Let T be a hyperbolic linear automorphism of $\mathbf{E}$ with skewness $a < 1$ (see Theorem 4.19) with respect to some norm $| \ |$ on $\mathbf{E}$. Basically, we are interested in showing the stability of T with respect to topological conjugacy under small smooth perturbations, or, more generally, under small Lipschitz perturbations. The central pillar of this theory is the theorem of Hartman [1], which we present below in a version strongly influenced by work of Moser [1] and Pugh [3].

It is immediate from the definition of hyperbolicity that the map $id - T$ is an automorphism. In fact, by the proof of Theorem 4.20 we have the further information that $|(id - T)^{-1}| \leq (1 - a)^{-1}$. The first statement implies that T has a unique fixed point (namely the origin), and we start by proving that this property is stable under Lipschitz perturbations of T.

(5.11) Lemma. *Let* $\eta: \mathbf{E} \to \mathbf{E}$ *be Lipschitz with constant* $\kappa < 1 - a$. *Then* $T + \eta$ *has a unique fixed point.*

Proof. The fixed point set of $T + \eta$ is precisely that of $(id - T)^{-1}\eta$, which is a contraction. Thus the contraction mapping theorem (C.5 of Appendix C) gives the result. □

(5.12) Exercise. Prove that the fixed point of $T + \eta$ depends in a C^r fashion on the map η regarded as an element of the map space $C^r(\mathbf{E})$ (see Appendix B). (*Hint:* Consider the function from $\mathbf{E} \times C^r(\mathbf{E})$ to $\mathbf{E}$ sending (x, η) to $(id - T)^{-1}\eta(x)$.)

(5.13) Exercise. Let B be the closed ball in $\mathbf{E}$ with centre 0 and finite radius $b > 0$. Prove that if $\eta: B \to \mathbf{E}$ is Lipschitz with constant $\kappa < 1 - a$ and if $|\eta|_0 \leq b(1 - a)$ then $T + \eta: B \to \mathbf{E}$ has a unique fixed point.

Recall (Appendix B) that $C^0(\mathbf{E})$ is the Banach space of all bounded continuous maps from $\mathbf{E}$ to $\mathbf{E}$, with the sup norm.

(5.14) Theorem. (*Hartman's linearization theorem*) *Let* $\eta \in C^0(\mathbf{E})$ *be Lipschitz with constant* $\kappa < \min\{1 - a, |T_s^{-1}|^{-1}\}$. *Then* $T + \eta$ *is topologically conjugate to* T.

It is technically convenient to prove a more detailed statement, as follows:

(5.15) Theorem. *Let* η, $\zeta \in C^0(\mathbf{E})$ *be Lipschitz with constant* κ *less than* $\min\{1 - a, |T_s^{-1}|^{-1}\}$. *Then there exists a unique* $g \in C^0(\mathbf{E})$ *such that*

(5.16) $(T + \eta)(id + g) = (id + g)(T + \zeta).$

Moreover, $id + g$ *is a homeomorphism, and is thus a topological conjugacy from* $T + \zeta$ *to* $T + \eta$.

Proof. By the Lipschitz inverse mapping theorem (Exercise C.11 of Appendix C), the condition $\kappa < |T_s^{-1}|^{-1}$ implies that $T + \zeta$ is a homeomorphism. Thus we may write (5.16) as

(5.17) $(T + \eta)(id + g)(T + \zeta)^{-1} = (T + \zeta)(T + \zeta)^{-1} + g.$

We define maps $\tilde{T}$ and $\tilde{\eta}$ of $C^0(\mathbf{E})$ into itself by

$$\tilde{T}(g) = Tg(T + \zeta)^{-1}$$

and

$$\tilde{\eta}(g) = \eta(id + g)(T + \zeta)^{-1} - \zeta(T + \zeta)^{-1}$$

so that (5.17) now reads $(\tilde{T} + \tilde{\eta})(g) = g$. Clearly $\tilde{\eta}$ is Lipschitz with constant κ. Equally clearly $\tilde{T}$ is a linear automorphism with inverse $g \mapsto T^{-1}g(T + \zeta)$. We also observe that $\tilde{T}$ is hyperbolic with skewness a, since it contracts and expands its invariant subspaces $C^0(\mathbf{E}, \mathbf{E}_s(T))$ and $C^0(\mathbf{E}, \mathbf{E}_u(T))$ in the correct proportions, and $C^0(\mathbf{E})$ is the direct sum of these subspaces. We now apply Lemma 5.11 and deduce the existence of a unique fixed point f of $\tilde{T} + \tilde{\eta}$. This gives the first part of the theorem.

Finally, let $g' \in C^0(\mathbf{E})$ be the unique map corresponding to g when η and ζ are interchanged in (5.15). Thus

$(T + \eta)(id + g)(id + g') = (id + g)(T + \zeta)(id + g') = (id + g)(id + g')(T + \eta).$

But also $(T + \eta)id = id(T + \eta)$. By the uniqueness in the first part of the theorem (applied with $\eta = \zeta$), $(id + g)(id + g') = id$. Similarly

$$(id + g')(id + g) = id.$$

Thus $id + g$ is a homeomorphism, since g and g' are continuous.   □

(5.18) Exercise. Show that the hypothesis that g is bounded is necessary for uniqueness in Theorem 5.15. (*Hint*: Put $\mathbf{E} = \mathbf{R}$, and $\eta = \zeta = 0$.)

(5.19) Exercise. Obtain an alternative proof of Hartman's theorem by considering the map from $C^0(\mathbf{E})$ to $C^0(\mathbf{E})$ taking $g = (g_s, g_u)$ to

$$((T_s g_s + \eta_s(id + g) - \zeta_s)(T + \zeta)^{-1}, T_u^{-1}(\zeta_u + g_u(T + \zeta) - \eta_u(id + g)),$$

where $C^0(\mathbf{E}) = C^0(\mathbf{E}, \mathbf{E}_s(T)) \times C^0(\mathbf{E}, \mathbf{E}_u(T))$.

It is natural to ask how smooth the conjugacy $id + g$ is in Theorem 5.15, and tempting to suppose that it is just as smooth as the maps $T + \eta$ and $T + \zeta$ that it conjugates. However, a moment's thought shows that this cannot always be the case. If η, ζ and g are C^1, we may differentiate (5.16) at the unique fixed point p of $T + \zeta$, and deduce that the differential of $T + \zeta$ at p is similar to the differential of $T + \eta$ at its fixed point $p + g(p)$. Thus we have a necessary condition for smoothness of g that is quite a heavy restriction on η and ζ. To see this, put $\zeta = 0$ and $\eta(0) = 0$, and observe that, however small we make η and its derivatives, the automorphism $T + D\eta(0)$ will not usually be similar to T. Moreover, this similarity condition is not sufficient for C^1 conjugacy. We return to this and similar questions in the appendix to this chapter.

Our main application of Hartman's theorem is to compare a diffeomorphism near a hyperbolic fixed point with its linear approximation at the point. The result is a local one, whereas, in Hartman's theorem as stated above, maps are defined on the whole of $\mathbf{E}$. Thus the proof of the result consists of extending local maps to global ones.

(5.20) Corollary. *A hyperbolic fixed point p of a C^1-diffeomorphism $f: X \to X$ is topologically conjugate to the fixed point 0 of $T_p f$.*

Proof. We can assume, after conjugation with an admissible chart at p, that we are working in the model space $\mathbf{E}$. We may suppose that $p = 0$ and that f is a C^1 diffeomorphism mapping some open neighbourhood U of 0 onto an open neighbourhood $f(U)$ of 0. We are given that the differential $Df(0)$ is hyperbolic. Let the skewness of $Df(0)$ be a with respect to some norm $|\ |$ on $\mathbf{E}$, and let κ be some positive number smaller than $\min\{1 - a, |Df(0)_s^{-1}|^{-1}\}$. Since f is C^1, there is some closed ball B centre 0 radius b on which $|Df(x) - Df(0)| \leqslant \frac{1}{2}\kappa$. This implies that $f - Df(0)$ is Lipschitz with constant $\frac{1}{2}\kappa$ on B. Define $\eta: \mathbf{E} \to \mathbf{E}$ by

$$\eta(x) = \begin{cases} f(x) - Df(0)(x) & \text{if } |x| \leqslant b, \\ f\left(\dfrac{bx}{|x|}\right) - Df(0)\left(\dfrac{bx}{|x|}\right) & \text{if } |x| \geqslant b. \end{cases}$$

We assert that η is Lipschitz with constant κ. We must prove that, for x and x' in $\mathbf{E}$, $|\eta(x) - \eta(x')| \leqslant \kappa|x - x'|$, where (as we may suppose) $|x'| \leqslant |x|$. The case $|x| \leqslant b$ is trivial, and the case $b < |x'|$ reduces to the case $|x'| \leqslant b \leqslant |x|$, since multiplying the vectors by $b/|x'|$ leaves the left-hand side unaltered and decreases the right-hand side. Suppose now that $|x'| \le b \le |x|$. Then

$$|\eta(x) - \eta(x')| = \left| \eta\left(\frac{bx}{|x|}\right) - \eta(x') \right|$$

$$\le \tfrac{1}{2}\kappa \left| \frac{bx}{|x|} - x' \right|$$

$$\le \tfrac{1}{2}\kappa (|x - x'| + |x| - b)$$

$$\le \tfrac{1}{2}\kappa (|x - x'| + |x| - |x'|)$$

$$\le \kappa |x - x'|,$$

as required. Now η is also bounded. By Hartman's theorem there is a topological conjugacy h, say, from $Df(0)$ to $Df(0) + \eta$. Let V be a neighbourhood of 0 such that $h(V) \subset B$. Then, on V,

$$fh = (Df(0) + \eta)h = hDf(0).$$

Thus f is conjugate at 0 to $Df(0)$ at 0.   □

(5.21) Corollary. *There are $4n$ topological conjugacy classes of hyperbolic fixed points that occur on n-dimensional manifolds.*

Proof. This follows immediately from Corollary 5.20 and the classification of hyperbolic linear homeomorphisms of $\mathbf{R}^n$ in Corollary 4.27.   □

We call a fixed point p of a diffeomorphism $f: X \to X$ *structurally stable* if for each sufficiently small neighbourhood U of p in X there is a neighbourhood V of f in Diff X such that, for all $g \in V$, g has a unique fixed point q in U and q is topologically conjugate to p. Even although we have not discussed the topology of Diff X in this chapter, we have assembled all the facts that are needed to prove that fixed points are structurally stable if (and, in finite dimensions, only if) they are hyperbolic. Since we are discussing local phenomena, we can work in the model space $\mathbf{E}$. The basic essentials are contained in the following exercise.

(5.22) Exercise. Let $f: U \to f(U)$ be a diffeomorphism of open subsets of $\mathbf{E}$ with a fixed point at 0. Suppose that 0 is hyperbolic, and that $Df(0)$ has skewness a with respect to the norm $|\ |$ on $\mathbf{E}$. Choose κ, with $0 < \kappa < 1 - a$, so small that, for all $T \in L(\mathbf{E})$ with $|T - Df(0)| \le \kappa$, T is hyperbolic and topologically conjugate to $Df(0)$. Let B be a closed ball in E with centre 0 and radius $b(<1)$ such that $|Df(x) - Df(0)| \le \kappa/2$ for all $x \in B$. Use Exercise 5.13 to prove that, for all C^1 maps $g: U \to E$ with $|g - f|_1 \le \kappa b/2$, g has a unique fixed point p in B. Show that $g|p$ is topologically conjugate to $f|0$, and hence that 0 is structurally stable under C^1-small perturbations of f.

Conversely, prove that if **E** is finite dimensional and if 0 is structurally stable under C^1-small perturbations of f then 0 is hyperbolic.

We have already discussed at some length in Chapter 4 the stability of linear automorphisms with respect to topological conjugacy under linear perturbations. As we commented there, it is possible to prove that hyperbolic automorphisms are stable in this sense (the result of Theorem 4.30) as a corollary of Hartman's theorem. This is just a matter of getting a local conjugacy at 0 between the original and the perturbed maps (one first doctors the perturbation to get it into $C^0(\mathbf{E})$) and then extending the local conjugacy to a global one, using the conjugacy relation much as in the proof of Theorem 4.24. We leave the details to the interested reader in the following pair of exercises.

(5.23) Exercise. Let $f: \mathbf{E} \to \mathbf{E}$ be Lipschitz with constant $\kappa > 0$. Given $\varepsilon > 0$, find a map $g \in C^0(\mathbf{E})$ such that g is Lipschitz with constant $\kappa + \varepsilon$ and $g = f$ on the unit ball in **E**. (*Hint*: Assume $f(0) = 0$, and choose $N > 0$ with $1/N < \varepsilon/2\kappa$. Prove that the map $\rho: \mathbf{E} \to \mathbf{R}$ defined by

$$\rho(x) = \begin{cases} 1 & \text{if } |x| \le 1 \\ (N+1-|x|)/N & \text{if } 1 \le |x| \le N+1 \\ 0 & \text{if } N+1 \le |x| \end{cases}$$

is Lipschitz with constant $1/N$. Deduce that $g = \rho \cdot f$ is locally Lipschitz with constant $\kappa + \varepsilon$, and hence Lipschitz with constant $\kappa + \varepsilon$.)

(5.24) Exercise. Let $T \in HL(\mathbf{E})$ have skewness a with respect to the norm $|\ |$ on **E**. Prove that any $T' \in L(\mathbf{E})$ with

$$|T' - T| < \min\{1 - a, |T_s^{-1}|^{-1}\}$$

is topologically conjugate to T.

III. HARTMAN'S THEOREM FOR FLOWS

Hartman's linearization theorem was independently discovered, in the flow context, by Grobman [1, 2]. We can readily deduce the global Banach space version for flows from its analogue for diffeomorphisms, as follows:

(5.25) Theorem. (*Hartman and Grobman*) *Let $T \in EL(\mathbf{E})$. For all Lipschitz maps $\eta \in C^0(\mathbf{E})$ with sufficiently small Lipschitz constant, there is a flow equivalence from T to $T + \eta$.*

Proof. Let $\phi: \mathbf{R} \times \mathbf{E} \to \mathbf{E}$ be the integral flow of T. Thus $\phi^1(= \exp T)$ is a hyperbolic automorphism. Suppose that it has skewness $a < 1$ with respect to some equivalent norm $|\ |$ on $\mathbf{E}$. We recall a few necessary facts from Chapter 3. Firstly, the vector field $T + \eta$ has an integral flow, ψ say (Theorem 3.39). Next, the map $\psi^1 - \phi^1$ is in $C^0(\mathbf{E})$ (Theorem 3.45). Finally, the Lipschitz constant of $\psi^1 - \phi^1$ can be made arbitrarily small by decreasing the Lipschitz constant of η (Theorem 3.46). Thus, for the latter constant sufficiently small, we may apply Hartman's theorem to the hyperbolic automorphism ϕ^1 and the perturbation $\psi^1 - \phi^1$, and obtain a topological conjugacy h from ϕ^1 to ψ^1 which (by Theorem 5.15) is unique subject to the condition $h - id \in C^0(\mathbf{E})$. We assert that $\psi^t h = h\phi^t$ for all $t \in \mathbf{R}$. Now, certainly,

$$\psi^1(\psi^t h \phi^{-t}) = \psi^t \psi^1 h \phi^{-t}$$
$$= \psi^t h \phi^1 \phi^{-t}$$
$$= (\psi^t h \phi^{-t})\phi^1.$$

Moreover

$$\psi^t h \phi^{-t} - id = (\psi^t - \phi^t)h\phi^{-t} + \phi^t(h - id)\phi^{-t}$$

is C^0-bounded. Thus, by the above mentioned uniqueness of h, $h = \psi^t h \phi^{-t}$, and h is a flow equivalence from ϕ to ψ. $\square$

Just as with diffeomorphisms, we may now deduce the theorem for hyperbolic fixed points of flows on manifolds. The proof, which we leave to the reader, is that of Corollary 5.20 with the obvious modifications.

(5.26) Corollary. *A hyperbolic zero p of a C^1 vector field on X is flow equivalent to the zero of its Hessian at p.* $\square$

(5.27) Corollary. *Flow equivalence and orbit equivalence coincide for hyperbolic zeros of C^1 vector fields on an n dimensional manifold X. There are precisely $n + 1$ equivalence classes of such points with respect to either relation.*

Proof. Immediate from Corollary 5.26 and Exercise 4.44. $\square$

(5.28) Exercise. Show that the flow equivalence h of Theorem 5.25 is unique subject to the condition $h - id \in C^0(\mathbf{E})$.

(5.29) Exercise. Let v be a C^1 vector field on $\mathbf{E}$ with a zero at p. Prove that if p is hyperbolic then it is stable with respect to flow equivalence under C^1-small perturbations of v. Prove, conversely, that if $\mathbf{E}$ is finite dimensional and if p is stable with respect to topological equivalence under C^1-small perturbations of v then p is hyperbolic.

It is worth emphasizing at this stage that hyperbolicity of fixed points is not an invariant of topological equivalence or of flow equivalence. (We might well have made a similar remark earlier in the context of topological conjugacy.) That is to say, it is perfectly possible for a non-hyperbolic fixed point to be topologically equivalent, or even flow equivalent, to a hyperbolic fixed point. For example, the non-hyperbolic zero 0 of the vector field $v(x) = x^3$ on $\mathbf{R}$ is clearly flow equivalent to the hyperbolic zero 0 of the vector field $w(x) = x$. Suppose that a fixed point p of a flow ϕ is topologically equivalent to a hyperbolic fixed point q of a flow ψ. There are three possibilities: (a) $(T_q\psi^1)_s = T_q\psi^1$, (b) $(T_q\psi^1)_u = T_q\psi^1$, and (c) neither (a) nor (b). We call p (and also, of course, q) in case (a) a *sink*, in case (b) a *source* and in case (c) a *saddle point* (see Figure 5.30). We use the same terminology for fixed points of a diffeomorphism that are topologically conjugate to hyperbolic fixed points.

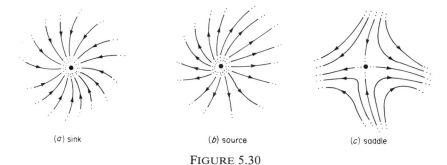

(*a*) sink   (*b*) source   (*c*) saddle

FIGURE 5.30

Any sink p has the property that any positive half orbit $\mathbf{R}^+ . x = \{t . x : t \geq 0\}$ starting at a point x near p stays near p and eventually ends at p, in the sense that $\omega(x) = p$. The point p is said to be *asymptotically stable in the sense of Liapunov*. In the language of differential equations, this type of stability is concerned with the way an individual solution of a system varies as the initial conditions are altered. This should not be confused with structural stability, which deals with the way the set of all solutions varies as we alter the system itself. We shall say some more about Liapunov stability in the appendix to this chapter, which also contains a section on the *index* of a fixed point. This last is an important integer valued invariant of local topological equivalence.

IV. HYPERBOLIC CLOSED ORBITS

Let p be a point on a closed orbit of a C^r flow ϕ ($r \geq 1$) on a manifold X. Suppose that the orbit $\Gamma = \mathbf{R} . p$ of ϕ through p has period τ. Then $\phi^\tau : X \to X$ is a C^r diffeomorphism with a fixed point at p. The tangent map $T_p\phi^\tau$ is a linear homeomorphism of $T_p X$. It keeps the linear subspace $\langle v(x) \rangle$ of $T_p X$ pointwise fixed, where $v(x)$ is the velocity of ϕ at x, because ϕ^τ keeps the orbit Γ pointwise fixed. We say that Γ is *hyperbolic* if $T_p X$ has a $T_p \phi^\tau$-invariant splitting as a direct sum

$$T_p X = \langle v(p) \rangle \oplus F_p$$

where $T_p \phi^\tau | F_p$ is hyperbolic. This definition is independent of choice of p on Γ, for if $q = \phi(t, p)$ then $T_q \phi^\tau = (T_p \phi^t)(T_p \phi^\tau)(T_p \phi^t)^{-1}$, so $T_q \phi^\tau$ and $T_p \phi^\tau$ are linearly conjugate.

Thus hyperbolicity for a closed orbit means that the linear approximation to ϕ^τ is as hyperbolic as possible, bearing in mind that it cannot be hyperbolic in the direction of the orbit. We can get a clear geometrical picture of what this means in terms of *Poincaré maps*. Let Y be some open disc embedded as a submanifold of X through p, such that $\langle v(x) \rangle \oplus T_p Y = T_p X$. We say that Y is *transverse* to the orbit at p, and call it a *cross section* to the flow there. Now $\phi(\tau, p) = p$. We assert that, for some small open neighbourhood U of p in Y, there is a unique continuous function $\rho : U \to \mathbf{R}$ such that $\rho(p) = \tau$ and $\phi(\rho(y), y) \in Y$. The function ρ is a *first return function* for Y. Any two such functions agree on the intersection of their domains; we could have phrased this in the language of germs (see Hirsch [1]). Intuitively $\rho(y)$ is the time that it takes a point starting at y to move along the orbit of the flow in the positive direction until it hits the section Y again. We call the point at which it does so $f(y)$ (see Figure 5.31). Thus $f : U \to Y$ is a map defined by

$$f(y) = \phi(\rho(y), y).$$

We call f a *Poincaré map* for Y. As with ρ, it is well defined up to domain.

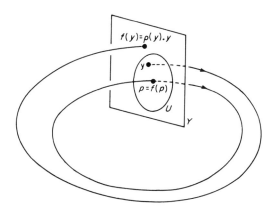

FIGURE 5.31

Note that these definitions do not require Γ to be hyperbolic. In fact, we now show that Γ is hyperbolic precisely when the Poincaré map for some section Y at p has a hyperbolic fixed point at p.

(5.32) Theorem. *For sufficiently small U, the first return function ρ is well defined and C^r, and the Poincaré map f is a well defined C^r diffeomorphism of U onto an open subset of Y. If T_pX has a $T_p\phi^\tau$-invariant splitting $\langle v(p)\rangle \oplus F_p$ then T_pf is linearly conjugate to $T_p\phi^\tau|F_p$. The orbit Γ is hyperbolic if and only if p is a hyperbolic fixed point of f.*

Proof. Notice that by Theorem 5.8 and Remark 5.10 we may identify p and some neighbourhood of it with the origin in the Banach space $\mathbf{E} = \mathbf{R} \times \mathbf{F}$ and some neighbourhood of it, with the flow given locally by

$$\phi(t, (s, y)) = (s + t, y).$$

This is because a C^r flow equivalence affects tangent maps at p only by a linear conjugacy. Under this identification the map ϕ^τ satisfies, near p,

$$\phi^\tau(s, y) = \phi^s(\phi^\tau(0, y)) = \phi^\tau(0, y) + (s, 0).$$

Thus if g and h are maps defined on some neighbourhood V of 0 in $\mathbf{F}$ by

$$\phi^\tau(0, y) = (g(y), h(y)) \in \mathbf{R} \times \mathbf{F}$$

then ϕ^τ is locally of the form

$$\phi^\tau(s, y) = (s + g(y), h(y)).$$

Notice that g and h are C^r and vanish at $y = 0$. Moreover if V is sufficiently small, h is a diffeomorphism of V onto an open subset of $\mathbf{F}$, by the inverse mapping theorem (Exercise C.11 of Appendix C).

We first prove the theorem for the particular cross section $\{0\} \times \mathbf{F}$ (or, rather, some neighbourhood of $p = (0, 0)$ in it). In this case, the first return function ρ is the map $(0, y) \mapsto \tau - g(y)$ and the Poincaré map f is $(0, y) \mapsto (0, h(y))$. If T_pX splits as in the statement of the theorem, we can suppose after a linear automorphism of $\mathbf{E}$ that $F_p = \{0\} \times \mathbf{F}$. Then the differential of ϕ^τ at $p = (0, 0)$ is given by

(5.33) $$D\phi^\tau(p)(s, y) = (s, Dh(0)(y)),$$

and it is clear that $T_p\phi^\tau|F_p$ equals T_pf. It follows immediately that if Γ is hyperbolic then p is a hyperbolic fixed point of f. Suppose conversely that p is a hyperbolic fixed point of f. It is sufficient to prove that T_pX splits as in the statement of the theorem. Now the differential of ϕ^τ at p is given by

(5.34) $$D\phi^\tau(p)(s, y) = (s + Dg(0)(y), Dh(0)(y)).$$

Since $Dh(0)$ is hyperbolic there is some $a > 1$ for which iterates $|Dh(0)^n(y)|$ grow as a multiple of $a^{|n|}|y|$, either for n positive or, if y is on the stable manifold of $Dh(0)$, for n negative. Thus (s, y) is in the eigenspace of $D\phi^\tau(p)$ corresponding to 1 if and only if $y = 0$. One easily checks directly that $D\phi^\tau(p)$ has the same resolvent as the map defined by (5.33), and so its spectrum is that of $T_p f$ together with the number 1. By spectral theory (see Corollary 4.58) $T_p X$ splits as required.

Finally consider an arbitrary cross section Y at p. Let π_1 and π_2 be the product projections of $\mathbf{R} \times \mathbf{F}$. Then π_2 maps some open neighbourhood of p in Y diffeomorphically onto an open neighbourhood of 0 in $\mathbf{F}$. A first return function for Y may be defined for sufficiently small y by the formula $y \mapsto \tau - g\pi_2(y) - \pi_1(y) + \pi_1(\pi_2|Y)^{-1}h\pi_2(y)$ and a Poincaré map by $y \mapsto (\pi_2|Y)^{-1}h\pi_2(y)$. Since the tangent map to the latter at p is linearly conjugate to $Dh(0)$, the proof is complete. □

(5.35) Remarks. The requirement in the statement of Theorem 5.32 that $T_p X$ should split, is not so very inconvenient. One may, in general, speed up or slow down the flow ϕ near p along individual orbits so that the formula (5.34) becomes (5.33). The new flow has the same orbits as the old one and in addition has the splitting property (see the proof of Theorem 5.40 below).

When X is finite dimensional it is clear from (5.34) that

$$\det (D\phi^\tau(p) - \lambda(id)) = (1 - \lambda) \det (Dh(0) - \lambda(id)).$$

Thus, whether or not the splitting occurs, the eigenvalues of $T_p\phi^\tau$ are those of $T_p f$, *with the correct multiplicity*, together with an extra eigenvalue 1. The eigenvalues of $T_p f$ are called the *characteristic multipliers* of Γ. Thus

(5.36) Corollary. *The orbit Γ is hyperbolic if and only if none of its characteristic multipliers lies on the unit circle in* $\mathbf{C}$. □

In order to analyse further the structure of a flow in the neighbourhood of a closed orbit, we recall the notion of the suspension of a diffeomorphism (Construction 1.23). The situation is now rather more complicated, as we wish to suspend diffeomorphisms which are only locally defined (in particular, Poincaré maps for cross sections of flows). Let Y be a manifold and let f be a C^r diffeomorphism $(r \geq 1)$ of a connected open subset U of Y onto an open subset $f(U)$ of Y, with a fixed point p. Let I be the open interval $]-\varepsilon, 1 + \varepsilon[$ for some small positive ε (certainly $\varepsilon < \frac{1}{2}$). Consider the quotient space of $\mathbf{R} \times Y$ under the equivalence relation $\sim$ given by $(s, y) \sim (s', y')$ if and only if $y \in U$, $y' = f(y)$ and $s = s' + 1$. Let V be the quotient subset $I \times U/\sim$ (see Figure 5.37). Then V is a differentiable manifold, and the standard unit vector field in the direction of the first component of $\mathbf{R} \times Y$ induces a vector field on V. The suspension $\Sigma(f)$ of f is the local integral of

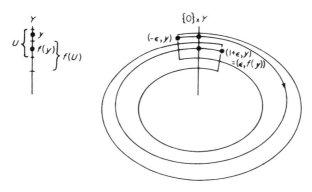

FIGURE 5.37

this vector field. Thus, if $[s, y]$ denotes the $\sim$ equivalence class of (s, y) and t is sufficiently small,

$$\Sigma(f)(t, [s, y]) = [s + t, y].$$

One identifies $y \in Y$ with $[0, y]$. Notice that $\Sigma(f)$ is C^r, and that the orbit through p is periodic of period 1. The set Y is a cross section of $\Sigma(f)$ with first return map the constant function with value 1, and f is a Poincaré map for Y.

Now let $f' : U' \to f'(U')$ be a diffeomorphism of open subsets of a manifold Y'. A *topological conjugacy from f to f'* is a homeomorphism $h : U \cup f(U) \to U' \cup f'(U')$ such that, for all $y \in U$,

$$hf(y) = f'h(y).$$

The following result considerably simplifies the problem of classifying suspensions.

(5.38) Proposition. *If two diffeomorphisms $f : U \to f(U)$ and $f' : U' \to f'(U')$ are topologically conjugate, then their suspensions are flow equivalent.*

Proof. Let h be a topological conjugacy from f to f'. Let $\Sigma(f)$ and $\Sigma(f')$ be the suspensions, local integrals on $V = (I \times U)/\sim$ and $V' = (I \times U')/\sim'$ respectively, where I is the interval $]-\varepsilon, 1+\varepsilon[$. Define a map $H : V \to V'$ by $H([s, y]) = [s, h(y)]'$ where $[\ ]'$ denotes the equivalence class with respect to $\sim'$. Provided that H is well defined (i.e. respects the identifications under $\sim$ and $\sim'$) it is clear that it is a flow equivalence from $\Sigma(f)$ to $\Sigma(f')$. But, for all $s \in]-\varepsilon, \varepsilon[$ and for all $y \in U$,

$$(s + 1, h(y)) \sim' (s, f'h(y) = (s, hf(y)),$$

and so the representatives $(s + 1, y)$ and $(s, f(y)$ of $[s + 1, y]$ both lead to the same value for H. Thus H is well defined. $\qquad\square$

(5.59) Remark. Topologically conjugate could be replaced by C^i conjugate above, in which case the flow equivalence H would be C^j ($1 \leqslant j \leqslant r$).

The main connection between the flow near a closed orbit Γ of a C^r flow ϕ on a manifold X and the Poincaré map at a cross section Y of the flow at the point p of Γ is made by the following theorem:

(5.40) Theorem. *Let $f: U \to f(U)$ be a Poincaré map at p for the cross section Y. Then there is a C^r orbit preserving diffeomorphism h from some neighbourhood of the orbit $\mathbf{R} \cdot p$ of the suspension $\Sigma(f)$ to some neighbourhood of the orbit Γ of ϕ such that $h(p) = p$.*

Proof. Let ε be as in the definition of $\Sigma(f)$ above, and let λ be a C^∞ real increasing function on the closed interval $[\varepsilon, 1 - \varepsilon]$ such that $\lambda(\varepsilon) = 0$, $\lambda(1 - \varepsilon) = 1$ and all derivatives of λ of order $\geqslant 1$ vanish at ε and $1 - \varepsilon$. Let A be the maximum value of the first derivative λ' on $[\varepsilon, 1 - \varepsilon]$. We may assume, by changing the time scale if necessary that Γ has period 1. We may also assume that, provided U is sufficiently small, the first return function ρ satisfies $\rho(y) > \max\{1 - 1/A, 2\varepsilon\}$ for all $y \in U \cup f(U)$. We define a C^r orbit preserving diffeomorphism h from the domain $V = (I \times U)/\sim$ of $\Sigma(f)$ to a neighbourhood of Γ in X by

$$h([s, y]) = \begin{cases} \phi(s, y) & \text{for } s \in \,]{-}\varepsilon, \varepsilon[, \, y \in U \\ \phi(s + (\rho(y) - 1)\lambda(s), y) & \text{for } s \in [\varepsilon, 1 - \varepsilon], \, y \in U \\ \phi(s + \rho(y) - 1, y) & \text{for } s \in \,]1 - \varepsilon, 1 + \varepsilon[, \, y \in U. \end{cases}$$

Notice that, since for $s \in \,]1 - \varepsilon, 1 + \varepsilon[$

$$\phi(s + \rho(y) - 1, y) = \phi(s - 1, f(y)),$$

h is well defined on V. $\square$

Now if Γ is hyperbolic, then any Poincaré map f has a hyperbolic fixed point at p. Since by Corollary 5.20 f is topologically conjugate at p to $T_p f$ at 0, we deduce from Proposition 5.38 that $\Sigma(f)$ is topologically equivalent to the suspension $\Sigma(T_p f)$ on some neighbourhood of its unique closed orbit. Putting all this together, we have:

(5.41) Corollary. *If Γ is hyperbolic then the flow ϕ is topologically equivalent at Γ to $\Sigma(T_p f)$ at its unique closed orbit.* $\square$

In the finite dimensional case, the classification of hyperbolic linear maps up to topological conjugacy in Corollary 5.21 yields immediately a classification of hyperbolic closed orbits:

(5.42) Corollary. *There are, up to topological equivalence, precisely* $4n$ *different hyperbolic closed orbits that can occur in a flow on an* $(n+1)$- *dimensional manifold* $(n \geqslant 1)$.

Proof. The different types are distinguished by the dimensions and orientability (or lack of it) of the pair of submanifolds whose α-set or ω-set is the closed orbit. These are the so-called *unstable* and *stable manifolds* of the closed orbit, and we shall discuss them further in the next chapter.   □

(5.43) Exercise. Visualize the above types of hyperbolic closed orbit for $n = 1$ and for $n = 2$.

Hyperbolic closed orbits are *structurally stable*, in the sense that, if Γ is such an orbit of a C^1 vector field v on X, and if w is a vector field on X that is C^1-close to v (see Appendix B), then for some neighbourhood U of Γ in X, w has a unique closed orbit in U, and this closed orbit is topologically equivalent to Γ. To construct a proof of this result, use Theorem 3.45 to show that, for a given cross section Y, the Poincaré map of w is C^1-close to the Poincaré map of v and then use the structural stability of hyperbolic fixed points of diffeomorphisms (see Exercise 5.22).

Appendix 5

I. SMOOTH LINEARIZATION

Recall that in Hartman's theorem we altered a hyperbolic linear homeomorphism T by a perturbation η and found a topological conjugacy $h = id + g$ from T to $T + \eta$. We pointed out at the time that h is not necessarily C^1 even when the perturbation η is C^∞, since differentiating the conjugacy relation would place algebraic restrictions on the first derivatives of $T + \eta$. The question now arises as to whether further differentiation places further restrictions on higher derivatives, and whether, even if these algebraic restrictions are satisfied, the smoothness of η has any effect on that of h. It turns out that, in finite dimensions at any rate, further restrictions are the exception rather than the rule, and that positive results on smoothness can be obtained. We state here, without proof, the major theorem to this effect, due to Sternberg [1, 2], and add, also without proof, two relevant theorems of Hartman [1]. We also set an exercise to show that Hölder continuity of the perturbation implies Hölder continuity of the conjugacy. This seems to be the only completely general result in which a property is transferred from the perturbation to the conjugacy.

(5.44) Theorem. (*Sternberg's Theorem*) *Let* $T \in L(\mathbf{R}^n)$ *have eigenvalues* $\lambda_1, \ldots, \lambda_n$ (*possibly complex or repeated*) *satisfying*

$$\lambda_i \neq \lambda_1^{m_1} \ldots \lambda_n^{m_n}$$

for all $1 \leq i \leq n$ *and for all non-negative integers* $m_1, \ldots, m_n$ *with* $\sum_{j=1}^{n} m_j \geq 2$. *Let* $\eta : U \to \mathbf{R}^n$ *be a* C^s *map* ($s \geq 1$) *defined on some neighbourhood* U *of* 0 *with* $\eta(0) = D\eta(0) = 0$. *Then* $(T + \eta)|0$ *is* C^r *conjugate to* $T|0$, *where, for given* T, r *depends only on* s *and tends to* ∞ *with* s. $\quad\square$

In particular, if η is C^∞, the maps T and $T + \eta$ are C^∞ conjugate at 0. Notice that the eigenvalue condition implies that $T \in HL(\mathbf{R}^n)$. There is a vector field version of Sternberg's theorem, where the above multiplicative eigen-value condition is replaced by the additive one

$$\lambda_i \neq m_1\lambda_1 + \cdots + m_n\lambda_n,$$

the conclusion being that, near 0, the vector field $T + \eta$ is induced (in the sense of Theorem 3.11) from T by some C^r diffeomorphism (provided s is sufficiently large) and that r tends to ∞ with s. See Nelson [1] for a good proof of this theorem and a more precise statement of how r depends on s.

(5.45) Theorem. (*Hartman*) *Let* $T \in L(\mathbf{R}^n)$ *be a contraction and* $\eta : U \to \mathbf{R}^n$ *be a* C^1 *map defined on some neighbourhood* U *of* 0, *with* $\eta(0) = D\eta(0) = 0$. *Then* $(T + \eta)|0$ *is* C^1 *conjugate to* $T|0$.   □

(5.46) Theorem. (*Hartman*) *Let* $T \in HL(\mathbf{R}^n)$, *where* $n = 1$ *or* 2, *and let* η *be as in Theorem 5.45. Then* $(T + h)|0$ *is* C^1 *conjugate to* $T|0$.   □

(5.47) Exercise. *A map* $g : \mathbf{E} \to \mathbf{E}$ *is Holder continuous with constant* $\lambda \, (>0)$ *and exponent* $\alpha > 0$ *if, for all* x, $x' \in \mathbf{E}$, $|g(x) - g(x')| \leq \lambda |x - x'|^\alpha$. *Prove that, for fixed* λ *and* α, *the subset* $\mathcal{H}(\lambda, \alpha)$ *of* $C^0(\mathbf{E})$ *consisting of all Hölder continuous maps with constant* λ *and exponent* α *is closed in* $C^0(\mathbf{E})$. *Let* $T \in HL(\mathbf{E})$, *and let* $\eta \in C^0(\mathbf{E})$ *be Lipschitz and Hölder continuous with exponent* $\alpha < 1$. *Prove that for any* $\lambda > 0$ *the map of* $C^0(\mathbf{E})$ *to itself defined in Exercise 5.19 (with* $\zeta = 0$, *for simplicity) takes* $\mathcal{H}(\lambda, \alpha)$ *into itself, provided that* α *and the Lipschitz and Hölder constants of* η *are sufficiently small. Deduce that in this case the map* g *of Theorem 5.15 is in* $\mathcal{H}(\lambda, \alpha)$. *(Note that, if we are only interested in the local behaviour of* $T + \eta$, *the Hölder condition on* η *is a consequence of the Lipschitz condition, since*

$$|x - x'| < |x - x'|^\alpha \qquad \text{for} \qquad |x - x'| < 1.)$$

In the negative direction, we give some counter examples, which emphasize the value of the foregoing theorems.

(5.48) Exercise. Prove that two linear automorphisms of $\mathbf{R}$ are Lipschitz conjugate at 0 (i.e. with local conjugacy a Lipeomorphism) only if they are equal.

(5.49) Exercise. Let $f : \mathbf{R}^2 \to \mathbf{R}^2$ be defined by $f(x, y) = (a^2x + y^2, ay)$ where a is fixed and positive. Prove that f is not C^2 conjugate to $Df(0)$ at 0. (*Hint*: Differentiate the relation $hf = Df(0)h$ twice at 0.)

(5.50) Exercise. Let $T : \mathbf{R}^3 \to \mathbf{R}^3$ be defined by $T(x, y, z) = (ax, acy, cz)$, where $a > 1 > c > a^{-1} > 0$ and let $\eta : \mathbf{R}^3 \to \mathbf{R}^3$ be defined, for some fixed positive ε, by $\eta(x, y, z) = (0, \varepsilon acxz, 0)$. Prove that there is no Lipschitz local conjugacy from T to $T + \eta$ at 0. (*Hint*: Show that, near 0, any local conjugacy h preserves the z-axis. Now suppose that h is Lipschitz. Prove that, near 0, h preserves the x-axis, and satisfies

$$c^{-n}h_2(x, 0, c^n z) - a^n h_2(a^{-n}x, 0, z) = n\varepsilon h_1(x, 0, c^n z)h_3(a^{-n}x, 0, z),$$

where $h = (h_1, h_2, h_3)$ and n is any positive integer. Obtain a contradiction by letting $n \to \infty$.

As we have seen the smoothness of the perturbation η is not always fully echoed by the smoothness of the conjugacy h from T to $T + \eta$. However, it is in other ways. For instance, the image under h of the stable manifold of T is a submanifold which is always as smooth as η: this is one of the main theorems of the next chapter. Another question that arises naturally is the dependence of h on η when both are regarded as points in map spaces. Here again the smoothness of η shows itself.

For example, let $\mathscr{B}^r$ denote the set of Lipschitz maps in $UC^r(\mathbf{E})$ (see Appendix B) with constant less than $\min\{1-a, |T_s^{-1}|^{-1}\}$. We may drop the U of UC^r if $r = 0$ or if $\mathbf{E}$ is finite dimensional. With the notations of Theorem 5.15, we then have:

(5.51) Theorem. *For fixed ζ, the map θ sending $\eta \in \mathscr{B}^0$ to the corresponding $g \in C^0(\mathbf{E})$ is Lipschitz, and its restriction to $\mathscr{B}^r$ is C^r.*

Proof. We define a map $\chi : \mathscr{B}^0 \times C^0(\mathbf{E}) \to C^0(\mathbf{E})$ by

$$\chi(\eta, g) = (id - \tilde{T})^{-1}((\eta(id + g) - \zeta)(T + \zeta)^{-1}),$$

where $\tilde{T}$ is as in the proof of Theorem 5.15. Then χ is a uniform contraction on the second factor, and uniformly Lipschitz on the first factor with constant $(1-a)^{-1}$. Thus, by Theorem C.7 of Appendix C, the fixed point map is Lipschitz, and this map is precisely θ. Now restrict χ to $\mathscr{B}^r \times C^0(\mathbf{E})$. The restricted map is C^r (see Theorem B.19 of Appendix B), and thus, by Theorem C.7 again, the restriction of θ to $\mathscr{B}^r$ is C^r. $\qquad\square$

(5.52) Exercise. Investigate the dependence of the map g of Theorem 5.15 on the pair (η, ζ) together.

(5.53) Exercise. Investigate the dependence of $h - id$ on η in (the proof of) Theorem 5.25.

II. LIAPUNOV STABILITY

Let p be a fixed point of a flow ϕ on a topological space X. We say that p is *stable in the sense of Liapunov*[†] (*or Liapunov stable*) if, given any neigh-

---

[†] The results of this section date back to Liapunov's doctoral thesis, written at the end of the last century. See Liapunov [1] for a French translation.

bourhood U of p, there is some neighbourhood V of p such that for all $x \in V$, $\mathbf{R}^+ . x \subset U$, where $\mathbf{R}^+ . x = \{t . x : t \geq 0\}$. If this is not the case then p is *unstable*. We say that p is a *asymptotically stable* if it is Liapunov stable and, in addition, for some neighbourhood W of p, $x \in W$ implies $t . x \to p$ as $t \to \infty$. We may similarly talk of *Liapunov stable* and *asymptotically stable* zeros of a vector field on X (see Exercise 3.44).

(5.54) Example. The flow $t . x = x e^{-t}$ on $\mathbf{R}$ has an asymptotically stable fixed point at 0. The flow $t . z = z e^{it}$ on $\mathbf{R}^2 (= \mathbf{C})$ has a fixed point 0 that is stable but not asymptotically stable. The flow $t . (x, y) = (x e^t, y e^{-t})$ on $\mathbf{R}^2$ has an unstable fixed point at $(0, 0)$. Figure 5.54 illustrates a flow on S^2 with a

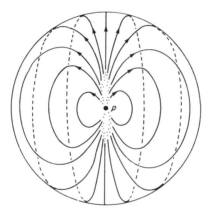

FIGURE 5.54

fixed point p that satisfies the additional condition for asymptotic stability without being Liapunov stable. It is the one-point compactification of the integral flow of a constant vector field on $\mathbf{R}^2$ (with p the point of compactification ∞). Of course one could start with $\mathbf{R}^n$ for any $n \geq 1$ and obtain a similar example.

Now let X be a *finite dimensional* smooth manifold. Let $f : X \to \mathbf{R}$ be a smooth function with an isolated critical point at p. Suppose that f has a (necessarily strict) maximum at p. Our experience of elementary several variable calculus leads us to expect that the level surfaces $\{x \in X : f(x) = \text{constant}\}$ of f near p should be, as in Figure 5.55, a sequence of smoothly embedded spheres of codimension 1 enclosing p. If ∇f is the gradient vector field of f (with respect to some Riemannian metric on X, see Examples 3.3 and A.57) then p is a zero of ∇f. Moreover the orbits of ∇f intersect the level surfaces of f orthogonally going inwards, the direction of increasing $f(x)$.

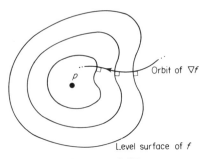

FIGURE 5.55

Thus p is asymptotically stable, as a zero of ∇f. (N.b. we shall shortly obtain a completely rigorous proof of this last statement.)

Suppose now that ϕ is the integral flow of a vector field v on X, and that, for all $x \neq p$ near p, $\langle \nabla f(x), v(x) \rangle > 0$. This condition says that $v(x)$ makes an acute angle with $\nabla f(x)$, and thus the orbit of v through x crosses the level surface at x in the same direction as the orbit of ∇f, namely inwards. Thus p is also asymptotically stable as a fixed point of ϕ (see Figure 5.56). If we

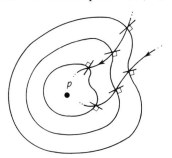

FIGURE 5.56

weaken the above inequality to $\geqslant$, we allow orbits of v to be tangential to the level surfaces. It is, of course, possible for an orbit to be tangential to a level surface at a point where it crosses over the surface. However it is difficult to see how an orbit can make much progress in an outwards direction if neither it *nor nearby orbits* may cross any level surface transversally outwards. We are led to suspect that, in this case, p is Liapunov stable as a fixed point of ϕ. This is, in fact, correct, but it is not easy to make the above approach rigorous. We leave it as a useful picture to have in mind, and start again.

Let $g : X \to \mathbf{R}$ be a smooth function†. Recall from the section on first integrals in Appendix 3 that the directional derivative $L_v g : X \to \mathbf{R}$ is the

---

† Actually, here, and in what follows, g need only be defined on some neighbourhood of p. However we do not wish to overcomplicate the notation.

continuous function defined by $Tg(v(x)) = (g(x), L_v g(x)) \in \mathbf{R}^2 = T(\mathbf{R})$. For all $t_0 \in \mathbf{R}$, $L_v g(t_0 . x)$ is the derivative with respect to t of $g(t . x)$ at t_0. With respect to any Riemannian metric $\langle , \rangle$ on X, $L_v g(x) = \langle v(x), \nabla g(x) \rangle$. We say that g is *positive definite* at p if it has a strict local minimum at p and $g(p) = 0$. We define *positive semi-definite* by dropping the word "strict", and *negative definite* and *semi-definite* in the obvious way. We say that g is a *Liapunov function for* v (*or for* ϕ) at p if g is positive definite at p and $L_v g$ is negative semi-definite at p. It is *strong* if $L_v g$ is negative definite. Notice that a strong Liapunov function for v at p has an isolated minimum at p, since $L_v g$ vanishes at any critical point of g. Thus, modulo a constant, g takes the place of $-f$ in the above geometrical description; this change of sign is traditional, and has no significance.

(5.57) Theorem. (*Liapunov's Theorem*) *If there exists a Liapunov function g for ϕ at p then p is Liapunov stable as a fixed point of ϕ. If further g is strong then p is asymptotically stable.*

Proof. Let g be a Liapunov function for ϕ at p, and let U be any neighbourhood of p. We wish to show that for some neighbourhood V of p, $x \in V$ implies $\mathbf{R}^+ . x \subset U$. By taking a smaller neighbourhood if necessary we may assume that U is compact and that, for all $x \in U$, $g(x) > 0$ and $L_v g(x) \leq 0$. The frontier ∂U of U is compact and non-empty (provided $U \neq X$; if $U = X$ then, of course, we put $V = X$). Thus the infimum, m say, of $g|\partial U$ is strictly positive. By continuity of g, there exists a neighbourhood V of p on which $g(x)$ is strictly less than m, and we assert that this V has the above property. This is because if, for any $x \in V$, $\mathbf{R}^+ . x$ leaves U, it does so at some point $t_0 . x$ of $\partial U (t_0 > 0)$, but since $g(t . x)$ is non-increasing while $t . x$ remains in U it can never attain a suitable value $g(t_0 . x) \geq m$. To be more precise, $\phi_x^{-1}(\text{int } U)$ is open in $\mathbf{R}$, and hence is a union of open intervals. If $t_0 < \infty$ is the end point of the interval containing 0 then $g(t_0 . x)$ is in ∂U but, since $d(g(t . x))/dt \leq 0$ on $]0, t_0[$, $g(t_0 . x) \leq g(x) < m$, which contradicts $m = \inf \{g(y): y \in \partial U\}$.

Now suppose that g is a strong Liapunov function for ϕ at p, and that U and V are neighbourhoods as above, with $L_v g(x) < 0$ for all $x \neq p$ in U. Then $g(t . x)$ decreases as t increases, and therefore tends to some limit $l \geq 0$ as $t \to \infty$. If $l > 0$, then, by continuity of g, $t . x$ never enters some open neighbourhood W of p. Thus, for all $t \geq 0$, $d(g(t . x))/dt \leq M < 0$, where M is the supremum of $L_v g$ on the compact set $U \backslash W$. However this inequality implies that $g(t . x) \to -\infty$ as $t \to \infty$, a contradiction. We deduce that $l = 0$. Now let $k = \inf \{g(y): y \in U \backslash W\}$. Since k is strictly positive, $g(t . x) < k$ for all sufficiently large t, and hence $t . x$ is eventually in W. Since this last argument holds equally well for any open neighbourhood W of p, $t . x \to p$ as $t \to \infty$, and hence p is asymptotically stable. $\square$

The attractive feature of Liapunov's theorem is that one does not need to integrate the vector field (i.e. solve the differential equations) in order to apply it. For this reason, it is sometimes called *Liapunov's direct method*. For example, if v is the vector field $v(x, y) = (-2y^3, x^3 - y^4)$ on $\mathbf{R}^2$ and g is the positive definite function $g(x, y) = x^4 + 2y^4$, then it is a trivial observation that

$$L_v g(x, y) = 4x^3(-2y^3) + 8y^3(x^3 - y^4) = -8y^4$$

is negative semi-definite, and so $(0, 0)$ is a Liapunov stable zero of v. The converse to Liapunov's theorem also holds (see Antosiewicz's survey [1] for details), and so we know that a Liapunov function (resp. strong Liapunov function) exists at any Liapunov stable (resp. asymptotically stable) fixed point. The snag about applying Liapunov's theorem is that in any specific case it may be very hard to recognize whether such a function does exist, and to find one if it does.

It would be rather a tall order to prove that a given fixed point is unstable by showing that no Liapunov functions exist there. The following instability theorem also due to Liapunov is sometimes useful:

(5.58) Theorem. *If there exists a C^1 function $h: X \to \mathbf{R}$ with $h(p) = 0$ such that $L_v h$ is positive definite at p and h is strictly positive on a sequence of points converging to p, then p is unstable as a fixed point of ϕ.*

Proof. Let U be a compact neighbourhood of p such that $L_v h(x) > 0$ for all $x \neq p$ in U, and let V be any neighbourhood of p. Then V contains a point $x \neq p$ such that $h(x) > 0$. Suppose that $\mathbf{R}^+ . X \subset U$. Then $h(t . x)$ is strictly increasing with t, so by continuity of h, there is some open neighbourhood W of p that contains no points of $\mathbf{R}^+ . x$. Since $U \backslash W$ is compact, the infimum of $L_v h$ on $U \backslash W$ is strictly positive, and thus $h(t . x) \to \infty$ as $t \to \infty$. But h is bounded on $U \backslash W$, so we have a contradiction. $\qquad \square$

(5.59) Exercise. (i) Use the function $g(x, y) = x^6 + 3y^2$ to prove that $(0, 0)$ is an asymptotically stable zero of the vector field $v(x, y) = (-3x^3 - y, x^5 - 2y^3)$ on $\mathbf{R}^2$. Prove that the vector field $v(x, y) = (3x^3 + y, x^5 - 2y^3)$ has an unstable zero at $(0, 0)$.

(ii) Prove that the function $f(x, y) = x^2 + 4y^2 + 2xy^2 + y^4$ is a Liapunov function for the vector field $v(x, y) = (-2xy^2, xy - 2y)$ at $(0, 0) \in \mathbf{R}^2$.

(5.60) Example. If a C^1 function $f: X \to \mathbf{R}$ has an isolated critical point at p, and this is a (necessarily strict) local maximum of f, then the gradient vector field ∇f (with respect to any Riemannian metric on X), has an asymptotically stable fixed point at p. For, let $g(x) = f(p) - f(x)$. Then g is positive definite at p, and $L_{\nabla f} g(x) = \langle -\nabla f(x), \nabla f(x) \rangle$ is negative unless $\nabla f(x) = 0$, which implies that x is a critical point of f. Thus $L_{\nabla f}g$ is negative definite at p, and so g is a strong Liapunov function for ∇f at p. Similarly if f has an isolated critical point at q and this is not a local maximum of f, then ∇f has an unstable fixed point at q by the instability theorem.

(5.61) Example. Recall (Example 3.49) that in conservative mechanics the Hamiltonian energy function $H = T + V$ is a first integral for the Hamiltonian vector field. Thus, rather trivially, H satisfies the condition that its directional derivative with respect to the vector field is negative semidefinite at any zero of the vector field. These zeros correspond to equilibrium positions of the system, and since the kinetic energy T is a positive definite function of the momenta, H is positive definite at an equilibrium position if and only if the potential energy V has a strict minimum there. For dissipative systems, H decreases along the orbits at points where the momenta are non-zero, and so the directional derivative is still negative semi-definite, and the above conclusion continues to hold.

Liapunov stability can be defined for orbits other than fixed points, and analogues of the above theorems can be obtained in this broader context. Equivalently one may discuss stability of a fixed point of a time dependent vector field, since stability of the solution γ of $x' = v(x)$ is equivalent to stability of the zero solution of $y' = v(y + \gamma(t)) - v\gamma(t)$. See, for example, Hale [1].

III. THE INDEX OF A FIXED POINT

Let v be a continuous vector field on an open subset V of $\mathbf{R}^2$, and let C be a simple closed curve in V. Suppose that v does not pass through any zero of v. Then we can associate with C an integer, called its *index with respect to v* which we may describe intuitively as follows. Consider a variable point x starting at some point $x_0 \in C$ and moving around C in the positive (anticlockwise) direction. The (anticlockwise) angle $\theta(x)$ that $v(x)$ makes with the positive x axis is only defined up to a multiple of 2π. However if we start, say, with $0 \leqslant \theta(x_0) < 2\pi$, we can choose a representative $\theta(x)$ that varies continuously with x. When we return to x_0 after one trip round C, $\theta(x)$ may not return to the original value $\theta(x_0)$: it takes up a value that differs from $\theta(x_0)$ by $2n\pi$, for some integer n. Thus $2n\pi$ is the *total angular variation* of the vector field around the curve C. The number n, which obviously does not depend upon the starting point x_0 and the speed with which x moves round C, is the index of C. In Figure 5.62, C has index 1 in (i)–(iii), but index 0 in (iv), index 2 in (v) and index -1 in (vi).

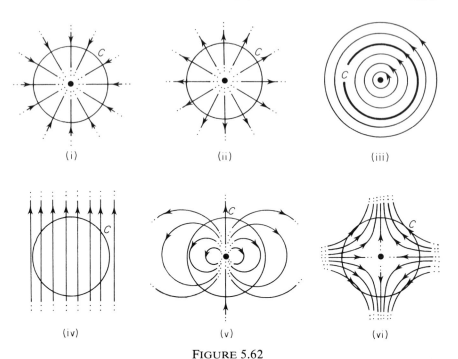

FIGURE 5.62

(5.63) Exercise. For any integer n, positive or negative, visualize a vector field and curve for which the index is n.

(5.64) Exercise. Let v be C^1 and let C be parametrized by the C^1 map $g:[0, \tau] \to \mathbf{R}^2$. Up to a multiple of π,

$$\theta g(t) = \tan^{-1}(v_2 g(t)/v_1 g(t)).$$

Prove that the index of C is

$$\int_0^\tau \langle vg(t), -iDv(g(t))(g'(t)) \rangle \, dt/2\pi |vg(t)|^2$$

(identifying $\mathbf{R}^2$ with $\mathbf{C}$ in the usual way). Use the formula to prove that the index of the unit circle with respect to $v(x, y) = (x, -y)$ is -1.

Suppose that C is deformed continuously into a curve C', through a family of curves none of which contains a zero of v. Then it is intuitively obvious that the total angular variation and hence the index, changes continuously, and, since the index is an integer, this can only mean that it remains constant. Similarly if the vector field v is deformed continuously into a new vector field v', then the index of C with respect to both these vector fields is the same provided that at no intermediate stage does the perturbed vector field have a zero on C. These two facts enable us to make some interesting observations. For example, C bounds a topological 2-dimensional ball B in $\mathbf{R}^2$. If B is in V and if B contains no zero of v, then C may be deformed until it lies in a small neighbourhood N of a given point x_0 of B. Since $v(x)$ is near $v(x_0)$ for $x \in N$, the total angular variation is now small, and hence zero, and hence the original curve C had index zero with respect to v (see Figure 5.65). Again, if

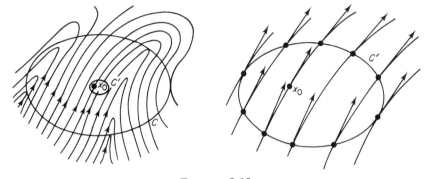

FIGURE 5.65

p is an isolated zero of v, and if C_1 and C_2 are circles with centre p, small enough to contain no other zeros of v and no points of $\mathbf{R}^2 \backslash V$, then C_1 has the same index as C_2, since we may deform the one radially to the other. Thus we may unambiguously define the *index of p with* respect to v to be the index of any sufficiently small circle with centre p. Moreover, by an argument familiar in contour integration and illustrated by Figure 5.66, the index of C is the sum of the indices of the zeros p_i in the interior domain B, provided that they are finite in number and that B is in V.

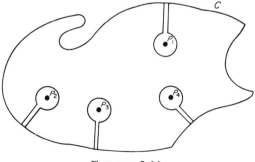

FIGURE 5.66

As an example of deforming the vector field, we may prove the *fundamental theorem of algebra*, which states that any complex polynomial $p(z)$ has a zero. For suppose that $p(z) = z^n + a_1 z^{n-1} + \cdots + a_n$, and let $r > 1$ be large enough for the inequality $r^n > |a_1| r^{n-1} + \cdots + |a_n|$ to hold. Then $p_t(z) = z^n + t(a_1 z^{n-1} + \cdots + a_n)$, for $0 \leq t \leq 1$, defines a deformation from the vector field $z \mapsto p(z)$ on $\mathbf{C} = \mathbf{R}^2$ to the vector field $z \mapsto p_0(z) = z^n$, and p_t has no zeros on the circle $z = r$. Using the formula of Exercise 5.64, the index of the circle with respect to p_0 is

$$\int_0^{2\pi} \langle r^n e^{int}, -inr^{n-1} e^{i(n-1)t} ir\, e^{it}\rangle \, dt/2\pi r^{2n},$$

which simplifies to n. Thus the index of the circle is n with respect to p. Since, by our above remarks the index would be zero if p had no zeros inside the circle, we deduce that p has zeros.

It is possible to generalize the concept of index to higher dimensions. Let us think of the Jordan curve C as being the image of a continuous map $g: S^1 \to V$ which preserves orientation (i.e. as x moves anticlockwise round S^1, $g(x)$ moves anticlockwise round C). Then the direction $vg(x)/|vg(x)|$ of the vector field at $g(x) \in C$ is a point $\theta(x)$ of S^1 (note that θ now has domain S^1 rather than C), so we have a continuous map $\theta: S^1 \to S^1$. The index of C is the number of times the new map θ wraps S^1 round itself (in the anticlockwise direction). At first sight this new approach is even vaguer than the previous one, but it may be made completely precise using homology theory. The index is just the *degree*, deg θ, of the map θ. For the definition of degree, and for other bits of algebraic topology used in this section, we recommend the reader to Greenberg [1], Hu [2] and Maunder [1]. Now suppose that v is a continuous vector field on an open subset V of $\mathbf{R}^n$ and that $g: M \to V$ is a continuous (not necessarily injective) map of a compact (connected) oriented $(n-1)$-dimensional manifold, and that no zeros of v lie on $g(M)$. Then we may define a continuous map $\theta: M \to S^{n-1}$ by $\theta(x) = vg(x)/|vg(x)|$ and the *index of g with respect to v*, $\mathrm{ind}_v\, g$, to be deg θ. Equivalently, we proceed as follows. The map $vg: M \to \mathbf{R}^n \backslash \{0\}$ induces a homology group homomorphism $(vg)_*: H_{n-1}(M) \to H_{n-1}(\mathbf{R}^n \backslash \{0\})$. Both groups are isomorphic to $\mathbf{Z}$, and the given orientation of M and the standard orientation of $\mathbf{R}^n \backslash \{0\}$ give generators which we identify with $1 \in \mathbf{Z}$. In this case, $\mathrm{ind}_v\, g$ is just $(vg)_*(1)$.

If g is homotopic to h by a homotopy that avoids the zero set Fix v of v, or if v is homotopic to w through vector fields v_t with no zeros on $g(M)$, then we obtain a homotopy from $vg: M \to \mathbf{R}^n \backslash \{0\}$ to vh or wg, as the case may be. Hence, $(vg)_* = (vh)_* = (wg)_*$, and:

(5.67) Proposition. $\mathrm{Ind}_v\, g = \mathrm{ind}_v\, h = \mathrm{ind}_w\, g.$ $\qquad\qquad\square$

In particular, since a constant map clearly has index 0 with respect to any vector field:

(5.68) Corollary. *If g is homotopic to a constant map in $V\backslash \mathrm{Fix}\, v$ then* $\mathrm{ind}_v\, g = 0$. $\qquad\square$

(5.69) Exercise. Prove that if $w: V \to \mathbf{R}^n$ is a vector field that never takes the opposite direction to v at any point of $g(M)$ (i.e. $\langle wg(x),\, vg(x)\rangle > -|wg(x)|\,.\,|vg(x)|$ for all $x \in M$) then $\mathrm{ind}_v\, g = \mathrm{ind}_w\, g$. Let $f \cdot \mathbf{R}^n \to \mathbf{R}^n$ be a continuous map taking the closed unit ball into itself. By putting $V = \mathbf{R}^n$, $M = S^{n-1}$, $v = id - f$, $w = id$ and $g = $ the inclusion, prove *Brouwer's theorem* that any continuous map of a closed ball into itself has a fixed point.

If g is a topological embedding of M in V, and $N = g(M)$, then $\mathbf{R}^n\backslash N$ has two connected components, one of which, D say, is bounded. If p is any point of D, and $g_*: H_{n-1}(M) \to H_{n-1}(\mathbf{R}^n\backslash\{p\}) \cong \mathbf{Z}$ is the induced map, then $g_*(1)$ is independent of p and takes the value ± 1. If it is $+1$ we say that g is *orientation preserving*, otherwise *orientation reversing*. If $g_1: M_1 \to V$ is another embedding of an oriented manifold M_1 with image N, then $gh = g_1$ defines a homeomorphism $h: M_1 \to M$, and, since the degree of the composite of two maps is the product of the degrees of the maps, $\mathrm{ind}_v\, g_1 = (\mathrm{ind}_v\, g)\,.\,(\deg h)$. Note that $\deg h = \pm 1$. It is $+1$ (i.e. h is orientation preserving) if and only if g and g_1 either both preserve or both reverse orientation. Thus we may define the *index of N with respect to v*, $\mathrm{ind}_v\, N$, to be $\mathrm{ind}_v\, g$ for any orientation preserving embedding $g: M \to V$ with image N. As in the $n = 2$ case, we may now define the *index of* an isolated zero p of v, $\mathrm{ind}_v\, p$, to be $\mathrm{ind}_v\, N$ for any sufficiently small $(n-1)$-sphere with centre p. Equivalently $\mathrm{ind}_v\, p = v_*(1)$, where $v_*: H_n(B, B\backslash\{p\}) \to H_n(\mathbf{R}^n, \mathbf{R}^n\backslash\{0\})$ is the induced homomorphism of relative homology groups, restricting v to some neighbourhood B of p that contains no other zeros of v. We make the same definition if p is a regular point of v. Since N is homologous to zero† in $N \cup D$, where D is the bounded component of $\mathbf{R}^n\backslash N$, we deduce:

(5.70) Proposition. *If $D \subset V$ and v has no zeros in D then* $\mathrm{ind}_v\, N = 0$. *If $D \subset V$ and v has only finitely many zeros $p_1, \ldots, p_r$ in D then* $\mathrm{ind}_v\, N = \sum_{i=1}^r \mathrm{ind}_v\, p_i$. $\qquad\square$

(5.71) Corollary. *If p is a regular point of v then* $\mathrm{ind}_v\, p = 0$. $\qquad\square$

(5.72) Exercise. Let $n = 2$, and let C be an integral curve of v. Prove that if C is a closed orbit then $\mathrm{ind}_v\, C = 1$. (*Hint*: Take $x \in C$ with minimal second coordinate x_2. Let τ be the period of C, and identify S^1 with $\mathbf{R}/\tau\mathbf{Z}$. Consider

---

† At least, this is clear if $N \cup D$ is polyhedral. More generally the proposition can be proved using Exercise H.2 of p. 361 of Spanier [1], together with Alexander duality (Theorem 16 of p. 296 of Spanier [1]).

the homotopy $G: S^1 \times [0, \tau] \to \mathbf{R}^2 \backslash \{0\}$ defined for all $t \in [0, \tau]$, by

$$G([t], u) = \begin{cases} \nu(v(x)) & \text{for } t = 0, \\ \nu((2t) \cdot x - x) & \text{for } 0 < t < u/2, \\ \nu(u \cdot x - x) & \text{for } t = u/2, \, 0 < u < \tau \\ -\nu(v(x)) & \text{for } t = u/2, \, u = \tau \\ \nu(u \cdot x - (2t - u) \cdot x) & \text{for } u/2 < t < u, \\ \nu(v(x)) & \text{for } t \geqslant u, \end{cases}$$

where $\nu : \mathbf{R}^2 \backslash \{0\} \to S^1$ is defined by $\nu(x) = x/|x|$. Prove that $[t] \to G([t], \tau)$ has degree 1.)

(5.73) Exercise. Let $V = \mathbf{R}^n$ and let v be a linear automorphism. Prove that $\text{ind}_v 0$ is 1 if $\det v > 0$ and -1 if $\det v < 0$. Thus $\text{ind}_v 0 = (-1)^m$, where m is the number of eigenvalues with negative real part, counting multiplicities.

We wish to prove that the index of a fixed point with respect to a vector field is a topological invariant. For this statement to make sense, we must be able to talk about orbits of the vector field, so we now assume that v integrates to give a partial flow on V (see Exercise 3.40). Actually there is no loss in assuming that the partial flow is a flow (see Exercise 3.44), so we make this assumption. We reformulate the definition of index in the context of flows. Let ϕ be any flow on V, let M be as above and let $g : M \to V$ be any continuous map such that $g(M)$ contains no fixed points of ϕ. Then $g(M)$ does not contain periodic points of arbitrarily small period, for if there were a sequence of periodic points in $g(M)$ whose periods converged to 0, then $g(M)$ would contain the limit of some subsequence, and this would necessarily be a fixed point. Let σ be any continuous positive valued function on M such that, for any $x \in M$ with $g(x) \in \text{Per } \phi$, $\sigma(x)$ is strictly less than the smallest positive period per $g(x)$ of $g(x)$. Let $\alpha : M \to \mathbf{R}^n \backslash \{0\}$ be given by $\alpha(x) = \sigma(x) \cdot g(x) - g(x)$. We define the *index of g with respect to ϕ*, $\text{ind}_\phi g$ to be the integer $\alpha_*(1)$, where $\alpha_* : H_{n-1}(M) \to H_{n-1}(\mathbf{R}^n \backslash \{0\})$ is the induced homomorphism of homology. It is easy to check that the definition is independent of choice of the map σ. For, if τ is another map with the above properties, then so also is the map $x \to (1 - u)\sigma(x) + u\tau(x)$ for all $u \in [0, 1]$. This gives us a homotopy from α to the map $\beta : M \to \mathbf{R}^n \backslash \{0\}$ defined by $\beta(x) = \tau(x) \cdot g(x) - g(x)$, and hence $\beta_*(1) = \alpha_*(1)$. Notice that if ϕ is an integral flow of v, then $v(y) = \lim (t \cdot y - y)/t$. Therefore, putting $\sigma(x) = t$, small and constant, gives us a map α such that $\text{ind}_\phi g = \alpha_*(1) = ((1/t)\alpha)_*(1)$, and, for all $u \in [0, 1]$, putting

$$\alpha_u(x) = \begin{cases} ((ut) \cdot g(x) - g(x))/ut & \text{for } 0 < u \leqslant 1 \\ v(g(x)) & \text{for } u = 0 \end{cases}$$

gives us a homotopy from vg to $(1/t)\alpha$. Thus $\text{ind}_v g = \text{ind}_\phi g$ in this case.

Let ψ be a flow on an open subset W of $\mathbf{R}^n$. In proving our topological invariance result, we need to assume that the domains V and W in question are homeomorphic to the open n-ball.

(5.74) Theorem. *Let V be a topological n-ball and let $h: V \to W$ be a topological equivalence from ϕ to ψ. Then $\text{ind}_\psi \, hg = \text{ind}_\phi \, g$ if h is orientation preserving and $\text{ind}_\psi \, hg = -\text{ind}_\phi \, g$ if h is orientation reversing.*

Proof. If $f: V \to \mathbf{R}^n$ is a homeomorphism, it is a topological equivalence from ϕ to the induced flow $t \cdot f(x) = f(t \cdot x)$ on $\mathbf{R}^n$. Since hf^{-1} and f^{-1} are topological equivalences with domain $\mathbf{R}^n$, and h factorizes as the composite $(hf^{-1})(f)$, we may assume from the start that $V = \mathbf{R}^n$.

Let $\sigma: M \to \mathbf{R}$ be as in the definition of $\text{ind}_\phi \, g$. By definition of topological equivalence, for each $y \in V$ there exists an increasing homeomorphism $\alpha_y : \mathbf{R} \to \mathbf{R}$ such that $h(t \cdot y) = \alpha_y(t) \cdot h(y)$ for all $t \in \mathbf{R}$. For all $x \in M$, put $\tau(x) = \alpha_{g(x)}(\sigma(x))$. Note that τ is positive valued. Moreover, since $\alpha_{g(x)}$ is bijective and $\sigma(x) < \text{per} \, g(x)$ if $g(x) \in \text{Per} \, \phi$, $\tau(x) < \text{per} \, hg(x)$ if $hg(x) \in \text{Per} \, \psi$.

We assert that τ is continuous. To see this, note that h maps the compact set $F = \{t \cdot g(x) : x \in M, t \in [0, \sigma(x)]\}$ homeomorphically onto the set $G = \{t \cdot hg(x) : x \in M, t \in [0, \tau(x)]\}$. Let $x_0 \in M$. Then, since $\sigma(x) \cdot g(x)$ tends to $\sigma(x_0) \cdot g(x_0)$ as $x \to x_0$ and h is continuous, $\tau(x) \cdot hg(x) \to \tau(x_0) \cdot hg(x_0)$ as $x \to x_0$. If τ is not continuous at x_0 then there exists some sequence $(x_r)_{r \geq 1}$ such that $x_r \to x_0$ as $r \to \infty$ and $\tau(x_r)$ is bounded away from $\tau(x_0)$. We may assume that either.

(i) $\tau(x_r) \to$ some limit $l < \tau(x_0)$, or

(ii) $\tau(x_r) > \tau(x_0)$ for all $r \geq 1$.

If (i) holds, then $\tau(x_r) \cdot hg(x) \to l \cdot hg(x_0)$ as $r \to \infty$, and thus $l \cdot hg(x_0) = \tau(x_0) \cdot hg(x_0)$, contrary to our above remark that $\tau(x) < \text{per} \, hg(x)$. If (ii) holds, we choose t with $\tau(x_0) < t < \inf \{\tau(x_r): r \geq 1\}$ and $t < \text{per} \, hg(x_0)$ if $hg(x_0) \in \text{Per} \, \psi$. Then $(t \cdot hg(x_r))_{r \geq 1}$ is a sequence in G converging to $t \cdot hg(x_0)$. Since $t \cdot hg(x_0)$ is not in G, this contradicts compactness of G. Thus τ is continuous at x_0.

The above properties of τ enable us to define $\text{ind}_\psi \, hg$ as $\gamma_*(1)$, where $\gamma: M \to \mathbf{R}^n \backslash \{0\}$ is given by $\gamma(x) = \tau(x) \cdot hg(x) - hg(x)$. Let $\alpha: M \to \mathbf{R}^n \backslash \{0\}$ be given by $\alpha(x) = \sigma(x) \cdot g(x) - g(x)$, so that $\text{ind}_\phi \, g = \alpha_*(1)$. For all $x \in M$, $\gamma(x) = h(\sigma(x) \cdot g(x)) - hg(x)$. Let $\gamma_u : M \to \mathbf{R}^n \backslash \{0\}$ be defined by

$$\gamma_u(x) = h(\sigma(x) \cdot g(x) - ug(x)) - h((1-u)g(x)).$$

This gives a homotopy from $\gamma_0 = \gamma$ to γ_1, where

$$\gamma_1(x) = h(\sigma(x) \cdot g(x) - g(x)) - h(0) = h\alpha(x) - h(0),$$

and so $\mathrm{ind}_\psi\, hg = (\gamma_1)_*(1)$. If h is orientation preserving, the homomorphism $h_*: H_{n-1}(\mathbf{R}^n\backslash\{0\}) \to H_{n-1}(W\backslash\{h(0)\})$ is just $id : \mathbf{Z} \to \mathbf{Z}$. So is the map back to $H_{n-1}(\mathbf{R}^n\backslash\{0\})$ induced from translation through $-h(0)$. Thus $\mathrm{ind}_\psi\, hg = \alpha_*(1) = \mathrm{ind}_\phi\, g$. Similarly, · if h is orientation reversing $h_* = -id$, and $\mathrm{ind}_\psi\, hg = -\alpha_*(1) = -\mathrm{ind}_\phi\, g$.   □

The corresponding result for the index of a map $g: M \to V$ with respect to a vector field is:

(5.75) Corollary. *Let V be a topological n-ball and let $h: V \to W$ be a topological equivalence from a C^1 vector field v on V to a C^1 vector field w on W. Then $\mathrm{ind}_w\, hg = \mathrm{ind}_v\, h$ if h orientation preserving, and $\mathrm{ind}_w\, hg = -\mathrm{ind}_v\, h$ if h is orientation reversing.*   □

(5.76) Exercise. If V and v are as shown in Figure 5.76 (i), find a map $g: S^1 \to V$ and a diffeomorphism $h: V \to W$ for which the conclusion of Corollary 5.75 does not hold when $w = (Th)vh^{-1}$. If V and v are as shown in Figure 5.76 (ii), explain how to find, for given integers r, s with $r - s$ even, a map $g: S^1 \to V$ and a diffeomorphism $h: V \to W$ such that $\mathrm{ind}_v\, g = r$ and $\mathrm{ind}_w\, hg = s$, where $w = (Th)vh^{-1}$.

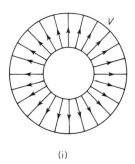
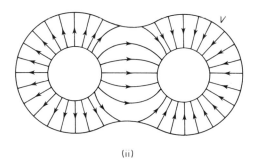

(i)   (ii)

FIGURE 5.76

If p is an isolated fixed point of a flow ϕ on V then we may, as for vector fields, define $\mathrm{ind}_\phi\, p$ to be $\mathrm{ind}_\phi\, g$ where $g: S^{n-1} \to V$ is an orientation preserving embedding with image a small sphere with centre p. In this case the index is preserved by any topological equivalence h. We do not get a change of sign when h is orientation reversing, since, although $\mathrm{ind}_\psi\, hg = -\mathrm{ind}_\phi\, g$, it is also true that $\mathrm{ind}_\psi\, h(p) = -\mathrm{ind}_\psi\, hg$, since $hg: S^{n-1} \to W$ is orientation reversing. That is to say:

(5.77) Corollary. *If an isolated fixed point p of a flow ϕ on V is topologically equivalent to a fixed point q of a flow ψ on W then $\mathrm{ind}_\phi\, p = \mathrm{ind}_\psi\, q$.*   □

Notice that if ϕ is the integral flow of the vector field v then $\text{ind}_\phi\, p = \text{ind}_v\, p$. Thus Corollary 5.77 also holds for vector fields.

We are now able to define the index $\text{ind}_\phi\, p$ of an isolated fixed point p of any flow ϕ (or C^1 vector field v) on any finite dimensional smooth manifold X by taking a chart $\xi: U \to U' \subset \mathbf{R}^n$ at p and defining $\text{ind}_\phi\, p$ to be the index of $\xi(p)$ with respect to the induced (partial) flow on U'. Since another chart η at p gives rise to a fixed point $\eta(p)$ that is topologically equivalent to $\xi(p)$, the integer $\text{ind}_\phi\, p$ does not depend on the choice of chart at p. Corollary 5.77 continues to hold when V and W are general finite dimensional manifolds. By Corollary 5.77, Exercise 5.73 and Corollary 5.20 we have:

(5.78) Proposition. *The index of a hyperbolic fixed point p of a flow ϕ is* $(-1)^m$ *where m is the dimension of the stable manifold at p.* □

If a flow ϕ on a finite dimensional manifold X has only finitely many fixed points, its *index sum* is $\sum \text{ind}_\phi\, p$ (summing over $p \in \text{Fix}\,\phi$). Similarly for vector fields. Thus, for example, the index sum of the north–south flow on S^n is 2 if n is even, and 0 if n is odd. If X is not compact then the index sum of ϕ may take any integer value, but if X is compact its topological structure puts stronger restrictions on the geometrical properties of the flows it can carry. In this case we have the celebrated theorem, due to Poincaré in dimension 2 and to Hopf in general, that the index sum is independent of the flow ϕ and depends only on the manifold X.

(5.79) Theorem. (*Poincaré–Hopf*) *The index sum of a flow ϕ (resp. C^0 vector field v) on a compact manifold X is independent of ϕ(resp. v), and equals the Euler characteristic of X.*

The proof of this theorem needs some transversality theory and it essentially contained in Hirsch [1]. Milnar [2] and Guillemin and Pollack [1] give nice versions of the proof for smooth v.

The whole of the above theory goes over perfectly well to diffeomorphisms $f: X \to X$, and even to continuous maps $f: X \to X$. If V is open in $\mathbf{R}^n$, $f: V \to \mathbf{R}^n$ is continuous, and $g: M^{n-1} \to V$ is a map of a compact oriented $(n-1)$-manifold whose image contains no points of Fix f, then the *index of g with respect to f*, $\text{ind}_f\, g$, is $\alpha_*(1)$, where $\alpha_*: H_{n-1}(M) \to H_{n-1}(\mathbf{R}^n \backslash \{0\})$ is induced by the map $\alpha(x) = fg(x) - g(x)$. So, in fact, $\text{ind}_f\, g = \text{ind}_v\, g$ where v is $f - id$ regarded as a vector field.

(5.80) Exercise. Prove that if $h: \mathbf{R}^n \to \mathbf{R}^n$ is a homeomorphism then $\text{ind}_{hfh^{-1}}\, hg = \text{ind}_f\, g$ if h is orientation preserving, and $-\text{ind}_f\, g$ if not.

The theory now develops as for flows. The *index of an isolated fixed* point p of f is more commonly known as the *Lefschetz number*, $\text{Lef}_p\,(f)$, *of f at p*. Thus if p is an isolated fixed point of the flow ϕ then $\text{ind}_\phi\, p = \text{Lef}_p\,(\phi^t)$ for any $t > 0$ with $t < \mathrm{per}\ q$ for all $q \in \mathrm{Per}\ \phi \cap S$, where S is a small sphere with centre p. The index sum of a continuous map $f : X \to X$ is called the Lefschetz number, Lef (f), of f. For compact X, the analogue of the Poincaré–Hopf theorem is not that this number is independent of f, but rather that it is an invariant of homotopy. This actually generalizes the Poincaré–Hopf theorem, since, for any two flows ϕ and ψ on X, ϕ^t and ψ^t are homotopic to the identity and hence to each other.

CHAPTER 6

Stable Manifolds

If we look at a picture of a saddle point (for example Figure 6.1), and try to analyse what qualitative features give it its characteristic appearance, we are bound to pick out the four special orbits that begin or end at the fixed point. These, together with the fixed point itself, form the *stable* and *unstable manifolds* of the dynamical system at the fixed point. We have noted in the last chapter the importance of hyperbolicity in the theory of dynamical systems, and a hyperbolic structure always implies the presence of such manifolds. If we know the "singular elements" of a system (in some sense which must include periodic points for diffeomorphisms, and fixed points and closed orbits for flows), and if we also know the way in which their stable and unstable manifolds fit together, then we have a pretty good hold on the orbit structure of the system. In this chapter we develop a theory for such manifolds, first for hyperbolic fixed points and then for more general invariant sets.

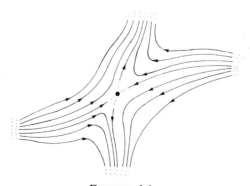

FIGURE 6.1

I. THE STABLE MANIFOLD AT A HYPERBOLIC FIXED POINT OF A DIFFEOMORPHISM

Let $f: X \to X$ be a diffeomorphism, and let p be a fixed point of f. The *stable set (or in-set) of f at p* is the set

$$\{x \in X : f^n(x) \to p \quad \text{as} \quad n \to \infty\}.$$

Notice that this set is always non-empty, since it contains p. The *unstable set (or out-set) of f at p* is the stable set of f^{-1} at p. (Thus results about stable sets can generally be restated in terms of unstable sets; we leave this to the reader.) Given any open neighbourhood U of p, we define the *local stable set of f|U at p* to be the set of all $x \in U$ such that $(f^n(x): n \geq 0)$ is a sequence in U converging to p. Notice that any point of the global stable set can be taken to a point of the local stable set by a suitable power of the diffeomorphism f. Our main theorem asserts that if p is hyperbolic then the global stable set is an immersed submanifold of X which is at least as smooth as the diffeomorphism f. We call it the *stable manifold* of f at p, $W_s(p)$. It need not be an embedded submanifold, however. For example, there is nothing to prevent the situation illustrated in Figure 6.2, where the stable and unstable manifolds of p coincide. Moreover, things look far worse when the stable and

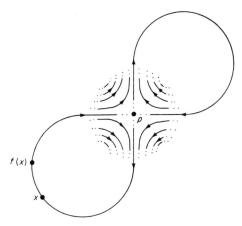

FIGURE 6.2

unstable manifolds at p have a point of transverse intersection elsewhere than at p. In such a situation, the point of intersection is said to be *transversally homoclinic*. This phenomenon is worth examining in more detail. First of all, we consider the situation (shown in Figure 6.3) where the

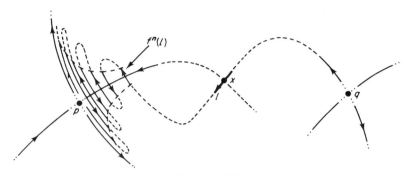

FIGURE 6.3

stable manifold at a hyperbolic fixed point p intersects the unstable manifold at another hyperbolic fixed point q transversally at some point x, say. The point x is then said to be *transversally heteroclinic*. The sequence $(f^n(x): n \geq 0)$ tends to p along the stable manifold at p, and is contained in a sequence $(f^n(l): n \geq 0)$ of images of a small segment l of the unstable manifold at q. The members of the latter sequence are transverse to the stable manifold at p and eventually "press themselves up against" any local unstable manifold at p. Thus any point of the local (and hence of the global) unstable manifold of p is a limit of points on the unstable manifold of q. Similarly every point of the stable manifold at q is a limit of points on the stable manifold at p. This is an interesting situation, but does not make any extreme demands on the topology of the submanifolds in question. However, if we now put $p = q$, then we have that each point of the stable manifold is a limit of points on other branches (that is, local connected components) of the stable manifold, and similarly for the unstable manifold. Thus these immersed submanifolds are not, globally, copies of the real line with the usual topology. The points of intersection of the stable and unstable manifold of p, other than p itself, are transversally homoclinic points. If the reader is sceptical as to whether such behaviour can actually occur, he should take another look at the toral automorphisms of Example 1.30.

We now return to the general theory. We shall deduce the global version of the stable manifold theorem from the local version. Roughly speaking, once we have planted a local stable manifold, we can easily grow it to obtain a global stable manifold; the difficulty lies in establishing it locally.

As usual, we take a chart at the point in question, and thus transfer the problem to the model space $\mathbf{E}$. Thus we obtain a local diffeomorphism with a hyperbolic fixed point at the origin, say, and the differential there is a linear approximation for the map itself nearby. We have been through this routine already in Corollary 5.20. The difference now is that we are interested in smoothness of maps, and whereas before we had no difficulty in extending a local Lipschitz map to a global one, a similar extension for smooth maps presents problems when $\mathbf{E}$ is an arbitrary Banach space. It seems a good idea, then, to give our basic results on perturbations of hyperbolic linear maps a local character.

We start out with a hyperbolic linear automorphism T of a Banach space $\mathbf{E}$. Thus $\mathbf{E}$ splits into stable and unstable summands

$$\mathbf{E} = \mathbf{E}_s(T) \oplus \mathbf{E}_u(T) = \mathbf{E}_s(T) \times \mathbf{E}_u(T).$$

We use the max $\{|\ |_s, |\ |_u\}$ norm on $\mathbf{E}$. Suppose that T has skewness a (see Theorem 4.19), and let κ be any number with $0 < \kappa < 1 - a$. Let $B = B_s \times B_u$ be the closed ball with centre 0 and radius b (possibly $b = \infty$) in $\mathbf{E}$. We have, for T, the picture on the left in Figure 6.4. We shall show that, after a

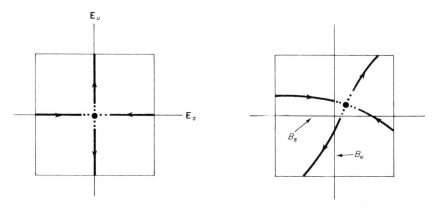

FIGURE 6.4

Lipschitz perturbation η with constant κ satisfying $|\eta|_0 \le b(1-a)$, the local stable set of $(T + \eta)|B$ is, as in the picture on the right, the graph of a map, h say, from B_s to B_u. Moreover the map h is as smooth as the perturbation η. By Exercise 5.13, $T + \eta$ has a unique fixed point in B, and by transferring the origin to this point we may assume, for simplicity, that $\eta(0) = 0$. We may now state:

(6.5) Theorem. (*Local stable manifold theorem*) *Let* $\eta : B \to \mathbf{E}$ *be Lipschitz with constant* κ, *where* $\kappa < 1 - a$. *Suppose* $\eta(0) = 0$. *Then there is a unique map* $h : B_s \to B_u$ *such that* graph h *is the stable set of* $(T + \eta)|B$ *at* 0. *Moreover the map* h *is Lipschitz, and* C^r *when* η *is* C^r.

Proof. The idea is to identify in $\mathscr{S}_0(B)$, which we recall (Appendix B) is a closed subset of the Banach space $\mathscr{S}_0(\mathbf{E})$ of sequences in $\mathbf{E}$ converging to 0, those members of the form $((T + \eta)^n(x) : n \ge 0)$ for some $x = (x_s, x_u) \in B$.

This is a good idea because the first members of such sequences are precisely the points of the local stable manifold. We show first that $\gamma \in \mathcal{S}_0(B)$ is of this form if and only if

(6.6) $$\gamma(n) = \chi(x_s, \gamma)(n)$$

where

(6.7) $$\chi(x_s, \gamma)(n) = \begin{cases} ((T+\eta)_s\gamma(n-1),\ T_u^{-1}(\gamma_u(n+1) - \eta_u\gamma(n))) & \text{for } n > 0 \\ (x_s,\ T_u^{-1}(\gamma_u(1) - \eta_u\gamma(0))) & \text{for } n = 0, \end{cases}$$

and $\eta = (\eta_s, \eta_u) : B \to \mathbf{E}_s \times \mathbf{E}_u$. This is clear, since the relation

$$\gamma_u(n) = T_u^{-1}(\gamma_u(n+1) - \eta_u\gamma(n))$$

holding on the unstable component for $n \geq 0$ may be rewritten as

$$\gamma_u(n+1) = (T+\eta)_u\gamma(n),$$

and this combines with a similar relation on the stable component to give

$$\gamma(n+1) = (T+\eta)\gamma(n)$$

for $n \geq 0$.

The useful feature of (6.6) is that it exhibits γ as the fixed point of a contraction of $\mathcal{S}_0(B)$. For, given γ and $\gamma' \in \mathcal{S}_0(B)$,

$$|\chi(x_s, \gamma) - \chi(x_s, \gamma')|$$

$$= \sup \{|(T+\eta)_s\gamma(n) - (T+\eta)_s\gamma'(n)|,$$

$$|T_u^{-1}(\gamma_u(n+1) - \gamma_u'(n+1) + \eta_u\gamma'(n) - \eta_u\gamma(n))| : n \geq 0\}$$

$$\leq \sup \{(|T_s| + \operatorname{Lip} \eta)|\gamma - \gamma'|,\ |T_u^{-1}|(1 + \operatorname{Lip} \eta)|\gamma - \gamma'|\}$$

$$\leq (a + \kappa)|\gamma - \gamma'|.$$

Since $a + \kappa$ is < 1, χ contracts the second factor uniformly. Putting $\gamma' = 0$ proves that $\chi(x_s, \gamma)$ takes values in B. A similar estimate on (6.7) shows that $\chi(x_s, \gamma)$ converges to 0.

We have shown that (6.7) defines a map $\chi : B_s \times \mathcal{S}_0(B) \to \mathcal{S}_0(B)$. Notice that χ is Lipschitz on the first factor, since

$$|\chi(x_s', \gamma) - \chi(x_s, \gamma)| = |x_s - x_s'|.$$

Thus by Theorems C.5 and C.7 of Appendix C, there is a unique map $g : B_s \to \mathcal{S}_0(B)$ such that, for all $x_s \in B_s$, (6.6) holds with γ replaced by $g(x_s)$. Correspondingly there is for each $x_s \in B_s$ a unique point $g(x_s)(0)$ of the local stable set with stable component x_s. Thus the map h of the theorem is given by $h(x_s) = (g(x_s)(0))_u$.

Notice that by Theorem C.7 the map g is Lipschitz. Now h is g composed with the map from $\mathscr{S}_0(B)$ to B that evaluates sequences at 0, followed by the product projection. Since the latter maps are both continuous linear, we deduce that h is Lipschitz.

Finally suppose that η is C'. We can split χ into the sum of two functions, one of which factors through B_s (this one takes (x_s, γ) to the sequence whose only non-vanishing term is $(x_s, 0)$ in the 0th place) and the other through $\mathscr{S}_0(B)$. The first of these functions is trivially C^∞, and the second is C', by Lemma B.4 and Theorem B.20 of Appendix B. This is because T_s, T_u^{-1}, product projections, and maps of $\mathscr{S}_0(B)$ that move sequences to the left or right are all continuous linear. Thus χ is C'. We deduce from Theorem C.7 that g, and hence h, is C'. $\qquad\square$

(6.8) Remark. The formula (6.7) defines equally well a contraction on $\mathscr{B}(B)$, the space of all bounded sequences in B, and one obtains a unique fixed point map : $B_s \to \mathscr{B}(B)$ which, by uniqueness, is the above map g. Thus the local stable manifold may be characterized as the set of all points x whose iterates $(T+\eta)^n(x)$ for $n \geq 0$ form a bounded sequence in B.

Notice that the stable manifold theorem works under substantially weaker hypotheses than Hartman's theorem. We do not necessarily assume η to be bounded when $b = \infty$, nor do we impose a condition to ensure that $T + \eta$ is a homeomorphism. In the appendix to this chapter we further investigate the dependence of h on η.

There are two other features of the local stable manifold that we would like to establish. Firstly the tangent space at 0 to the local stable manifold is the stable summand of the tangent map T_0f, where we are now writing $f = T + \eta$. Secondly, iterates under f of points of the local stable manifold do not drift in gently towards 0; they approach it, and one another, exponentially.

(6.9) Theorem. (*i*) *If η is C^1 in Theorem 6.5, then the tangent to graph h at 0 is parallel to the stable manifold of the hyperbolic linear map $T + D\eta(0)$.*

(*ii*) *The maps h and $f|$graph h are Lipschitz with constant λ, where $f = T + \eta$ and $\lambda = a + \kappa < 1$. For any norm $\| \ \|$ on $\mathbf{E}$ equivalent to $| \ |$, there exists $A > 0$ such that, for all $n \geq 0$ and for all x and $y \in$ graph h,*

$$\|f^n(x) - f^n(y)\| \leq A\lambda^n \|x - y\|.$$

Proof. (i) Differentiating the relation $g(x_s) = \chi(x_s, g(x_s))$ at 0 gives $Dg(0) = D\chi(0, 0)(id, Dg(0))$. We compute that, for all $(x_s, \gamma) \in B_s + \mathscr{S}_0(B)$, $D\chi(0, 0)(x_s, \gamma)$ is the sequence

$$n \mapsto \begin{cases} ((T+D\eta(0))_s\gamma(n-1),\ T_u^{-1}(\gamma_u(n+1) - D\eta_u(0)\gamma(n))) & \text{for } n > 0 \\ (x_s,\ T_u^{-1}(\gamma_u(1) - D\eta_u(0)\gamma(0))) & \text{for } n = 0. \end{cases}$$

Thus $Dg(0): \mathbf{E}_s \to \mathscr{S}_0(\mathbf{E})$ is the fixed point map for χ when η is replaced by $D\eta(0)$. Thus the stable manifold map for $T + D\eta(0)$ is $x_s \mapsto (Dg(0)(x_s)(0))_u$. But this is the map $Dh(0)$.

(ii) Recall that we are working with the norm
$$|x| = \max\{|x_s|, |x_u|\}.$$
Let x and $y \in$ graph h, with $x \neq y$. Suppose that $|x - y| = |x_u - y_u|$. Consider
$$f(x) - f(y) = (T_s(x_s - y_s) + \eta_s(x) - \eta_s(y), T_u(x_u - y_u) + \eta_u(x) - \eta_u(y)).$$

The norm of the first component is no greater than $(a + \kappa)|x - y|$, while that of the second component is at least $((1/a) - \kappa)|x - y|$. Since $a + \kappa < 1$ and
$$\frac{1}{a} - \kappa > \frac{1}{a} - \frac{\kappa}{a(a + \kappa)} = \frac{1}{a + \kappa} > 1,$$
the norm of the whole expression is the norm of its second component. Thus, by induction, for all $n \geq 0$,
$$|f^n(x) - f^n(y)| \geq (a + \kappa)^{-n}|x - y|.$$
Since $(a + \kappa)^{-n} \to \infty$ as $n \to \infty$, the sequences $(f^n(x))$ and $(f^n(y))$ are not both bounded, which contradicts the fact that x and y are on the local stable manifold. We deduce that $|x_u - y_u| < |x_s - y_s|$. But now observe that, since $f(x)$ and $f(y) \in$ graph h, $|f(x) - f(y)|$ is also the norm of its first component, and hence, as above, not greater than $(a + \kappa)|x - y|$. Thus $f|$graph h is Lipschitz with constant λ. The given inequality follows immediately by changing the norm in
$$|f^n(x) - f^n(y)| \leq \lambda^n|x - y|.$$
Finally, if $|x_u - y_u| > \lambda|x_s - y_s|$, then $|f_u(x) - f_u(y)| > ((\lambda/a) - \kappa))|x - y|$, which is a contradiction, since $(\lambda/a) - \kappa = 1 + \kappa(1 - a)/a > 1 > a + \kappa$. Thus h is Lipschitz with constant λ. $\qquad\square$

(6.10) Exercise. Prove an *unstable manifold theorem* for $T + \eta$ by using the formula
$$\chi(x_u, \gamma)(n) = \begin{cases} ((T + \eta)_s \gamma(n + 1), T_u^{-1}(\gamma_u(n - 1) - \eta_u\gamma(n))) & \text{for } n > 0, \\ ((T + \eta)_s \gamma(1), x_u) & \text{for } n = 0. \end{cases}$$

(6.11) Exercise. We call T α-*hyperbolic* ($\alpha > 0$) if the circle of radius α in $\mathbf{C}$ does not intersect the spectrum of T. Thus $\alpha^{-1}T$ is hyperbolic, and we deduce from Theorem 4.19 that there is a decomposition $\mathbf{E} = \mathbf{E}_s \oplus \mathbf{E}_u$ into T-invariant summands and an equivalent norm $|\ | = \max\{|\ |_s, |\ |_u\}$ on $\mathbf{E}$ with respect to which the restrictions T_s and T_u satisfy $|T_s| < \alpha$ and $|T_u^{-1}| < 1/\alpha$. Let B be a ball with centre 0 in $\mathbf{E}$ (if $\alpha > 1$ we require $B = \mathbf{E}$), and let $\eta : B \to \mathbf{E}$ be a Lipschitz map with constant $\kappa < \max\{\alpha - |T_s|, |T_u^{-1}|^{-1} - \alpha\}$ and with $\eta(0) = 0$. Replace $\mathscr{S}_0(B)$ by $\tilde{\alpha}(\mathscr{S}_0(B))$ (see Exercise B.22 of

Appendix B) in the proof of Theorem 6.5, and thus prove that there is a unique map $h: B_s \to B_u$ such that, for all $x \in B$, $x \in$ graph h if and only if $\{\alpha^{-n}(T + \eta)^n(x): n \geq 0\}$ is well defined and converges to 0. Prove that h is Lipschitz. Prove also that if $\alpha \leq 1$† and η is C^r then h is C^r. We call graph h the *local α-stable manifold of $T + \eta$ at 0*. Convince yourself that this result is not an immediate application of Theorem 6.5 to the hyperbolic automorphism $\alpha^{-1}T$.

We now prove the stable manifold theorem for fixed points of a diffeomorphism of a smooth manifold X. We start with the local version. The proof is completely straightforward, but we give it in full to establish some notation.

(6.12) Theorem. *Let p be a hyperbolic fixed point of a C^r diffeomorphism $(r \geq 1)$ f of X. Then, for some open neighbourhood U of p, the local stable set of $f|U$ at p is a C^r embedded submanifold of X, tangent at p to the stable summand of $T_p f$.*

Proof. Let $\xi: V \to V'$ be an admissible chart at p, with $\xi(p) = 0$. Let g be defined, on some small neighbourhood W of 0 by $g(y) = \xi f \xi^{-1}(y)$. Then g is C^r and has a hyperbolic fixed point at 0. Pick a norm $|\ |$ on $\mathbf{E}$ with respect to which $Dg(0)$ has skewness $a < 1$. Then on some open ball B with centre 0, $|Dg(y) - Dg(0)| \leq \kappa < 1 - a$. Thus $\eta = g - Dg(0): B \to \mathbf{E}$ is Lipschitz with constant κ. Let $h: B_s \to B_u$ be the corresponding C^r stable manifold map, given by Theorem 6.5. Then $\xi^{-1}(id, h)$ is a C^r embedding of B_s in X with image the local stable manifold of $f|U$ at p, where $U = \xi^{-1}(B)$. Tangency follows from Theorem 6.9. □

(6.13) Theorem. (*Global stable manifold theorem*) *Let p be a hyperbolic fixed point of C^r diffeomorphism f $(r \geq 1)$ of X. Then the global stable set Y of f at p is a C^r immersed submanifold of X, tangent at p to the stable summand of $T_p f$.*

Proof. We continue with the notation of the proof of Theorem 6.12. We shall give the set Y the structure of a C^r manifold by constructing a C^r atlas on Y, and then prove that the inclusion of Y in X is a C^r immersion. Let Y_0 denote the local stable manifold of $f|U$ at p, so that $Y_0 = \xi^{-1}$ (graph h). For all integers $i \geq 0$, let $Y_i = f^{-i}(Y_0)$, and let $\xi_i: Y_i \to B_s$ be defined by $\xi_i(y) = \pi_s \xi f^i(y)$, where $\pi_s: B \to B_s$ is projection onto the stable summand. Then the family $\{Y_i: i \geq 0\}$ covers Y. For all i, j with $i \geq j \geq 0$, $\xi_i(Y_i \cap Y_j) = B_s$ and $\xi_i(Y_i \cap Y_j) = \pi_s g^{i-j}(\text{graph } h)$. Moreover the coordinate transformation

$$\xi_{ij} = \xi_i \xi_j^{-1} : \xi_i(Y_i \cap Y_j) \to \xi_i(Y_i \cap Y_j)$$

---

† The reason for the failure of this approach in the case $\alpha > 1$ is that Exercise B.22 does not cope successfully with smoothness. However there are some positive results in this case; see, for example, Irwin [2] and Hirsch, Pugh and Shub [1].

and its inverse ξ_{ji} are the C^r diffeomorphisms given by $\xi_{ij}(x) = \pi_s g^{i-j}(x, h(x))$ and $\xi_{ji}(x) = \pi_s g^{j-i}(x, h(x))$. Notice that B_s is open, and hence, by the inverse function theorem $\pi_s g^{i-j}$ (graph h) is open. Hence $\{\xi_i : i \geq 0\}$ is a C^r atlas on Y.

Let $y \in Y_i$. To show that the inclusion is a C^r immersion at y, it suffices to show that its composite with f^i is an immersion at y (since f^i is a C^r diffeomorphism). But the representative of this composite, with respect to the charts ξ_i at y and ξ at $f^i(y)$ is the C^r embedding (id, h). $\qquad \square$

The nature of the charts ξ_i in the above theorem strongly suggests that the global stable manifold of f at p is an immersed copy of the stable manifold of the linear approximation. This is borne out by the following exercise.

(6.14) Exercise. By extending the map η (in the proof of Theorem 6.12) to the whole of $\mathbf{E}$, construct a locally Lipschitz bijection of the Banach space $\mathbf{E}_s(Dg(0))$ onto the stable manifold of f. Show that if f is C^r and if $\mathbf{E}$ admits C^r bump functions then this can be done so that the bijection gives a C^r immersion of $\mathbf{E}_s(Dg(0))$ in X.

One may extend the above theory from fixed points to periodic points of a diffeomorphism with no extra effort. If p is a periodic point of a diffeomorphism $f: X \to X$, then p is a fixed point of the diffeomorphism f^k, where k is the period of p. Notice that $d(f^m(x), f^m(p)) \to 0$ as $m \to \infty$ (where f is some admissible distance function on X) if and only if $f^{nk}(x) \to p$ as $n \to \infty$. Thus, if we define the stable set of f at p to be

(6.15) $\qquad \{x \in X : d(f^m(x), f^m(p)) \to 0 \text{ as } m \to \infty\},$

and say that p is *hyperbolic* if it is a hyperbolic fixed point of f^k, then Theorem 6.13 becomes:

(6.16) Theorem. *Let p be a hyperbolic periodic point of period k of a C^r diffeomorphism f of X, where $r \geq 1$. Then the stable set of f at p is a C^r immersed submanifold of X tangent at p to the stable summand of $T_p f^k$.* $\qquad \square$

Once X is given a distance function d, our new definition of stable set makes sense for *any* point $p \in X$. We shall shortly be proving a stable manifold theorem of a more general nature that involves non-periodic as well as periodic points.

II. STABLE MANIFOLD THEORY FOR FLOWS

The stable manifold theorem for a hyperbolic fixed point of a flow is a simple corollary of the corresponding theorem for diffeomorphisms. We first give a definition of stable set which will serve for any point of the manifold. Let ϕ be a flow on a manifold X, and let d be an admissible distance function on X. The *stable set of ϕ at $p \in X$* is the set

$$\{x \in X : d(\phi(t, x), \phi(t, p)) \to 0 \text{ as } t \to \infty\}$$

and the *unstable set of ϕ at p* is the set

$$\{x \in X : d(\phi(t, x), \phi(t, p)) \to 0 \text{ as } t \to -\infty\}$$

The connection between stable manifolds of diffeomorphisms and flows is made by recalling the result of Exercise 1.39, that if p is a fixed point of ϕ then $\phi(t, x) \to p$ as $t \to \infty$ if and only if $\phi(n, x) \to p$ as $n \to \infty$, where $t \in \mathbf{R}$ and $n \in \mathbf{Z}$. Thus the stable set of the flow ϕ at p is precisely the stable set of the diffeomorphism ϕ^1 at p, or, equally, of ϕ^t for any other $t > 0$). We deduce:

(6.17) Theorem. (*Global stable manifold theorem*) *Let p be a hyperbolic fixed point of a C^r flow ($r \geq 1$) ϕ on X. Then the stable set of ϕ at p is a C^r immersed submanifold of X, tangent at p to the stable summand of $T_p\phi^1$.* ☐

A local version is only less trivial to deduce in that, if U is an open neighbourhood of p then the local stable manifold of the flow ϕ in U is not necessarily the whole of the local stable manifold of ϕ^1 in U because orbits may leave U at non-integer values of t and then come back again. Nevertheless the former is an open subset of the latter, and thus a submanifold of it. One can get a Lipschitz version of the theorem, corresponding to part of Theorem 6.5, and also a local version when a C^r vector field does not have an integral flow defined for all t. Both are completely straightforward to prove.

The theory for a hyperbolic closed orbit $\Gamma = \mathbf{R} \cdot p$ is slightly more complicated. Let Γ have period τ. It is an easy consequence of continuity of ϕ and compactness of Γ that $d(\phi(t, x), \phi(t, p)) \to 0$ as $t \to \infty$ in $\mathbf{R}$ if and only if $\phi(n\tau, x) \to \phi(n\tau, p) = p$ as $n \to \infty$ in $\mathbf{Z}$ (see Exercise 1.39). Thus we are interested in the stable set of the diffeomorphism ϕ^τ at its fixed point p. Now p is not a hyperbolic fixed point of ϕ^τ; we have the splitting $T_pX = \mathbf{E}_s \oplus \langle v \rangle \oplus \mathbf{E}_u$, where $\mathbf{E}_s$ and $\mathbf{E}_u$ are the stable and unstable summands at p and v is tangent to Γ at p. If β is the spectral radius of $T_p\phi^\tau | \mathbf{E}_s$, we may choose any α with $\beta < \alpha < 1$ and deduce from Exercise 6.11 the existence of a C^r local α-stable manifold $W_s^\alpha(p)$ of ϕ^τ at p. This submanifold is independent of the choice of α, by uniqueness, and is, as above, contained in the stable set of ϕ at p. We may extend it backwards into a global C^r immersed submanifold $W_s(p)$ modelled on $\mathbf{E}_s$, just as in the proof of Theorem 6.13. The images under ϕ^t of $W_s(p)$, for all $t \in \mathbf{R}$, give a family of submanifolds, one through each point $q = \phi^t(p)$ of Γ. We write $W_s(q)$ for $\phi^t(W_s(p))$ if $q = \phi^t(p)$, and $W_s(\Gamma)$ for $\{x \in W_s(q) : q \in \Gamma\}$. Note that, since $\phi(n\tau, x) \to p$ as $n \to \infty$ for all

$x \in W_s(p)$, $\phi(n\tau, x) \to q$ as $n \to \infty$ for all $x \in W_s(q)$ (see Figure 6.18, where the double arrows indicate orbits of the diffeomorphism ϕ^τ). Notice also that $W_s(\Gamma)$ is determined by the small portion of it at p that is the image of a small neighbourhood of $(0, 0)$ in $\mathbf{R} \times \mathbf{E}_s$ under the map $\phi(id \times h)$, where h is the local α-stable manifold function at p. Since $\langle v \rangle$ does not lie in $\mathbf{E}_s$, which is tangent to the α-stable manifold at p, this map is a C^r immersion, and it follows that globally $W_u(\Gamma)$ is a C^r immersed submanifold modelled on $\mathbf{R} \times \mathbf{E}_s$ and C^r foliated (see Appendix A) by the family $\{W_s(q): q \in \Gamma\}$.

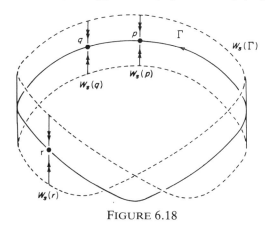

FIGURE 6.18

We call $W_s(\Gamma)$ the *stable manifold* of ϕ at Γ. We emphasize that we have not yet ruled out the possibility that

(i) $W_s(q)$ depends on choice of p, or that

(ii) for some point x, $\phi^{n\tau}(x) \to p$ as $n \to \infty$, but so slowly that x is not on the α-stable manifold for any $\alpha < 1$, or that

(iii) for some point x, the distance from $\phi(t, x)$ to Γ tends to 0 as $t \to \infty$ without $\phi^{n\tau}(x)$ tending to any particular point q of Γ.

If any of these phenomena were actually to occur, we would have points outside $W_s(\Gamma)$ that would nevertheless have a good claim to being called members of the stable set of Γ. However it is easy to see, in finite dimensions at any rate, that (i)–(iii) cannot occur. For, let dim $\mathbf{E}_s = d$, and take a cross section Y to the flow at p. Let $f: U \to Y$ be a Poincaré map at p, and let W be the (d-dimensional) local stable manifold of $f|U$ at p. If $\rho: U \to \mathbf{R}$ is the first return function, then the *positive orbit* $\mathbf{R}^+ . W = \{t . y: t \geq 0, y \in W\}$ is a $(d + 1)$-dimensional submanifold consisting of all points x in the neighbourhood $V = \{t . y: y \in U, 0 \leq t \leq \rho(y)\}$ of Γ whose positive orbits $\mathbf{R}^+ . x$ are wholly in V. All points w of $\mathbf{R}^+ . W$ have the property that the distance from $t . w$ to Γ tends to 0 as $t \to \infty$, and any orbit whose points have this property contains points of $\mathbf{R}^+.W$. But now observe that $A = \{t.y: y \in W_s^\alpha(p), 0 \leqslant t \leqslant \tau\}$ is a $(d+1)$-dimensional submanifold with $\mathbf{R}^+.A = A$ and, provided $W_s^\alpha(p)$ is small enough, with $A \subset V$. We deduce that A is contained in $\mathbf{R}^+.W$, and hence, by the dimensions, that A is a neighbourhood of Γ in $\mathbf{R}^+.W$. Therefore $W_s(\Gamma)$ is the set of points $x \in X$ such that the distance from $t.x$ to Γ tends to 0 as $t \to \infty$, the *stable set of ϕ* at Γ. This effectively rules out possibilities (i), (ii) and (iii). In particular $W_s(p)$ is precisely the stable set of ϕ at p. In the infinite dimensional case, we are still able to prove that A is a neighbourhood of Γ in $\mathbf{R}^+.W$. It is not hard to show that the two submanifolds have the same tangent space $\mathbf{E}_s + \langle v \rangle$ at p, and the result follows from this. Our theorem, then, may be stated as follows:

(6.19) Theorem. (*Stable manifold theorem for closed orbits*) *Let $\Gamma = \mathbf{R}.p$ be a hyperbolic closed orbit of a C^r flow ϕ ($r \geqslant 1$) on a manifold X with distance function d. Then the stable set $W_s(p)$ of ϕ at p is a C^r immersed submanifold tangent at p to the stable summand of $T_p X$ with respect to $T_p \phi^\tau$, where τ is the period of Γ. If $q = \phi^t(p)$, then $W_s(q) = \phi^t(W_s(p))$. The stable set $W_s(\Gamma)$ of ϕ at Γ is a C^r immersed submanifold which is C^r foliated by $\{W_s(q): q \in \Gamma\}$.* □

The submanifold $W_s(p)$ is called the *stable manifold of ϕ at p*. Since by the above remarks $W_s(p)$ is also the stable manifold of the diffeomorphism ϕ^τ at p, it is independent of the distance function d, as also is $W_s(\Gamma) = \bigcup_{q \in \Gamma} W_s(q)$.

III. THE GENERALIZED STABLE MANIFOLD THEOREM

Let X be a finite dimensional Riemannian manifold (see Appendix A) and let $f: X \to X$ be a diffeomorphism. We have already commented that our definition (6.15) of the stable set of f at a point p is valid for any $p \in X$, and we have seen that, when p is periodic and has a hyperbolic structure with respect to f, its stable set is a manifold. The question now arises as to whether we can extend our notion of hyperbolicity to non-periodic points p and get a stable manifold theorem for such points. The difficulty, of course, is that $T_p f^n$ does not map $T_p X$ to itself for $n > 0$, and so we cannot define hyperbolicity in terms of eigenvalues of this map. If we concentrate for the moment on the expanding and contracting properties of hyperbolic maps rather than their eigenvalues, we can see how a definition might go, but it will involve the whole orbit of p (as indeed does the definition of stable set). In fact, we may as well define the term hyperbolic structure for an arbitrary invariant subset of X, since many applications of stable manifold theory need this degree of generality.

Let Λ be any invariant subset of X, and let $T_\Lambda M$ be the tangent bundle of X over Λ (that is, $\{T_x X: x \in \Lambda\}$). We say that Λ has a *hyperbolic structure* (with respect to f) if there is a continuous splitting of $T_\Lambda X$ into the direct sum of Tf-invariant subbundles E_s and E_u such that, for some constants A and λ and for all $v \in E_s$, $w \in E_u$ and $n \geq 0$,

(6.20) $|Tf^n(v)| \leq A\lambda^n |v|, \qquad |Tf^{-n}(w)| \leq A\lambda^n |w|,$

where $0 < \lambda < 1$. Thus one may say that Tf is eventually contracting on E_s and eventually expanding on E_u. A *hyperbolic subset* of X (with respect to f) is a closed invariant subset of X that has a hyperbolic structure.

There is a temptation to make "eventually" into "immediately" by putting $A = n = 1$ in the above definition, but the result would not be invariant under differentiable conjugacy. If Λ is compact, one may always introduce a new Riemannian metric on X with respect to which (6.20) does hold with $A = n = 1$. Such a metric is said to be *adapted to f*. The proof, which resembles that of Theorem 4.47, is due to Mather [1], and also appears in Shub [4], Hirsch and Pugh [1] and Nitecki [1]. However we shall not need this result.

The idea behind the generalized stable manifold theorem is as follows. If Λ is a compact hyperbolic subset, the space $C^0(\Lambda, X)$ of all continuous maps from Λ to X may be given the structure of a smooth manifold modelled on the Banach space $\Gamma_\Lambda^0(X)$ of C^0 sections of $T_\Lambda(X)$, given the sup norm derived from the Riemannian Finsler $|\ |_x$ on X (see Appendix A). One considers the map $f^\#: C^0(\Lambda, X) \to C^0(\Lambda, X)$ defined by $f^\#(h) = fh(f|\Lambda)^{-1}$. Clearly this has a fixed point at ι (the inclusion of Λ in X). It turns out that the hyperbolic structure of Λ makes ι a hyperbolic fixed point. Thus $f^\#$ has a stable manifold at ι, consisting of $h \in C^0(\Lambda, X)$ such that $(f^\#)(h) \to \iota$ as $n \to \infty$. For such h, $f^n h f^{-n}(x) \to x$ as $n \to \infty$ for all $x \in \Lambda$, or, equivalently, by compactness of Λ, $d(f^n h(y), f^n(y)) \to 0$ as $n \to \infty$ for all $y \in \Lambda$. Here d is the distance function on X derived from the Riemannian metric (see Appendix A). So if we evaluate h at $y \in \Lambda$, we get a point on the stable set of y. If we do so for all h near ι, we get the local stable manifold at y, and properties (such as smoothness) of the stable manifold at ι yield, after evaluation at y, properties of the stable manifold at y. Moreover, if we let y vary, we can get information about how the stable manifold at y varies with y.

There is no need for us to give a full proof of all the statements in the previous paragraph. Since we are only applying the stable manifold theorem at a single point of $C^0(\Lambda, X)$ we can, provided we have a chart at ι, remove the problem to the model space $\Gamma_\Lambda^0(X)$, and, indeed, doing so makes some of our arguments easier to apply. For the manifold structure of $C^0(\Lambda, X)$ and related theory we refer the reader to Franks [2], Eliasson [1], Foster [1]. Notice, however, that for $\sigma \in \Gamma_\Lambda^0(X)$ near the origin, composing with the exponential map exp: $TX \to X$ gives an element h of $C^0(\Lambda, X)$ near the inclusion, and that all such h come about in this way. Reversing the process gives us a chart at ι. Moreover the tangent map $Tf^\#$ is, in terms of this chart, the map $\sigma \mapsto (Tf)\sigma(f|\Lambda)^{-1}$. The decomposition

$$\Gamma_\Lambda^0(X) = C^0(E_s) \oplus C^0(E_u)$$

is $T_\iota f^\#$-invariant, and, if we write T_s and T_u for the restrictions of $T_\iota f^\#$ to the two summands, we have that $|T_s^n| \le A\lambda^n$ and $|T_u^{-n}| \le A\lambda^n$. Thus, by our remarks about the spectral radius in the appendix to Chapter 4, the spectra of T_s and T_u are separated by the unit circle, and so ι is hyperbolic, as asserted.

We now make some definitions needed for the statement of our main theorem. Let $B(x, a)$ be the open ball in X with centre x and radius a, with respect to the Riemannian distance function d, and let $\Sigma(x, b)$ denote the set $\{y \in X : d(f^n(x), f^n(y)) < b \text{ for all } n \ge 0\}$. For $b \ge a \ge 0$ we define the *stable set of size* (b, a) of f at x to be $B(x, a) \cap \Sigma(x, b)$. We say that a map of an open subset of the total space of a vector bundle into a manifold is F^r (r times continuously fibre differentiable) if, with respect to admissible atlases, all partial derivatives in the fibre direction up to order r exist and are continuous as functions on the total space. One often says, in this case, that the images of the fibre vary C^r-continuously. Some of the theory of such maps is given in Appendix B. We can now state:

(6.21) Theorem. (*Generalized stable manifold theorem*) *Let f be a C^r diffeomorphism of X, and let Λ be a compact hyperbolic subset of X, with associated decomposition $T_\Lambda X = E_s \oplus E_u$. Then there exists an open neighbourhood W of the zero section in E_s and an F^r map $h : W \to X$, such that, for some $b \ge a \ge 0$ and for all $x \in \Lambda$, g restricted to the fibre W_x over x is a C^r embedding with image $W_s^{loc}(x)$, the stable set of size (b, a) at x. The tangent space to $W_s^{loc}(x)$ at x is $(E_s)_x$.*

Proof. For some neighbourhood P of the zero section in $T_\Lambda X$, we have a map $\tilde{\phi} : P \to T_\Lambda X$ defined by

$$\tilde{\phi}(v) = (\exp_{f(x)})^{-1} f \exp(v),$$

where v is in the fibre P_x. We may define a map $\phi : C_b(P) \to C_b(T_\Lambda X)$ by

$$\phi(\rho) = \tilde{\phi}\rho(f|\Lambda)^{-1},$$

where $C_b(T_\Lambda X)$ is the Banach space of bounded sections of $T_\Lambda X$ (using the sup norm) and $C_b(P)$ is the subset of sections with values in P. We observe that the map taking ρ to $(Tf)\rho(f|\Lambda)^{-1}$ is continuous linear, and that the map on sections induced by the F^r map $\tilde{\phi}(Tf^{-1})$ is C^r (see Remarks B.26 of Appendix B). Thus ϕ is C^r.

The differential of ϕ at $0 \in C_b(P)$ is the linear automorphism $\rho \mapsto (Tf)\rho(f|\Lambda)^{-1}$ of $C_b(T_\Lambda X)$. As indicated earlier, this automorphism is hyperbolic, with stable summand $C_b(E_s)$ and unstable summand $C_b(E_u)$. We may, by Theorem 4.19, give those subspaces norms (equivalent to the previous ones) with respect to which the stable component of $D\phi(0)$ and the inverse of its unstable component have operator norm <1. We may then apply Theorem 6.5, with $\eta = \phi - D\phi(0)$, and obtain a C^r stable manifold map $\psi: B_s \to B_u$, for some small balls B_s and B_u of equal radii (see Theorem 6.9 (ii)) and with centre 0 in $C_b(E_s)$ and $C_b(E_u)$ respectively. This ψ is the unique map with the property that $\tau = \psi(\sigma)$ if and only if $\phi^n(\sigma, \tau) \in B_s \times B_u$ for all $n \geq 0$.

To get back to the original norm on $C_b(T_\Lambda X)$, choose balls $B(0, a)$ and $B(0, b)$ with respect to that norm, and balls D_s and D_u with respect to the new norm, all with centre 0, such that

$$B(0, a) \subset D_s \times D_u \subset B(0, b) \subset B_s \times B_u.$$

We have a stable manifold function: $D_s \to D_u$ which is the restriction of $\psi: B_s \to B_u$. Let G denote its graph $\{(\sigma, \psi(\sigma)): \sigma \in D_s\}$. If $\rho \in B(0, a) \cap G$, then $\phi^n(\rho)$ stays in $D_s \times D_u$, and hence in $B(0, b)$ for all $n \geq 0$, whereas if $\rho \in B(0, a)\backslash G$, then $\phi^n(\rho)$ leaves $B_s \times B_u$, and hence leaves $B(0, b)$, for some $n > 0$. Let V denote the image of $B(0, a) \cap G$ in $C_b(E_s)$ by the product projection. Clearly V is open in $C_b(E_s)$.

Now consider the restriction $\psi: V \to C_b(E_u)$. It is the unique map with the property that $\tau = \psi(\sigma)$ if and only if $(\sigma, \tau) \in B(0, a)$ and $\phi^n(\sigma, \tau) \in B(0, b)$ for all $n \geq 0$. Equivalently, $\tau = \psi(\sigma)$ if and only if, for all $x \in \Lambda$, $\exp(\sigma(x), \tau(x))$ is in the stable manifold of size (b, a) at x. The first thing that this characterization shows is that $\tau(x)$ is a function of $\sigma(x)$ rather than the whole section σ. By this we mean that, if two elements of V had the same value at $x \in \Lambda$ but their images under ψ had different values at x, we could interchange the latter values without disturbing the stable manifold property (here we use the fact that our sections are merely bounded). This offends uniqueness of ψ. We conclude that if any $\sigma \in V$ has $\sigma(x) = v$ for some $x \in \Lambda$, then the section σ_v defined by

$$\sigma_v(x) = v, \ \sigma_v(y) = 0 \quad \text{for } y \neq x$$

is in V, and $\psi(\sigma)(x) = \psi(\sigma_v)(x)$. Thus, if W denotes the open subset $\{\sigma(x): x \in \Lambda, \sigma \in V\}$ of E_s, we may define a map $g: W \to E_u$ by

$$g(v) = \psi(\sigma_v)(x),$$

where $v \in W_x$, such that ψ is induced by g. That is to say, $V = C_b(W)$ and, for all $\sigma \in V$, $\psi(\sigma) = g\sigma$. In the notation of Appendix B, $\psi = g_*$.

The map h of the theorem is defined by

$$h(v) = \exp(v, g(v)).$$

Thus if g is F^r, so is h. We would like to deduce the result for g from the fact that g_* is C^r. This is not immediate. However if we now restrict our attention to the subspace $C^0(T_\Lambda M)$ of all continuous sections of $T_\Lambda M$, we obtain as before a C^r stable manifold map from $C^0(W) = V \cap C^0(E_s)$ to $C^0(E_u)$ which agrees with $\psi: W \to C_b(E_u)$ by uniqueness of the latter. Thus the C^r map $\psi: C^0(W) \to C^0(E_u)$ is induced from g, and this fact, by Theorem B.27 of Appendix B, ensures that g is F^r.

Now that we know that g is F^r, we may work with bounded sections again. By Theorem 6.9, the differential $D\psi(0)$ of $\psi: V \to C_b(E_u)$ is the zero map from $C_b(E_s)$ to $C_b(E_u)$. The fibre differential of g at $v \in W_x$ is the linear map from $(E_s)_x$ to $(E_u)_x$ whose value at w is the value at x of $D\psi(\sigma_v)(\sigma_w) \in C_b(E_u)$, where σ_v is the section defined above (see Remarks B.26 of Appendix B). Thus the fibre differential at $0 \in W_x$ is the zero map from E_s to E_u, and this gives the required tangency property. $\qquad\square$

(6.22) Corollary. *There exists constants A and λ, with $0 < \lambda < 1$, such that, for all $x \in \Lambda$, for all y and $z \in W_s^{\mathrm{loc}}(x)$ and for all $n \geqslant 0$,*

$$d(f^n(y), f^n(z)) \leqslant A\lambda^n d(y, z).$$

For all x and $x' \in \Lambda$, $W_s^{\mathrm{loc}}(x) \cap W_s^{\mathrm{loc}}(x')$ is open in $W_s^{\mathrm{loc}}(x)$.

Proof. The inequality follows easily from Theorem 6.9 (ii) applied to the stable manifold of the map ϕ in the above proof. Now let $y \in W_s^{\mathrm{loc}}(x) \cap W_s^{\mathrm{loc}}(x')$, and let $z \in W_s^{\mathrm{loc}}(x)$. By the first part

$$d(f^n(z), f^n(x')) \leqslant A\lambda^n d(z, x')$$

$$\leqslant A\lambda^n(d(z, x) + d(x, y) + d(y, x'))$$

$$\leqslant 3aA\lambda^n,$$

and so, for all $n \geqslant$ some n_0, $d(f^n(z), f^n(x')) < b$ for all $z \in W_s^{\mathrm{loc}}(z)$. But, for z sufficiently near y, $d(y, z) < a - d(y, x')$ and, by continuity of f, $d(f^n(y), f^n(z)) < b - d(f^n(y), f^n(x'))$ for all n with $0 \leqslant n \leqslant n_0$. Thus $z \in W_s^{\mathrm{loc}}(x')$. $\qquad\square$

We shall not give a detailed account of the generalized stable manifold theorem for flows. We could obtain a version by following through the proof of Theorem 6.21, and replacing stable manifolds by α-stable manifolds (for $\alpha < 1$) as in the proof of Theorem 6.19. There would remain the difficulty of proving that the α-stable manifolds so obtained are precisely the stable sets of the points of the invariant set Λ. The generalized stable manifold theorem for flows is actually a corollary of an even more general theorem, Theorem 6.1 of Hirsch, Pugh and Shub [1], which is the reference for further reading on the subject.

Appendix 6

I. PERTURBED STABLE MANIFOLDS

We extend the proof of the local stable manifold theorem to give information about the dependence of the stable manifold function h upon the perturbation η. We continue with the notations of Theorem 6.5. In addition, let N^r ($r \geq 0$) denote the subset of $C^r(B, \mathbf{E})$ consisting of maps η that are Lipschitz with constant κ and satisfy $\eta(0) = 0$.

(6.23) Theorem. *For all $\eta \in N^r$, the stable manifold function h^η of $T + \eta$ is C^r-bounded. The map $\theta: N^r \to C^r(B_s, B_u)$ sending η to h^η is continuous at η_0 if η_0 is uniformly C^r.*

Proof. We regard η as another independent variable on the right-hand side of (6.7), so that χ becomes a map from $N^r \times B_s \times \mathscr{S}_0(B)$ to $\mathscr{S}_0(B)$. We obtain a fixed point map $f: N^r \times B_s \to \mathscr{S}_0(B)$. Using Corollary B.11 of Appendix B, $\chi^\eta - \chi^0$ is C^0-bounded and $D\chi^\eta$ is C^{r-1}-bounded. Thus by Theorem C.10 of Appendix C, $g^\eta - g^0$ is C^r-bounded, and hence $h^\eta - h^0$ is C^r-bounded. Since h^0, the stable manifold function of T, is zero, h^η is C^r-bounded.

Now let $\eta_0 \in N^r$ be uniformly C^r, and let $\eta \in N^r$. Then the map from N^r to $C^r(B_s \times \mathscr{S}_0(B), \mathscr{S}_0(\mathbf{E}))$, taking η to $\chi^\eta - \chi^{\eta_0}$ is continuous at η_0 (this is essentially Corollary B.12). Using Theorem B.20, we see that the map χ^{η_0} is uniformly C^r. Hence, by Theorem C.10, the map from N^r to $C^r(B_s, \mathscr{S}_0(B))$ taking η to $g^\eta - g^{\eta_0}$ is continuous at η_0. Thus so is θ, which is this map composed with $(\pi_u ev^0)_*$, where $\pi_u: \mathbf{E} \to \mathbf{E}_u$ is the product projection. $\square$

CHAPTER 7

Stable Systems

We have, with the generalized stable manifold theorem, reached the border between elementary theory and recent research. In this final chapter we describe one direction in which the subject has developed. We have already aired the general philosophy behind this development in earlier chapters. The guiding idea is to find a family of dynamical systems which, on the one hand, contains "almost all" of them and, on the other hand, can be classified in some qualitatively significant fashion. For some years it was conjectured that structurally stable systems would fulfil both requirements. Although this turned out not to be the case except in very low dimensions, structural stability is such a natural property, both mathematically and physically, that it still holds a central place in the theory.

The two ingredients of structural stability are the topology given to the set of all dynamical systems and the equivalence relation placed on the resulting topological space. The latter is topological equivalence for flows, and topological conjugacy for diffeomorphisms. The former is the C^r topology (where $1 \leqslant r \leqslant \infty$). This topology is explained in detail in Appendix B, but the idea is clear enough. For example, two diffeomorphisms are C^r-close when their values, and also the values of corresponding derivatives up to order r, are close at every point. One may think of the derivatives as being defined via charts on the manifolds. Since the derivatives depend on the charts there is some ambiguity, but, for compact manifolds at any rate, this is unimportant provided we work with suitable finite atlases. Once we have defined the C^r topology we may be more specific about what we mean by "almost all" systems. Certainly an open dense subject of this topological space would be admirable for our purpose. Even the intersection of countably many such sets (called a *Baire* or *residual* subset†; think of the irrationals as a subset of the reals) would satisfy us.

† Such a subset would itself be dense, since the C^r-topology makes the space of dynamical systems a Baire space.

Let Diff^r X $(1 \leq r \leq \infty)$ denote the set of all C^r diffeomorphisms of the smooth manifold X, provided with the C^r topology. We say that $f \in$ Diff^r X is *structurally stable* if it is in the interior of its topological conjugacy class[†]. That is to say, f is structurally stable if and only if any C^r-small perturbation takes it into a diffeomorphism that is topologically conjugate to f. Similarly, let $\Gamma^r X$ denote the set of all C^r vector fields on X topologized with the C^r topology. The vector field $v \in \Gamma^r X$ (or the corresponding integral flow ϕ) is *structurally stable* if it is in the interior of its topological equivalence class. The definition of structural stability is due to Andronov and Pontrjagin [1].

I. LOW DIMENSIONAL SYSTEMS

As the simplest possible example, consider a vector field v on S^1, with no zeros. Then there is precisely one orbit (exercise: why?) and this is periodic. We may identify TS^1 with $S^1 \times \mathbf{R}$. Let $\tilde{v} : S^1 \to \mathbf{R}$ be the principal part of v. Since $\tilde{v}$ is continuous and S^1 is compact $|\tilde{v}(x)|$ is bounded below by some constant $a > 0$. Any perturbation of v with C^0-size less than a does not introduce any zeros. Thus the perturbed vector field still has only one orbit, and is topologically equivalent to v. So v is structurally stable (C^0 structurally stable, in fact).

Now suppose that v has finitely many zeros, all of which are hyperbolic. We call such a vector field on S^1 *Morse–Smale* (a definition for higher dimensions follows later). Then there is an even number $2n$ of zeros, with sources and sinks alternating around S^1, as in Figure 7.1. The hyperbolic zeros are individually structurally stable (see Exercise 5.22), so a sufficiently

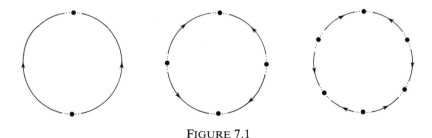

FIGURE 7.1

[†] Notice that this means that we have (apparently) more than one notion of structural stability for f, since we may also regard it as in Diff^s X for any $s < r$. One ought, really, to talk of C^r structural stability, and we occasionally will. At the time of writing it is not known which, if any, of these notions are equivalent.

C^1-small perturbation of v leaves the orbit configuration unaltered on some neighbourhood of the zeros. But a sufficiently C^0-small perturbation introduces no further zeros outside this neighbourhood. Thus the orbits flow between the components of the neighbourhood as before, and the perturbed vector field is topologically equivalent to the original one. Thus v is C^1 structurally stable.

It is important to note that there are vector fields v which are not Morse–Smale but which nevertheless have the above orbit structures. In this case the principal part $\tilde{v}$ has zero derivative at one or more of the zeros, and the inward or outward motion near the zero is given by "higher order terms". One may also have vector fields on S^1 with "one-way" zeros, as in Figure 7.2. However, provided that v has only finitely many zeros and is not

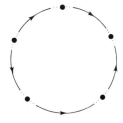

FIGURE 7.2

Morse–Smale, one can easily alter its orbit configuration. Choose a non-hyperbolic zero p. By a C^∞-small perturbation of the vector field near p, alter p to a source if it was not originally one, or to sink if it was originally a source. Since we leave all the other original zeros of v unaltered, we have changed the number of sources in either case. We have shown, then, that structurally stable vector fields on S^1 with finitely many zeros are Morse–Smale. However, a stronger result is true:

(7.3) Theorem. *A vector field on S^1 is C^1 structurally stable if and only if it is Morse–Smale. Morse–Smale vector fields form an open, dense subset of $\Gamma^r S^1$* $(1 \leqslant r \leqslant \infty)$.

Proof. We have proved sufficiency above. Let v be a vector field on S^1. Suppose that v is not the zero vector field. The tangent bundle TS^1 is isomorphic to the trivial bundle $S^1 \times \mathbf{R}$, so the image of v may be pictured as a curve on the cylinder (see Figure 7.3). The zeros of v are the points where the curve crosses the zero section, identifying the latter, as usual, with S^1. The zero is hyperbolic provided the curve is not tangent to the horizontal there.

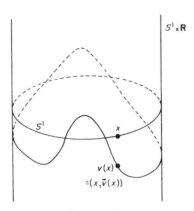

FIGURE 7.3

Now the image of the principal part $\tilde{v}$ of v is a closed subinterval $[a, b]$ with $a < b$. The critical points of $\tilde{v}$ (over which the curve is tangent to the horizontal) may form an infinite subset of S^1 but, by Sard's theorem†, their image under $\tilde{v}$ (the set of critical values) has measure zero in $\mathbf{R}$. Thus arbitrarily near $0 \in \mathbf{R}$ is a number c which is not a critical value of $\tilde{v}$. We perturb $\tilde{v}(x)$ to $\tilde{v}(x) - c$ and get a new vector field with only hyperbolic zeros. These, being isolated, are finite in number. Thus the perturbed vector field is Morse–Smale. We have proved the density part of the theorem, as the perturbation is arbitrarily C^∞-small. Note that if v were the zero vector field, we would first make a C^∞ small perturbation to get something else.

Now suppose that v is structurally stable. Then v is certainly not the zero vector field. Since v can be made Morse–Smale by an arbitrarily small perturbation, it is topologically equivalent to a Morse–Smale diffeomorphism. But we showed in the run-up to the theorem that this implies that v is Morse–Smale.   □

The above theorem is a simple illustration of a fruitful approach to the classification problem. One attempts to prove that certain properties are *open and dense* (i.e. hold for systems in an open, dense subset of $\mathrm{Diff}^r X$ or $\Gamma^r X$) or *generic* (the same thing with "Baire" replacing "open, dense"), using *transversality* theory. This is a theory that investigates the way submanifolds of a manifold cross each other, and how a map of one manifold to another throws the first across a submanifold of the second. Sard's theorem is the opening shot in this theory; for further details see Hirsch's

† See Hirsch [1]. The proof in the present case is by dividing S^1 into n equal subintervals and letting $n \to \infty$. By Taylor's Theorem, if a subinterval contains a critical point of $\tilde{v}$ then its image under $\tilde{v}$ has a length which is bounded by a fixed multiple of the square of the length of the subinterval. Thus the total length of all such images tends to zero as $n \to \infty$.

book [1]. If a property is dense, then automatically a stable system is equivalent to a system with the property. One then anticipates, and attempts to prove, that the stable system itself has the property (since the property will often be one that is not shared by all equivalent systems). For example, the property of having only hyperbolic zeros is open and dense for flows on a compact manifold of any dimension, and any structurally stable flow possesses this property. The proof is essentially that for S^1 given above. Similarly for diffeomorphisms of a compact manifold X the following properties are generic, and satisfied by structurally stable diffeomorphisms:

(i) all periodic points are hyperbolic,

(ii) for any two periodic points x and y, the stable manifold of x and the unstable manifold of y intersect *transversally*.

(This means that the tangent spaces to the two submanifolds at any point p of intersection generate the whole tangent space of the manifold X at p: $T_pW_s(x) + T_pW_u(y) = T_pX$. See Figure 7.4. We denote this by the standard notation $W_s(x) \pitchfork W_u(y)$.) The above result is known as the Kupka–Smale theorem (Kupka [1], Smale [3], Peixoto [1]).

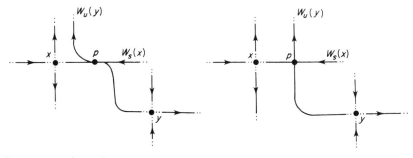

Non-transverse intesection at p Transverse intersection at p

FIGURE 7.4

Theorem 7.3 also holds for vector fields on compact orientable 2-manifolds and for diffeomorphisms of S^1, a result due to Peixoto [2]. Of course we must extend our definition of the Morse–Smale property to cover these cases. In fact, we give a definition for dynamical systems on manifolds of any dimension. We say that a dynamical system is *Morse–Smale* if

(i) its non-wandering set is the union of a finite set of fixed points and periodic orbits, all of which are hyperbolic, and

(ii) if $G.x$ and $G.y$ are any two orbits of the non-wandering set then $W_s(G.x) \pitchfork W_u(G.y)$.

Thus in the case of a 2-dimensional flow the non-wandering set consists of at most five types of orbit, viz. hyperbolic sources, sinks and saddle points, and hyperbolic expanding and contracting closed orbits. The second condition is vacuous unless both x and y are saddle points, and, when they are, it reduces to the condition that $W_s(x)$ and $W_u(y)$ do not have a common orbit (apart from $\{x\}$ in the case $x = y$). One says informally "no orbit joins x to y". So the configurations in Figure 7.5 are not allowed in Morse–Smale flows. It is

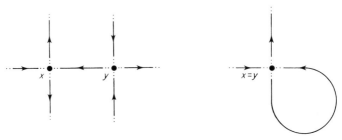

FIGURE 7.5

easy to get some idea of why condition (ii) should be necessary for structural stability. If it does not hold, we can take some point p on the offending orbit, and by making a slight (downwards in Figure 7.6) deflection of the vector

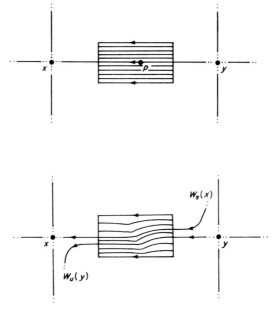

FIGURE 7.6

field in a thin flow tube containing p, we can divert $W_s(x)$ and $W_u(y)$ into different paths. Thus we have altered the orbit configuration. Or have we? It certainly looks as though we have in our picture, but it is dangerous to lean too heavily on pictures. For example we might, in our alteration of the vector field, join up two perfectly harmless orbits which happended to pass through the tube, and thus create a new orbit joining some other pair of saddle points, x' and y' say. In this case the new system might possibly be equivalent to the old one by a homeomorphism taking x' to x and y' to y. Or again, suppose that condition (i) were not in force and that there were infinitely many orbits joining pairs of saddle points. It is far from obvious that obliterating one of them would alter the topology of the picture as a whole. In fact we have to be a bit careful, and we have to assume that condition (i) holds. We take a neighbourhood U of the non-wandering set sufficiently small not to contain p. Since there are only finitely many *separatrices* (as we call orbits that begin or end at a saddle point) and since each is in U for all but a bounded period of time, each will pass through a flow tube about p in the complement of U only finitely many times. By making the tube thinner, we can ensure that $\mathbf{R}.p$ passes through the tube only once and the other separatrices miss it altogether. If we perform the perturbation in this tube we do indeed decrease the number of saddle point connections by one, and hence alter the orbit configuration topologically.

To show that a 2-dimensional Morse–Smale vector field v is structurally stable, one builds up a topological equivalence between it and any small perturbation of it first on a neighbourhood U of the non-wandering set of v (on which v is structurally stable by hyperbolicity), then on the components outside U of the separatrices (these have not been moved far by the perturbation and so still join corresponding components of U) and finally on the homogeneous looking blocks of orbits that remain. We shall not embark on the details of this programme, nor deal with the density problem. This is rather technical and special to two dimensions, and we refer the reader to Peixoto's paper [2]. We just quote the result:

(7.7) Theorem. *A vector field on a compact orientable 2-manifold X (resp. a diffeomorphism of S^1) is C^1 structurally stable if and only if it is Morse–Smale. Morse–Smale systems are open and dense in $\Gamma^r(X)$ (resp. $\mathrm{Diff}^r S^1$) for $1 \leqslant r \leqslant \infty$.*

The sufficiency of the conditions in the case when X is a 2-dimensional disc, was first suggested by Andronov and Pontrajagin [1] and proved for an analytic vector field by De Baggis [1]. In a recent paper [3] Peixoto has developed his theorem into a classification of structurally stable systems. The theorem for non-orientable 2-manifolds is still an open problem, except for some cases which have been proved by Gutierrez [1].

II. ANOSOV SYSTEMS

For some time it was hoped that the above results might generalize to higher dimensions. Unfortunately neither the characterization nor the density of structurally stable systems goes over. It is, then, a relief to find that, for all dimensions:

(7.8) Theorem. *A Morse–Smale system on a compact manifold is C^1 structurally stable.*

In fact, it is doubly welcome, because, since every compact manifold admits a Morse–Smale system[†], every compact manifold admits a structurally stable system. This was an early conjecture which turned out to be hard to prove, and survived until Palis and Smale's paper [1].

By the time this paper appeared it was known that there were other structurally stable systems besides Morse–Smale systems. The first of these to emerge were the toral automorphisms (see Example 1.30). These are structurally stable, but their non-wandering sets are the whole of the tori on which they are defined, so they certainly fail (i) of Morse–Smale (in fact, they are the opposite extreme to Morse–Smale systems in this sense!). Toral automorphisms were first put forward by Thom as a counterexample to the density of Morse–Smale diffeomorphisms. The justification of this is by proving them structurally stable, and this was first done by Anosov. His proof was for a wider class of systems which we now define.

A diffeomorphism $f: X \to X$ of a manifold X is *Anosov* if X has a hyperbolic structure with respect to f. Recall that this means that the tangent bundle TX splits continuously into a Tf-invariant direct-sum decomposition $TX = E_s \oplus E_u$ such that Tf contracts E and expands E_u (with respect to some Riemannian metric on X). Trivially hyperbolic linear maps f possess this property, since one has the identification $T\mathbf{R}^n = \mathbf{R}^n \times \mathbf{R}^n$ and $Tf(x, v) = (f(x), f(v))$, so that the splitting of $\mathbf{R}^n$ into the stable and unstable manifolds of f, W_s and W_u say, gives a splitting of $T\mathbf{R}^n$ as $(\mathbf{R}^n \times W_s) \oplus (\mathbf{R}^n \times W_u)$. In the case of toral automorphisms this splitting is carried over to the torus when we make the identification, and so toral automorphisms are Anosov. Similarly a vector field on X is *Anosov* if X has a hyperbolic structure with respect to it, and as examples of such we have all suspensions of Anosov diffeomorphisms. The main result due originally to Anosov [1, 2] is:

(7.9) Theorem. *Anosov systems on compact manifolds are C^1 structurally stable.*

[†] Take the unit time map of the gradient flow of almost any smooth real function on the manifold (see Smale [1]).

The diffeomorphism case of the theorem has a nice functional analytic proof due to Moser [1] (see also Franks [1] or Nitecki [1]). Recall, from the preamble to Theorem 6.21, that the set $C^0(X, X)$ of all C^0 maps of the compact manifold X to itself has the structure of a Banach manifold modelled on $\Gamma^0(X)$. Let $f : X \to X$ be an Anosov diffeomorphism and let $g : X \to X$ be C^1-near f. As in Chapter 6, one considers the map $f^\# : C^0(X, X) \to C^0(X, X)$ sending h to fhf^{-1}. Let $G : C^0(X, X) \to C^0(X, X)$ be the C^1-nearby map sending h to ghf^{-1}. Since, as in Chapter 6, $f^\#$ has a hyperbolic fixed point at the identity id_X, G has a fixed point, h say, near id_X. So $h = ghf^{-1}$, or, equivalently, $hf = gh$. Now the map f is *expansive*, meaning that the orbits of different points of X are never close together. More precisely there is a number $a > 0$ such that $d(f^n(x), f^n(y)) < a$ for all $n \in \mathbf{Z}$ implies $x = y$. This is clear from the fact that the map $\phi : C_b(P) \to C_b(TX)$ defined as in the proof of Theorem 6.21 has a hyperbolic fixed point at 0, so that only iterates of 0 under ϕ stay near 0. Suppose that $h(x) = h(y)$. Then $hf^n(x) = g^n h(x) = g^n h(y) = hf^n(y)$ for all $n \in \mathbf{Z}$. Since h is near id_X, $f^n(x)$ is near $f_n(y)$ for all $n \in \mathbf{Z}$, and thus since f is expansive, $x = y$. This shows that h is injective. But it is a standard result of algebraic topology that a continuous injection $h : X \to X$ of a compact manifold is a homeomorphism. Thus f is structurally stable.

It is worth while getting as much insight as possible into the effects of hyperbolic structure, and so we give a more down-to-earth description of the existence of h and the expansiveness of f (due to Robinson [4]). The existence of a hyperbolic structure on X implies that, for any $x \in X$, $T_{f^{-1}(x)}f$ throws a product ε-disc neighbourhood $D_s \times D_u$ of 0 in $T_{f^{-1}(x)}X$ across a similar ε-disc neighbourhood of 0 in $T_x X$ as shown in Figure 7.10. This

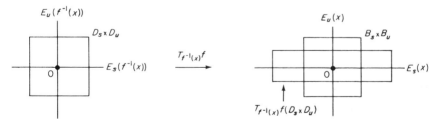

FIGURE 7.10

behaviour is echoed by f in the manifold X, provided that ε is small; it throws $D_{f^{-1}(x)} = \exp_{f^{-1}(x)}(D_s \times D_u)$ across $D_x = \exp_x(B_s \times B_u)$, as shown in Figure 7.11. We have similar product neighbourhoods D_y for all $y \in X$. We observe that f throws $D_{f^{-2}(x)}$ across $D_{f^{-1}(x)}$ as shown in Figure 7.11, and hence that $\bigcap\{f^n(D_{f^{-n}(x)}) : 0 \leq n \leq 2\}$ is as shown (shaded). It is, of course, possible that

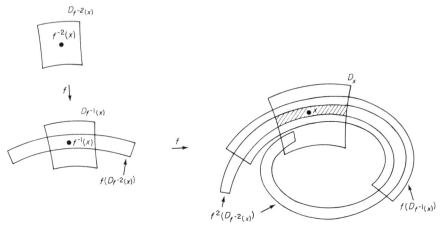

FIGURE 7.11

$f^2(D_{f^{-2}(x)})$ has further intersections with D_x, and we have attempted to indicate this in the diagram. If g is C^1-close to f then we have a similar picture replaced by $g^n(D_{f^{-n}(x)})$. It is fairly clear from the way that the $(s-)$ height of the sets $D_x^m(g) = \bigcap \{g^n(D_{f^{-n}(x)}): 0 \leqslant n \leqslant m\}$ decreases, that $D_x^\infty(g)$ is a disc stretching across D_x in the u-direction, as shown in Figure 7.12.

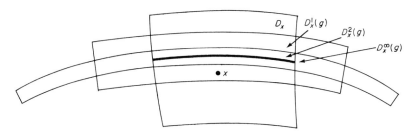

FIGURE 7.12

Since $D_x^\infty(g)$ in the set of points of D_x whose negative g-iterates stay near their corresponding f-iterates, g^{-1} maps $D_{f(x)}^\infty(g)$ into $D_x^\infty(g)$. Moreover it contracts it, since the discs are in the u-direction. Thus $\bigcap \{g^{-n}(D_{f^n(x)}^\infty(g)): n \geqslant 0\}$ is a single point, which we denote $h(x)$. We observe that $h(x)$ is the unique point of D_x such that every g-iterate of it stays near the corresponding f-iterate of x. Expansivity of f follows by putting $g = f$. In general, since $g^n(gh(x)) \in D_{f^{n+1}(x)} = D_{f^n f(x)}$ for all $n \in \mathbf{Z}$, we deduce that $hf(x) = gh(x)$.

There are, at the time of writing, several unsolved problems about Anosov diffeomorphisms. For example, is their non-wandering set always the whole manifold? Do they always have a fixed point? Not all manifolds admit Anosov diffeomorphism (in contrast to Morse–Smale diffeomorphisms). Do all n-dimensional manifolds which do admit them have $\mathbf{R}^n$ as universal covering space? One positive thing that we can say is that Anosov diffeomorphisms of tori are well understood; they are all topologically conjugate to the toral automorphisms (see Manning [1]).

III. CHARACTERIZATION OF STRUCTURAL STABILITY

The main problem now facing us is how to characterize structural stability, bearing in mind that we have to reconcile such apparently dissimilar systems as Morse–Smale and Anosov systems. The link comes when one recognizes, as Smale did, that if we generalize the Morse–Smale definition by replacing the term "closed orbits" by "basic sets" (definition shortly) then Anosov systems, and others as well, are allowed in. For example, toral automorphisms have only one basic set, the whole torus, so they trivially satisfy the new conditions. We say that a dynamical system has an Ω-*decomposition* if its non-wandering set Ω is the disjoint union of closed invariant sets Ω_1, $\Omega_2, \ldots, \Omega_k$. If the system is *topologically transitive* on Ω_i (that is, Ω_i is the closure of the orbit of one of its points) for all i, we say that $\Omega_i \cup \ldots \cup \Omega_k$ is a *spectral decomposition*, and that the Ω_i are basic sets. One could define the concept of a basic set in isolation by saying that a closed invariant set, $\Lambda \subset \Omega$ is *basic* if the system is topologically transitive on Λ but Λ does not meet the closure of the orbit of $\Omega \backslash \Lambda$. Note that a basic set is *indecomposable*, in that it is not the disjoint union of two non-empty closed invariant sets. In general there is no reason why the non-wandering set of a system should have a spectral decomposition. For example in the flow illustrated in Figure 7.13, the figure eight consisting of two orbits joining a saddle point is an

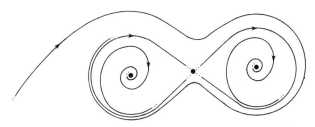

FIGURE 7.13

isolated part of the non-wandering set, but there is no way of splitting it into basic sets. However, Smale [5, 6] made the following definition:

A dynamical system is (or satisfies) *Axiom A* if its non-wandering set
  (a) has a hyperbolic structure, and
  (b) is the closure of the set of closed orbits of the system†,
and proved a fundamental theorem:

(7.14) Theorem. (*The Spectral Decomposition Theorem*) *The non-wandering set of an Axiom A dynamical system on a compact manifold is the union of finitely many basic sets.*

Proof. To get some idea of how Axiom A brings about this decomposition, let us examine the local structure of the non-wandering set Ω of an Axiom A diffeomorphism f of a compact manifold X. We know that Ω has a generalized stable manifold system, and the first point to realize is that Ω is given locally as the set of vertices of a sort of grid formed by the family of stable manifolds crossing the family of unstable manifolds (this is the so-called *local product structure* of Ω). To see this, note that if x and y are nearby points of Ω then, locally, by continuity of the unstable and stable manifold systems, $W_s(x)$ intersects $W_u(y)$ transversally in a single point z, and similarly $W_u(x)$ intersects $W_s(y)$ transversally in a single point t. We assert that z and t are in Ω (a slightly more general result is known as the *cloud lemma*). By Axiom A(b), continuity of the stable manifolds system and the fact that Ω is closed, we may assume that x and y are periodic, or even fixed (since periodic points are fixed by some power of f). Under positive iterates of f, any neighbourhood U of z presses up against and is spread out along $W_u(x)$, until it intersects $W_s(y)$ (see Figure 7.14). It is then pressed against $W_u(y)$ and spreads out along it until it eventually intersects U.

Also note that if N is a small open neighbourhood of x in Ω and if U is a non-empty open subset of N then the orbit of U is dense in N. For, given

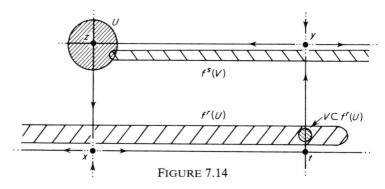

FIGURE 7.14

---

† It was for several years conjectured that A(a) implied A(b). This conjecture is now known to be true for diffeomorphisms of two-dimensional manifolds (Newhouse and Palis [1]) but false for diffeomorphisms of all higher dimensional manifolds (Dankner [1]).

$p \in N$, we may, as above, assume p fixed, and also assume that there is a fixed point q in U. Then $W_u(q)$ intersects $W_s(p)$ in a point z which is, by local product structure, in Ω. Some negative power of f takes z into U, and some positive power takes it arbitrarily near p. Thus p is in the closure of the orbit of U.

We now define a relation $\sim$ on Ω by $x \sim y$ *if and only if y is in the closure of the orbit of every neighbourhood of x in* Ω. We assert that $\sim$ is an equivalence relation. Reflexivity is trivial. Suppose $x \sim y$. Let P be any open neighbourhood of y in Ω, and let N be a small open neighbourhood of x in Ω (i.e. small enough for the density property mentioned in the previous paragraph to hold). Since $x \sim y$, P intersects some $f^r(N)$, and hence $N \cap f^{-r}(P)$ is a non-empty set, U say. Since the orbit of U is dense in N, so is the orbit of P, and thus $y \sim x$. See Figure 7.15. Similarly, suppose that $x \sim y$ and $y \sim z$. Let

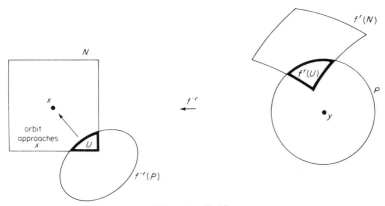

FIGURE 7.15

P be any neighbourhood of x in Ω, and let N_1 and N_2 be small neighbourhoods in Ω of y and z respectively. Then since $y \sim z$, some $f^r(N_1)$ intersects N_2, and hence $N_1 \cap f^{-r}(N_2)$ is a non-empty open subset, V say. Also, since $x \sim y$, some $f^s(P)$ intersects N_1 in a non-empty open set U_1, say. Since the orbit of U_1 is dense in N_1, some $f^t(U_1)$ intersects V. Thus $f^{r+t}(U_1)$ intersects N_2 in a non-empty open subset U_2. Since the orbit of U_2 is dense in N_2, so is the orbit of P. Thus $x \sim z$. See Figure 7.16.

Let Ω_i be the equivalence classes of Ω under $\sim$. By the proof of symmetry, we get a typical Ω_i by taking any open set N_i on Ω small enough to have the density property and taking the closure of its orbit. Thus there are only finitely many Ω_i, because we can take a finite covering of the compact set Ω by small enough open sets N_i, and, of course, the associated Ω_i, some of which may coincide, cover Ω. The Ω_i are, by the above description, closed

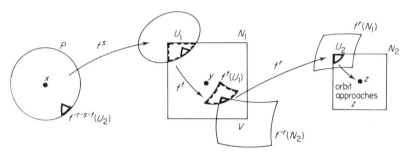

FIGURE 7.16

invariant subsets. We need finally to show that each Ω_i is topologically transitive, and this is a standard deduction from the fact that it contains an open set U_i with dense orbit. For, let V_j be a countable basis for the topology of Ω_i, and let O_j be the orbit of V_j. Then O_j is open and dense in Ω_i (because Ω_i is an $\sim$-equivalence class), and hence $\bigcap_j O_j = D$ (say) is residual, and hence dense in Ω_i. Let $x \in D$. Then the orbit of x is dense in Ω_i, because any open subset V of Ω_i contains some V_j, and $x \in$ some $f^n(V_j)$, so $f^{-n}(x) \in V_j \subset V$. ☐

To visualize a general Axiom A system, then, one thinks of a system with finitely many fixed points and periodic orbits, all hyperbolic (such as the gradient of the height function on the torus, or a Morse–Smale system, for example Figure 7.17 for S^2), but in higher dimensions and with the fixed

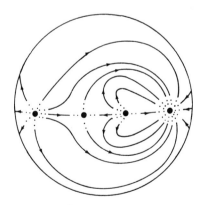

FIGURE 7.17

points and periodic orbits replaced by more general basic sets. The basic set could, for example, be a torus on which the system (a diffeomorphism) restricts to a toral automorphism. For example, if one takes the product of a toral automorphism on $S^1 \times S^1$ with the (North pole)–(South pole) diffeomorphism of S^1, one gets an Axiom A diffeomorphism of $S^1 \times S^1 \times S^1$ with two basic sets, $S^1 \times S^1 \times \{$North pole$\}$ and $S^1 \times S^1 \times \{$South pole$\}$. There is, however, no need for a basic set to be a submanifold. We shall describe two examples for which it is not. Such basic sets are usually termed *strange* or *exotic*, and a good deal of work has been done on their structure (see, for example, Williams [1, 3], and Sullivan and Williams [1]).

The first of these is the *expanding attractor* described by Smale [5]. An *attractor* of a diffeomorphism is a subset Λ that has a compact neighbourhood N such that $f(N) \subset \text{int } N$ and $\bigcap_{r \geqslant 0} f^r(N) = \Lambda$. For a flow ϕ the corresponding conditions are $\phi^t(N) \subset N$ for all $t \supset 0$ and $\bigcap_{t \geqslant 0} \phi^t(N) = \Lambda$. We visualize a map f of the solid torus $S^1 \times B^2$ into itself, stretching the S^1 factor to approximately twice its length, contracting the B^2 factor into something under half its width, and wrapping it twice around the target copy, as shown in Figure 7.18. We insist that each cross-sectional 2-disc $\{[\theta]\} \times B^2$

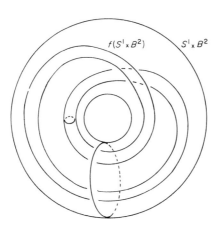

FIGURE 7.18

is mapped into $\{[2\theta]\} \times B^2$. Imagine the domain $S^1 \times B^2$ as embedded in a 3-ball B^3 in some 3-manifold X. Certainly f extends to a diffeomorphism of X. One could even produce a deformation of X, fixed outside B^3, which would slide $S^1 \times B^2$ into $f(S^1 \times B^2)$ so that x finishes up at $f(x)$. Thus the phenomenon that we are about to describe is by no means pathological. It may be met with locally in diffeomorphisms of any manifold of dimension greater than two.

Consider $f^2(S^1 \times B^2)$. This is a thinner longer tube which winds twice around $f(S^1 \times B^2)$. Similarly $f^3(S^1 \times B^2)$ is thinner and longer still, and winds twice around $f^2(S^1 \times B^2)$ (so eight times around $S^1 \times B^2$). Any point x of $S^1 \times B^2$ that is outside $f^r(S^1 \times B^2)$ for some $r > 0$ has a neighbourhood U (in $S^1 \times B^2$) that is also outside $f^r(S^1 \times B^2)$. Thus $f^s(U)$ does not intersect U for $s \geq r$, and hence x is wandering. Thus, if we are interested in the non-wandering set of f in $S^1 \times B^2$, we need only look at $\Lambda = \bigcap_{r \geq 0} f^r(S^1 \times B^2)$. If we look at the intersections of $f^r(S^1 \times B^2)$ with a single cross section $\{[\theta]\} \times B^2$ (Figure 7.19), we notice a striking resemblance to the well-known

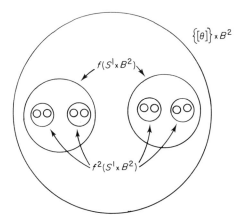

FIGURE 7.19

construction of the Cantor set from an interval by repeatedly deleting middle thirds. Thus it is no surprise to find that $\Lambda \cap (\{[\theta]\} \times B^2)$ is homeomorphic to the Cantor set. Technically speaking, Λ is a fibre bundle over S^1 with projection the product projection of $S^1 \times B^2$ and fibre the Cantor set.

It is not hard to see that Λ is, in fact, the non-wandering set of f on $S^1 \times B^2$. Let $x \in \Lambda$, and let U be any neighbourhood of x. It is clear that any segment $[a, b] \times B^2$ of $S^1 \times B^2$, however short, will, under a sufficiently large iterate f^s, wrap once round $S^1 \times B^2$ (i.e. its image will intersect every disc $\{[\theta]\} \times B^2$). A similar remark holds for any of the image tubes $f^r(S^1 \times B^2)$. We may choose a segment V of some such tube with $x \in V \subset U$. By the remark, for some $s > 0$, $f^s(V)$ intersects every cross-sectional disc of V. Thus $f^s(U) \cap U$ is non-empty, which proves that x is non-wandering. A similar argument shows that any open subset of Λ has its orbit dense in Λ, which implies that Λ is topologically transitive. Since Λ is locally like the product of a real interval and the Cantor set, we can talk of the 1-manifold part of Λ at x, meaning the fibre of the projection onto the Cantor set. The hyperbolic structure of Λ is clear: the stable subspace at $x = ([\theta], p)$ is the tangent to the cross-sectional 2-disc at x, and the unstable manifold is the tangent to the 1-manifold part at x. This expanding behaviour on the 1-manifold part is the reason we call the attractor *expanding*.

We now describe our second exotic basic set, the *Smale Horseshoe* (Smale [5]). Again the example is for a diffeomorphism, but would yield a basic set of a flow by suspension. It is fundamentally a 2-dimensional phenomenon, but is, perhaps, rather harder to grasp than the expanding attractor. This time we take a solid 2-dimensional square S, and four subrectangles, two vertical, R_0 and R_1, and two horizontal, $_0R$ and $_1R$ as shown (Figure 7.20).

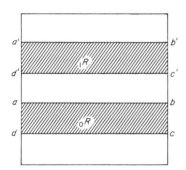 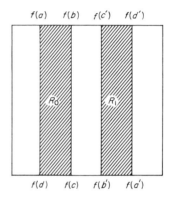

FIGURE 7.20

We map the vertices of $_0R$ to those of R_0 as shown, and extend linearly for a map f of $_0R$ to R_0. Similarly for $_1R$. *Note the half turn* in this second case. The map f so far defined is clearly hyperbolic. We now extend f to the whole square S, mapping it onto the shaded horseshoe shape in Figure 7.21 (whence the name of the example). We make sure that $f(S)$ is contained in the interior of $C = D_0 \cup S \cup D_1$, where D_0 and D_1 are half discs on the horizontal edges of S as shown. We then extend f over D_0 and D_1, mapping them into int D_0. We may actually arrange for $f|D_0$ to be a contraction, so that it has a unique attracting fixed point p. At this stage, f has a basic set of the type that we wish to consider. Notice again that, since we could produce f by deforming some neighbourhood of C in the plane, we can consider it as embedded locally in any 2-manifold X, as part of a diffeomorphism $f: X \to X$. We suppose this done.

We are interested in locating $\Omega(f) \cap S$. We observe that if a point $s \in S$ is not in $_0R$ or $_1R$ then it is not in $\Omega(f)$. This is because some neighbourhood U of it in S gets mapped into D_0 or D_1 by f, so into D_0 by f^2, and so into D_0 by

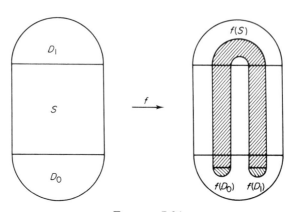

FIGURE 7.21

any higher power of f. Hence $U \cap f^r(U)$ is empty for $r > 0$. Similarly if x is not in $R_0 \cup R_1$, then some neighbourhood of it came from outside C under f, and its iterates under f^{-1} are all outside C, since $f(C) \subset C$. Summing up, if $x \in \Omega(f) \cap S$ then $x \in {}_iR_j = {}_iR \cap R_j$, for some i, j. Let $S_1 = \bigcup_{ij} R_j$.

We now make a similar observation for f^2. Let $R_{ij} = f({}_iR_j)$ and ${}_{ij}R = f^{-1}({}_iR_j)$ (see Figure 7.22). Note that $R_{ij} \subset R_i$ and ${}_{ij}R \subset {}_jR$. If $x \in S$ is outside

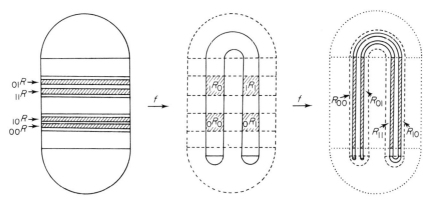

FIGURE 7.22

$\bigcup {}_{ij}R$, then so is some neighbourhood of x in S, f^2 takes the neighbourhood into $D_0 \cap D_1$ and further iterates of f keep it there. Hence $x \notin \Omega(f)$. Similarly $x \in S \backslash \bigcup R_{ij}$ is not in $\Omega(f)$. Hence $\Omega(f) \cap S \subset \bigcup({}_{ij}R \cap R_{kl}) = S_2$.

Now define ${}_iR_{jk} = {}_iR \cap R_{jk}$ and ${}_{ij}R_k = {}_{ij}R \cap R_k$. Notice that $f({}_{ij}R_k) = {}_iR_{jk}$. Define ${}_{ijk}R = f^{-1}({}_{ij}R_k)$ and $R_{ijk} = f({}_iR_{jk})$. Notice that $R_{ijk} \subset R_{ij}$ and ${}_{ijk}R \subset$

$_{jk}R$. Repeat the above argument, showing that $\Omega(f) \cap S \subset \bigcup(_{ijk}R \cap R_{lmn}) = S_3$ (see Figure 7.23). It is clear that, with a bit of care, we could define S_r

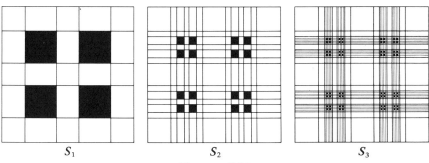

$$S_1 \qquad\qquad S_2 \qquad\qquad S_3$$

FIGURE 7.23

inductively, for all $r > 1$, and, hence, $S_\infty = \bigcap_{r \geqslant 1} S_r$ with the property that $\Omega(f) \subset S_\infty$. In fact S_∞ is homeomorphic to the Cantor set. There is a standard representation A of the Cantor set as infinite bisequences $a = \ldots a_{-2}a_{-1}a_0 . a_1 a_2 a_3 \ldots$ where each a_i is 0 or 1. If $U_n(a)$ is the set of all such bisequences that agree with a for n terms on either side of the decimal point then $\{U_n(a): n \geqslant 1\}$ is a basis of open neighbourhoods of a in A. We have a homeomorphism from A to S_∞ which sends the bisequence a to the (unique) point in $\bigcap_{n \geqslant 0}(_{a_{-n+1}\ldots a_0}R \cap R_{a_1\ldots a_n})$. Identifying S_∞ with A by this homeomorphism, it is clear that $f|S_\infty$ is the well known *shift* automorphism of A, which moves the decimal point one place to the left (i.e. $f(a)_i = a_{i-1}$ for all $i \in \mathbf{Z}$). Notice that $U_n(a)$ contains many periodic points of f, for example the one whose entries are the $2n$ terms $a_{-n+1} \ldots a_n$ repeated as a block time after time. This shows that $S_\infty \subset \overline{\mathrm{Per}\, f}$, and hence that $S_\infty \subset \Omega(f)$, so that $S_\infty = \Omega(f) \cap S$. Also there is a bisequence c which contains every possible finite sequence as a block of consecutive entries, so that, for every $a \in A$ and $n > 0, f^r(c) \in U_n(a)$ for some $r \in \mathbf{Z}$. Thus the orbit of c is dense in S_∞. Finally note that the hyperbolic structure for S_∞ is induced from $f|(_0R \cup _1R)$; the stable summands are horizontal and the unstable summands vertical.

Returning to the question of characterizing structural stability, and following Smale, we make another definition. We say that a system is *AS* if it satisfies both Axiom A and the *strong transversality* condition: for all x and y in the non-wandering set of the system, the stable manifold of the orbit of x intersects the stable manifold of the orbit of y transversally. This latter condition is, then, just the general version of the second condition in the definition of Morse–Smale systems. The best set of criteria of structural stability that has so far emerged is due to Robbin [1] (for C^2 diffeomorphisms) and, later, to Robinson [2, 3, 4] (for C^1 diffeomorphisms and flows).

They prove

(7.24) Theorem. *Any AS system is C^1-structurally stable.*

It seems likely that the converse of this theorem is also true, which would characterize structural stability very satisfactorily. At the time of writing the nearest approach to this is due to Franks [3] (see also Robbin [2]). A diffeomorphism $f : X \to X$ is *absolutely C^1-structurally stable* if for some C^1-neighbourhood N of f, there is a map σ (called a *selector*) associating with $g \in N$ a homeomorphism $\sigma(g)$ of X such that

   (i) $\sigma(f) = id_X$,
   (ii) for all $g \in N$, $g = hfh^{-1}$, where $h = \sigma(g)$,
   (iii) σ is Lipschitz at f with respect to the C^0 metric d (i.e. for some $\kappa > 0$ and all $g \in N$, $d(\sigma(g), id_X) \leqslant \kappa d(g, f)$).

Then:

(7.25) Theorem. *Any diffeomorphism is absolutely C^1 structurally stable if and only if it is AS.*

It is known that structural stability is equivalent to AS when $\Omega(f)$ is finite (Palis and Smale [1]) and that structural stability and Axiom A imply strong transversality (Smale [5]). It is also known that a structurally stable system is *weak Axiom A* (which is Axiom A with the hyperbolic structure condition relaxed for non-periodic points of $\Omega(f)$); this uses Pugh's celebrated C^1 closing lemma (Pugh [1, 2]).

IV. DENSITY

It is in attempting to generalize the second part of Peixoto's theorem, which deals with the density of Morse–Smale systems, that things go disastrously wrong. Obviously Morse–Smale systems are not dense in higher dimensions, since there are, as we have seen, other types of structurally stable systems. One might, nevertheless, have hoped that structurally stable systems are dense. This is not the case in higher dimensions, and we are now in a position to see why. The two ingredients for a counterexample are an exotic basic set (because such a set contains arbitrarily close but topologically distinct orbits, viz. periodic and non-periodic orbits) and a failure of the strong transversality condition. The idea is to construct a system where the unstable manifold of a fixed point has a point of tangency with a member of the stable manifold system of an exotic basic set Λ. Roughly speaking, if the dimensions are right, nearby systems exhibit the same tangency, but the stable manifold in question neither consistently contains nor consistently avoids periodic points of Λ. More precisely, one ensures that the stable manifolds of periodic points of Λ are dense. Thus systems for which they are tangent to the unstable manifold of the fixed point are dense near the given system, whereas, by the Kupka–Smale theorem, the stable and unstable manifolds of periodic points intersect transversally for a structurally stable system.

The above idea was originally due to Smale [4], who constructed examples for diffeomorphisms of compact 3-manifolds and flows on compact 4-manifolds. Peixoto and Pugh [1], by similar methods, showed that structurally stable systems are not dense on any non-compact manifold of dimension $\geqslant 2$. Finally Williams [2] completed the picture by reducing the dimensions of Smale's examples by 1. Thus, for instance, structurally stable diffeomorphisms are not dense on the two-dimensional torus T^2. One starts with a toral automorphism f on T^2, and modifies it in a neighbourhood of the fixed point 0 by composing with a diffeomorphism that is the identity outside some smaller neighbourhood, preserves the y-coordinate (the unstable manifold direction) but near 0 expands strongly in the x-direction (the stable manifold direction). This has the effect of turning 0 into a source and introducing two new saddle points, one each side of it. One might almost imagine that the old unstable manifold had been split in two lengthwise and the unstable manifold of a source grafted into the hole (see Figure 7.26).

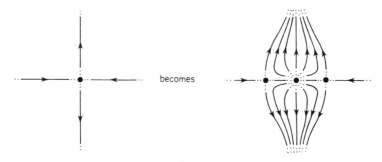

FIGURE 7.26

At this stage the system is called a *DA* (*derived from Anosov*) system. It has two basic sets, the source 0 and an exotic 1-dimensional expanding attractor Λ. The stable manifold system is still the stable manifold system of the original Anosov diffeomorphism (horizontal lines in Figure 7.26) except that one of them is broken in two at 0. We now remove a neighbourhood of the source at 0, and replace it by a plug consisting of two sources and a saddle point with the latter at 0 (see Figure 7.27). This modifies the stable manifold

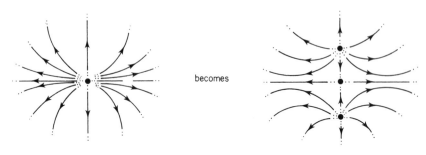

becomes

FIGURE 7.27

system of Λ to give Figure 7.28. At the moment the unstable manifold of the saddle point 0 is horizontal. We now perturb the diffeomorphism (call it g now) slightly in the neighbourhood of some point p of the unstable manifold

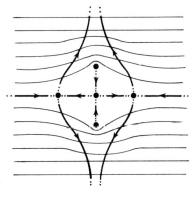

FIGURE 7.28

to produce a kink in it at $g(p)$ as shown in Figure 7.29. The diagram is oversimplified, as we do not attempt to reproduce the further kinks in $W_u(0)$ (at $g^n(p)$, $n > 1$) and in $W_s(\Lambda)$ (at and near $g^{-n}(p)$, $n \geq 0$). However we now have the tangency that we are aiming for.

There are two obvious courses of action open to us, as a response to the non-density of structural stability. We can either alter the equivalence relation on the space of all dynamical systems, in the hope that stability with respect to the new relation might be dense, or we can ask for something less than density in the given topology. We shall examine another equivalence relation in the next section. We finish this section by giving a positive result in the second direction.

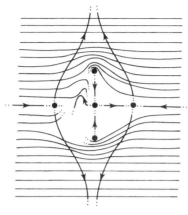

FIGURE 7.29

A natural question to ask is the following. Given an arbitrary dynamical system, can we deform it into a structurally stable system? If we can, how small a deformation do we need to make? We cannot make it arbitrarily C^1-small, since this would imply C^1-density of structural stability. But we might be able to do it by an arbitrarily C^0-small deformation. Notice that we are talking only about the *size* of the deformation needed to produce structural stability; the smoothness of the maps involved and the definition of structural stability are still as before. These questions are answered by the following theorem of Smale [9] and Shub [1] (see also Zeeman [1] and Franks [4]).

(7.30) Theorem. *Any C^r diffeomorphism $(1 \leq r \leq \infty)$ of a compact manifold is C^r isotopic to a C^1 structurally stable system by an arbitrarily C^0-small isotopy*[†].

The idea behind the proof is to triangulate the manifold X (see Munkres [1]) and hence to get a handlebody decomposition (Smale [10]; see also Mazure [1], who calls them differentiable cell decompositions). The technique then adopted is illustrated in two dimensions by Figure 7.31. The shaded picture on the left, part of the handlebody decomposition of X is originally mapped into X by f as shown. We first isotope f to a map g with the property that the image of every i-handle is contained in a union of j-handles for various $j \leq i$. This effect may be produced on the whole of X, using an inductive argument.

---

† Thus structural stability is dense in Diff^r X with respect to the C^0-topology. Of course it is no longer open if we use this topology. The analogous theorem for flows has been announced by D'Oliviera [1].

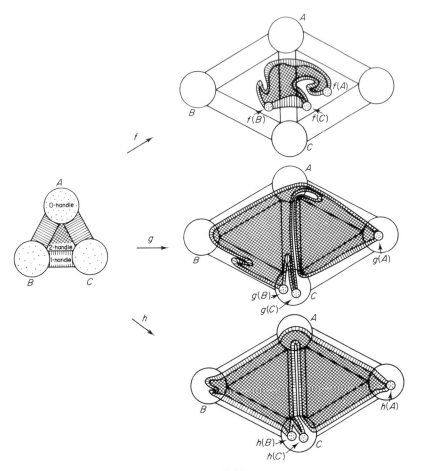

FIGURE 7.31

We then isotope g to a new map h, such that (i) the 0-handles are contracted, (ii) the 2-handles are expanded at all points of h^{-1}(2-handles), and (iii) the 1-handles are mapped in a nice linear fashion, expanding lengthwise and contracting radially, at all points of h^{-1}(1-handles). Notice the horseshoe that appears along the handle AC. With a little care, we have achieved a structurally stable map h, the Ω-set consisting of periodic sources associated with 2-handles, periodic sinks associated with 0-handles and saddle-type basic sets associated with 1-handles which may very well be exotic, for example the above horseshoe. The size of the isotopy involved depends on the fineness of the triangulation of X, and so can be made arbitrarily small, in the C^0-sense, by taking a fine enough triangulation.

V. OMEGA STABILITY

The failure of structural stability to be dense in higher dimensions led to consideration of other equivalence relations on the pace of all systems. Of these, the one that has so far aroused the most interest is Ω-*stability*. We have already introduced the notion of Ω-equivalence at the end of Chapter 2, and Ω-stability, is, of course, stability with respect to this equivalence relation. Thus, for example, a diffeomorphism $f: X \to X$ is Ω-*stable* (in the C^r sense, $1 \leqslant r \leqslant \infty$) if, for any nearby $g \in \text{Diff}^r X$, there exists a topological conjugacy between $f|\Omega(f)$ and $g|\Omega(g)$. At first sight it may seem rather an extreme measure to ignore all behaviour off the Ω-set. There are two answers to this criticism. Firstly, it is only in defining the equivalence relation that the non-Ω behaviour is ignored; it is very important in determining stability with respect to the relation. Secondly, it is arguable that in applications of the theory to mathematical modelling the Ω-sets are the only parts of the system with real physical significance. Certainly if a dissipative system in dynamics gives rise to a single dynamical system then only the attractors correspond to phenomena that are physically observable.

In order to get some feeling for the way in which the non-wandering set of a system may alter when the system is perturbed, we examine a few simple examples. First consider a "one-way" zero of a flow on S^1 (Figure 7.32). By

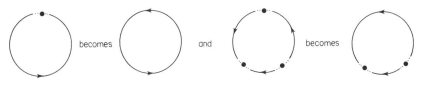

FIGURE 7.32

making a small local perturbation (adding a small velocity in the one-way direction) we eliminate the zero, and, in the first case, introduce a periodic orbit. Thus we have *exploded* the Ω-set from a single point to the whole circle. In the second case, we *implode* the Ω-set from three points to two. If we have an isolated non-hyperbolic zero which is "two-way" then we cannot entirely eliminate it by a local perturbation. However, for any non-hyperbolic zero, we can, by a C^1-small perturbation produce an interval, centred on the point, consisting entirely of zeros (the size of the interval varying with the size of the perturbation, of course), or introduce other more complicated behaviour on such an interval (Figure 7.33). Here the explosion in the Ω-set is local; its size is limited by the size of the perturbation allowed, in contrast to the previous explosion.

can become, for example,

or

FIGURE 7.33

A more spectacular explosion occurs if we start with the first picture in Figure 7.34, which represents the Ω-set of a diffeomorphism f of S^2, though, as usual, we draw flow-like orbits for easier visualization. The Ω-set consists of two sources y and t, a sink z and a saddle point x, together with an invariant arc l joining x to itself as shown (so that $l \subset W_s(x) \cap W_u(x)$). Notice that l cannot have a hyperbolic structure (Exercise: Why not?).

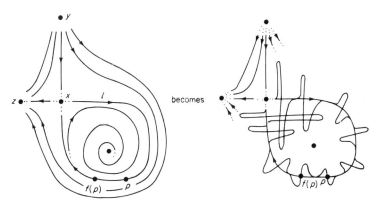

becomes

FIGURE 7.34

We perturb f on a small neighbourhood U of p, so that the resulting map g takes $l \cap U$ transversally across $l \cap g(U)$ at $f(p) = g(p)$. The unstable manifold of x with respect to g, $W_u(x)$, is the same as it was before the perturbation as one travels around it from x to p. However, after passing p moving to the left, it has a series of kinks which get larger and larger as they approach x. They press themselves up against $W_u(x)$, and eventually spread out past p into the kinks already formed, and so on. Similarly as $W_s(x)$ passes p going right, it develops a series of kinks which press themselves up against earlier parts of $W_s(x)$. Of course we have just described again the transverse homoclinic point phenomenon of Chapter 6. The point is that $W_s(x)$ now intersects $W_u(x)$ in an extremely complicated and rather random way, and, by the sort of "cloud lemma" argument used in the proof of Theorem 7.14, any such intersection is in $\Omega(g)$. We leave the reader to prove the intuitively obvious fact that $\Omega(f)$ is topologically different from $\Omega(g)$.

At first site, it appears that Ω-explosions are caused by a lack of hyperbolicity, so that we have, in Axiom A, the condition to eliminate them. This turns out to be true, in a sense, for local Ω-explosions, but Axiom A does not prevent global explosion. To see this consider the example (Smale [6]) of a diffeomorphism of S^2 with Ω-set consisting of six hyperbolic fixed points, of which two are sinks, two sources and two saddles, laid out as in Figure 7.35. To see that the Ω-set is as described note that any point outside the closed curve has a neighbourhood that either falls into the sink c under positive iterates of f, or into the source a under negative iterates. Similarly for points inside the closed curve. A small neighbourhood of a point such as p, under positive iterates of f, presses up against the underside of $W_u(x)$ and the right-hand side of the portion of $W_u(y)$ going to d. It does not come back and re-enter U, so p is wandering. Note that under negative iterates of f, U presses itself against the topside of $W_u(x) \cap W_s(y)$ and so positive and negative iterates of U come arbitrarily close together. Thus if we, by local perturbation at q on $W_u(x) \cap W_s(y)$, bring $W_u(x)$ above $W_s(y)$, we cause the positive and negative iterates $f^n(U)$ and $f^{-n}(U)$ to intersect for large n (Figure 7.35). Hence p becomes non-wandering, and similarly, any other

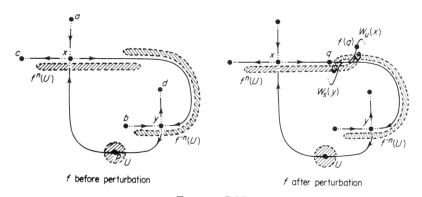

f before perturbation   f after perturbation

FIGURE 7.35

point of $W_s(x) \cap W_u(y)$. On the other hand the points of $W_u(x) \cap W_s(y)$ are wandering, just as they were before. Hence the new Ω-set is the six points together with the arc $W_s(x) \cap W_u(y)$. Note that, again, we cannot have a hyperbolic structure. If we make another perturbation so that $W_s(x)$ and $W_u(y)$ cross transversally, we cause another explosion. We have forced intersections $W_s(x) \cap W_u(x)$ and $W_s(y) \cap W_u(y)$ to appear and these join the new versions of $W_s(x) \cap W_u(y)$ and $W_s(y) \cap W_u(x)$ and the six fixed points to give the Ω-set of the new map. The set is similar to, but even harder to visualize than the one in the last example.

We now give an example of a system that is Ω-stable but not structurally stable. Consider the gradient flow of the height function on the torus T^2 already described in Example 3.3 and illustrated again in Figure 7.36. A

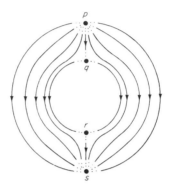

FIGURE 7.36

slight perturbation of the vector field will make the separatrices beginning at q end at s instead of r. However for all C^1-small perturbations the orbit structure is the same near the zeros, and elsewhere the orbits continue to move downwards, which precludes any global recurrence. Thus the Ω-set is still the four zeros, and so the system is Ω-stable.

Since this last example does not have strong transversality, it is clear that this property is not a relevant consideration when trying to characterize Ω-stability. We must examine the above examples to see what causes the explosions. We can distinguish two types, local explosions and global explosions. Local explosions are ones where the old and new Ω-sets are close together, the degree of closeness depending on the size of the perturbation. Global explosions are ones where arbitrarily small perturbations produce Ω-sets that are not close to the original ones†. The examples in Figures 7.33 and 7.34 are local, those in Figures 7.32 and 7.35 are global. As we have already hinted, local explosions are due to a lack of hyperbolicity. The basic sets of an Axiom A diffeomorphism f are locally stable. That is to say, given any basic set Λ, there is a neighbourhood U of Λ such that, for all g

---

† In some references (e.g. Shub [4]) the term "Ω-explosion" is only applied to global explosions.

sufficiently C^1-near f, $U \cap \Omega(g)$ has a subset Λ' such that

  (i) Λ' contains every g-invariant subset of U, and

  (ii) $g|\Lambda'$ is topologically conjugate to $f|\Lambda$.

Thus there may be other points of $\Omega(g)$ in U, but their orbits go outside U. This stability property is a close relative to the stability theorem for Anosov diffeomorphisms, and, not surprisingly its proof (see Smale [6] and Hirsch and Pugh [1]) has a lot in common with that of Theorem 7.9 sketched above.

Several times in this chapter, in proving that a point is non-wandering, we have used arguments that involve open sets spreading out from basic set to basic set until they get back near where they started. Any two consecutive basic sets in the route have to be linked together, or the open set cannot make it across from the one to the other. The link consists of an unstable manifold of the one and a stable manifold of the other with non-empty intersection. If the manifolds meet transversally the open set can always get across; if not then it may still be lucky or it may not (see Figure 7.37).

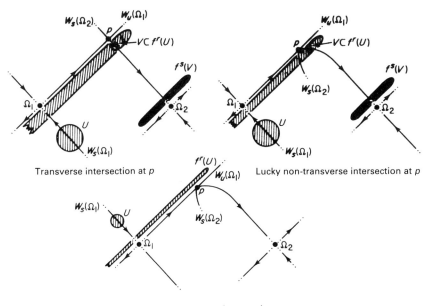

Transverse intersection at p

Lucky non-transverse intersection at p

Unlucky non-transverse intersection at p

FIGURE 7.37

Another unlucky situation, illustrated in the first picture in Figure 7.35 occurs when the open set reaches the first basic set of a link but cannot get near the intersection point of the link. This can happen when, due to an earlier lack of transversality, the open set does not intersect the stable manifold of the basic set ($f^n(U)$ does not intersect $W_s(y)$ in the picture). These unlucky situations, since they are all due to non-transversality, may be eliminated by arbitrarily small perturbations producing transversality. Thus a recipe for a global Ω-explosion of an Axiom A system is to start with a collection of basic sets, linked up to form a circular route, but with one or more unlucky situations incorporated (of course—we could not have separate basic sets otherwise). Then, by a small perturbation, we can remove all the bad luck situations, simultaneously if we do not insist on the perturbation being a local one, and we have our explosion.

The above description motivates the following definition (Rosenberg[1]). An *n-cycle* of an Axiom A dynamical system is a sequence of basic sets $\Omega_0, \Omega_1, \ldots, \Omega_n$, with $\Omega_0 = \Omega_n$ and $\Omega_i \neq \Omega_j$ otherwise, and such that $W_u(\Omega_{i-1}) \cap W_s(\Omega_i)$ is non-empty for all i, $1 \leqslant i \leqslant n$. An Axiom A dynamical system satisfies the *no-cyclic condition* if it has no *n*-cycles for all $n \geqslant 1$. The main theorem on Ω-stability, due to Smale [6] in the diffeomorphism case, and to Pugh and Shub [1] in the flow case is:

(7.38) Theorem. *If an Axiom A system on a compact manifold has the no-cycle property then it is Ω-stable.*

The idea of the proof is to show first, as indicated above, that Axiom A rules out local explosions. We then have a situation rather similar to that illustrated in Figure 7.36. If $W_u(\Omega_i) \cap W_s(\Omega_j)$ is non-empty, we can think of Ω_i as being situated above Ω_j (the no-cycle condition ensures that we do not find, paradoxically, that Ω_i is above itself) and the movement between basic sets as being downwards. It seems quite clear that C^1-small perturbations cannot produce an upwards motion and so global recurrence cannot occur in the perturbed system. Technically the proof uses a *filtration* (see Shub and Smale [1]). This is an increasing sequence of compact manifolds with boundary $X_1 \subset \cdots \subset X_r = X$, such that dim $X_i = $ dim X, $f(X_i) \subset $ int X_i and each $X_i \backslash X_{i-1}$ contains a single basic set. We picture a suitable filtration for the Figure 7.36 example in Figure 7.39. As with structural stability, it is conjectured that we have here a complete characterization; that Ω-stability implies Axiom A and no cycles. Palis [1] has the partial result that Axiom A and Ω-stability imply no cycles. There is, again, a notion of *absolute* stability, and this is characterized by Axiom A and no cycles (see Franks [3], Guckenheimer [1]).

Unfortunately Ω-stability is no more successful than structural stability as far as density is concerned. Abraham and Smale [1] took the product of a toral automorphism and a horseshoe diffeomorphism on S^2,, and by making the former *DA* over one of the two fixed points $(\ldots 00 \cdot 00 \ldots$ and $\ldots 11 \cdot 11 \ldots)$ of the horseshoe, they constructed a

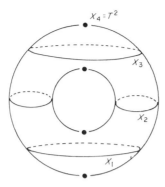

FIGURE 7.39

diffeomorphism of $T^2 \times S^2$ that has a C^1-neighbourhood all of whose members are neither Axiom A nor Ω-stable. Finally, Newhouse [1], by modifying the horseshoe diffeomorphism, constructed a counterexample to C^2-density of Axiom A and of Ω-stability for diffeomorphisms of any surface.

VI. BIFURCATION

Several other notions of stability have been put forward in the hope that they might prove to be generic, for example *future stability* (Shub and Williams [1]), *tolerance stability* (Takens [1], White [1]), *finite stability* (Robinson and Williams [1]) and *topological Ω-stability* (Hirsch, Pugh and Shub [1]). Some have survived longer than others; none has yet been wholly successful. In the course of time, a feeling has developed that perhaps it is too optimistic to expect to find a single natural equivalence relation with respect to which stability is dense (see, for example, Smale [8]). Recently, more attention has been focused on the interesting and important question of bifurcation of systems.

Up to now we have not devoted much time to systems such as the one-way zero in Figure 7.32. Our feeling has been that the most important systems are ones which can be used to model the dynamics of real life situations. But no real life situation can ever be exactly duplicated, and we should expect this to lead to slight variations in the model system. Thus a theory making use of qualitative features of a dynamical system is not convincing unless the features are shared by nearby systems. That is to say, good models should possess some form of qualitative stability. Hence our contempt for the one-way zero, which is extremely unstable. Now this argument has some force, but it is too simple-minded. Firstly, there may, in a physical situation, be factors present that rule out certain dynamical systems as models. Conservation laws have this effect, and so has symmetry in the physical situation. In this case, the subset of those dynamical systems that *are* allowable as models may be nowhere dense in the space of all systems, and so the stable systems that we have considered in this book are totally irrelevant. One has to consider afresh which properties are generic in the space of admissible systems. This has been done for Hamiltonian systems (see Robinson [1]). Secondly, even if the usual space of systems is the relevant one, the way in which a stable system loses its stability as it is gradually perturbed may be of importance, since the model for an event may consist not of a single system but of a whole family of systems. In his theory of morphogenesis, Thom [1] envisages a situation where the development of an organism, say, is governed by a collection of dynamical systems, one for each point of space time. The dynamical systems are themselves controlled by a number of parameters, which could for example be quantities like temperature and concentrations of chemicals in the neighbourhood of the point in question. A *catastrophe* is a point where the form of the organism changes discontinuously, and it corresponds to a topological change in the orbit structure of the dynamical systems. We say that the family of dynamical systems *bifurcates* there.

We shall give some examples of bifurcations of flows. These particular ones are local changes that can take place on any manifold. By taking a suitable chart we may work in Euclidean space. First consider a vector field v_α on **R** given by

$$v_\alpha(x) = \alpha + x^2.$$

Here α is a single real parameter, so for each $\alpha \in \mathbf{R}$ we have a vector field on *R*. We are interested in how the orbit structure varies with α. The analysis is very easy, and we find that

(i) for $\alpha > 0$ we have no zeros, and the whole of **R** is an orbit, oriented positively,

(ii) for $\alpha = 0$ we have a single zero at $x = 0$, which is a one way zero in the positive direction,

(iii) for $\alpha < 0$, we have two zeros, a sink at $-\sqrt{-\alpha}$ and a source at $\sqrt{-\alpha}$.

Thus the bifurcation occurs at $\alpha = 0$. Figure 7.40 illustrates the bifurcation.

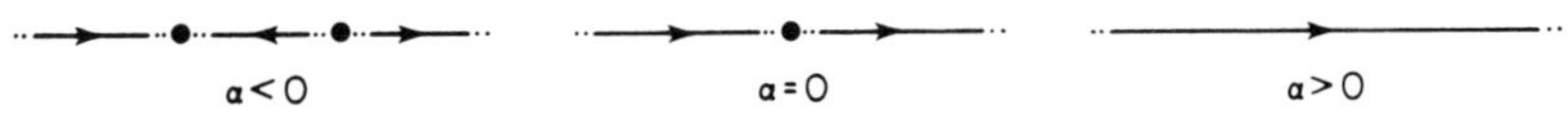

$a < 0$ $a = 0$ $a > 0$

FIGURE 7.40

If one takes the product of v_α with a fixed (i.e. independent of α) vector field on $\mathbf{R}^{n-1}$ having a hyperbolic fixed point at 0, one obtains a bifurcation of the resulting vector field on $\mathbf{R}^n$. Of course we have essentially added nothing to the original bifurcation in doing this. All such bifurcations are known as *saddle-node bifurcations*, because of the picture that one gets for $n = 2$ (Figure 7.41). A saddle point and a node come together, amalgamate and cancel each other out.

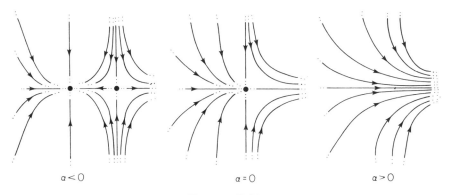

FIGURE 7.41

Saddle-node bifurcations are important because (when properly defined) they are stable as bifurcations of one-parameter families. Roughly speaking, then, one-parameter families near to a family with a saddle-node bifurcation also exhibit something that is topologically like a saddle node bifurcation near (in the positions of the amalgamating zeros, for example) the original one. One speaks of them as being *codimension one* bifurcations; one is then visualizing the set of systems exhibiting zeros of the $\alpha = 0$ type above as (in some sense) a submanifold of codimension one in $\Gamma^r(X)$, and the one-parameter family as being given by an arc in $\Gamma^r(X)$ crossing the submanifold transversally. Notice that the bifurcation illustrated in Figure 7.42 (a node bifurcating into two nodes and a saddle point) is *not* stable for one-parameter families. It can be perturbed slightly so that there is a saddle-node bifurcation near to but not at the original node.

The saddle-node bifurcation is the typical bifurcation obtained when the sign of a real eigenvalue of (the differential at) a zero is changed by varying a single parameter governing the system. There is, similarly, a typical co-dimension one bifurcation which comes about when the sign of the real part of a complex conjugate pair of eigenvalues is changed by varying a single parameter. This is known as the *Hopf bifurcation*, since it was first described by E. Hopf [1]. Consider the vector field v_α on $\mathbf{R}^2$ given by

$$v_\alpha(x, y) = (-y - x(\alpha + x^2 + y^2), \ x - y(\alpha + x^2 + y^2)).$$

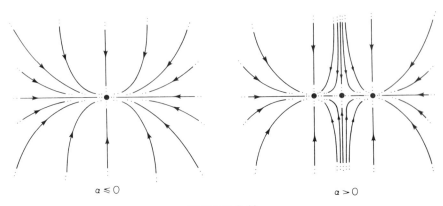

$\alpha \leqslant 0$ $\alpha > 0$

FIGURE 7.42

Again α is a single real parameter. For all $\alpha \in \mathbf{R}$, v_α has a zero at the origin, and the linear terms make this a spiral source for $\alpha < 0$ and a spiral sink for $\alpha > 0$. For $\alpha = 0$ the linear terms would give a centre (recall that this is a zero surrounded by closed orbits), but the cubic terms make the orbits spiral weakly inwards. The interesting feature of the bifurcation is not, however, the zero, but the unique closed orbit that we get at $x^2 + y^2 = -\alpha$, for each $\alpha < 0$. The bifurcation is illustrated in Figure 7.43. The periodic attractor

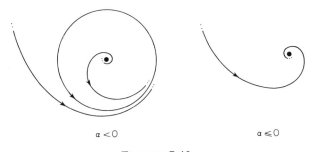

$\alpha < 0$ $\alpha \leqslant 0$

FIGURE 7.43

decreases in size until it amalgamates with the spiral source to form a spiral sink. The reverse bifurcation, in which a spiral sink splits into a spiral source and a periodic attractor, is intriguing because of the feeling one has that from something inert and dead (the source) one has created something pulsating and alive (the periodic orbit). This makes it very popular as a component in mathematical models.

It is not possible at present to give a coherent account of bifurcation theory for dynamical systems, since the subject is still in an early stage of development. Papers of Sotomayor [1, 2, 3] give some general principles for tackling the problem and prove some generic properties of bifurcations of one parameter families of vector fields (see also Guckenheimer [3]). Newhouse and Palis [2, 3] have examined the situation after one passes the first bifurcation of a Morse–Smale vector field, and found that it may be very complicated. Newhouse and Peixoto have investigated conditions under which pairs of structurally stable vector fields may be connected by one-parameter families with only finitely many stable bifurcations (see Newhouse and Peixoto [1] and Newhouse [3]). Takens [2, 3, 4] has written papers on various aspects of bifurcations of singularities of dynamical systems. Guckenheimer [2] has shown that the analysis of bifurcation of functions on manifolds is fundamentally different from and simpler than the corresponding theory for gradient vector fields.

The above discussion has mainly been for bifurcations in the context of topological equivalence of flows. The theory for topological conjugacy of diffeomorphisms is less well developed, and seems to differ in some respects. Also, of course, any other equivalence relation on dynamical systems has an associated bifurcation theory. Very little work has been done so far on such bifurcations. It seems likely, though, that for any interesting relation it will be a difficult task to establish a satisfactory general theory.

Theory of Manifolds

In this appendix we assemble most of the definitions and theorems about manifolds and maps that are needed in the course of the book. Our main purpose is to establish notations and terminology, and to collect useful examples. On the whole, therefore, arguments are omitted or summarized when they are already available in easily accessible texts†. We do not make substantial use of infinite dimensional manifolds in the book, apart from the trivial example of an open subset of an infinite dimensional Banach space. Nevertheless, we shall give the infinite dimensional theory here, since it is useful for further reading in the subject, and the generalization requires only a modicum of care.

I. TOPOLOGICAL MANIFOLDS

Let **E** be a real Banach space, finite or infinite dimensional. A *topological manifold modelled* on **E** is a topological space X such that, for each $x \in X$, there is an open neighbourhood U of x and a homeomorphism $\xi: U \to U'$ onto an open subset U' of **E**. It is usual to place further restrictions on the topology of X in order to avoid pathological examples. *We shall always assume that X is Hausdorff with a countable basis of open sets.* Together with the manifold hypothesis, these conditions imply that X is normal, metrizable and paracompact. *In addition we shall suppose, unless otherwise stated, that X is connected.*

---

†For example, Hirsch [1] is an excellent reference for differential topology, backed up by Lang [1] for infinite dimensional manifolds and Helgason [1] for Riemannian metrics. There are also very readable introductions to differential topology by Guillemin and Pollack [1] and Chillingworth [1]. Chillingworth's book is particularly relevant since it contains a very nice account of dynamical systems theory.

The homeomorphism ξ is called a *chart* at x. A set of charts whose domains cover X is called an *atlas* for X. If $\mathbf{E}$ is n-dimensional ($n < \infty$) then X is said to have *dimension* n, and to be a (topological) *n-manifold*. Since two finite dimensional real normed vector spaces are homeomorphic if and only if their dimensions are the same (see, for example, § 1 of Chapter 4 of Hu [2]), the dimension of X is uniquely defined (with the trivial exception of $X = $ empty set) and we may take $\mathbf{E} = \mathbf{R}^n$. If $\mathbf{E}$ is infinite dimensional, we say that X is infinite dimensional. The term *Banach manifold* commonly carries the connotation of infinite dimension.

Examples

(A.0) The topological space consisting of a single point is a 0-dimensional manifold.

(A.1) Any connected open subset of $\mathbf{E}$ is a manifold modelled on $\mathbf{E}$. The inclusion is a chart, and the single chart is an atlas. Any connected open subset V of a manifold X modelled on $\mathbf{E}$ is a manifold modelled on $\mathbf{E}$. If $\xi: U \to U'$ is a chart on X, $\xi|U \cap V: U \cap V \to \xi(U \cap V)$ is a chart on V.

(A.2) Let $L(\mathbf{E}, \mathbf{F})$ be the Banach space of continuous linear maps from the Banach space $\mathbf{E}$ to the Banach space $\mathbf{F}$. We write $L(\mathbf{E})$ for $L(\mathbf{E}, \mathbf{E})$, and denote by $GL(\mathbf{E})$ the set of (topological) linear automorphisms of $\mathbf{E}$. (Here *topological* means that the automorphisms and their inverses are continuous.) Then $GL(\mathbf{E})$, which is called the general linear group of $\mathbf{E}$, is open in $L(\mathbf{E})$ (Lemma 7.6.1 of Dunford and Schwartz [1]), and hence is a (not necessarily connected) manifold modelled on $L(\mathbf{E})$. For example, $GL(\mathbf{R}^n)$ has two components (automorphisms with positive determinant and automorphisms with negative determinant) which are n^2-manifolds.

(A.3) If $p: \mathbf{E} \to X$ is a covering map then, by definition, every $x \in X$ has an open neighbourhood U such that $p^{-1}(U)$ is a disjoint union of subsets U'_i each of which is mapped homeomorphically onto U by p. Thus, for any i, $(p|U'_i)^{-1}$ is a chart at x, and so X is a manifold modelled on $\mathbf{E}$. In many simple applications, X is the quotient $\mathbf{E}/G$ of a discrete group G acting on $\mathbf{E}$, and p is the quotient map. For example, the *circle* S^1 is the quotient $\mathbf{R}/\mathbf{Z}$ where the action is given by $n \cdot x = x + n$. Points of S^1 are equivalence classes $[x]$ under the equivalence relation on $\mathbf{R}$ given by $x \sim x'$ if and only if $x - x' \in \mathbf{Z}$, and p is defined by $p(x) = [x]$. We shall give an alternative definition of S^1 shortly. Some examples of 2-manifolds covered by $\mathbf{R}^2$ are (i) the *torus* $T = \mathbf{R}^2/\mathbf{Z}^2$ where $(m, n) \cdot (x, y) = (x + m, y + n)$, (ii) the *Klein bottle* $\mathbf{R}^2/\mathbf{Z}^2$ where $(m, n) \cdot (x, y) = ((-1)^n x + m, y + n)$ and (iii) the *Möbius band* $\mathbf{R}^2/\mathbf{Z}$ where $n \cdot (x, y) = (x + n, (-1)^n y)$. See Figure A.3.

Similarly, if $p: X \to Y$ is any covering of a space Y by a manifold X modelled on $\mathbf{E}$, then Y is a manifold modelled on $\mathbf{E}$.

(A.4) The *connected sum $X \# Y$* of manifolds X and Y modelled on $\mathbf{E}$ is obtained by removing a ball from each and gluing the remnants together along their spherical boundaries. More precisely, let $B_r(0) = \{x \in \mathbf{E} : |x| \leq r\}$ and $S_r(0) = \{x \in \mathbf{E} : |x| = r\}$. Take charts $\xi: U \to U'$ on X and $\eta: V \to V'$ on Y

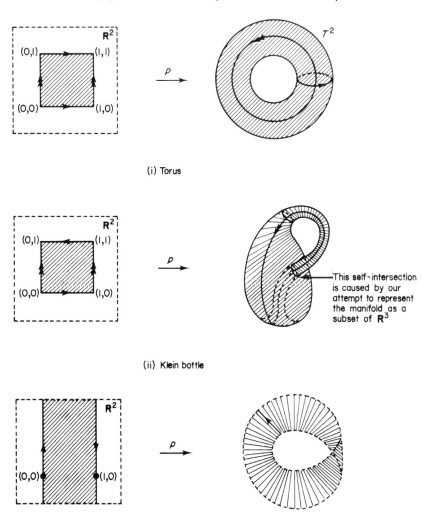

(i) Torus

(ii) Klein bottle

This self-intersection is caused by our attempt to represent the manifold as a subset of $\mathbf{R}^3$

(iii) Möbius band

FIGURE A.3

with $B_r(0) \subset U' \cap V'$. Let

$$X^- = X \backslash \xi^{-1}(\text{int } B_{r/2}(0)) \quad \text{and} \quad Y^- = Y \backslash \eta^{-1}(\text{int } B_{r/2}(0))$$

and put

$$X \# Y = X^- \cup Y^- / (\xi^{-1}(x) = \eta^{-1}(x) : x \in S_{r/2}(0)).$$

The definition is, up to homeomorphism, independent of choice of charts. For example, if $X = Y = T^2$ then $X \# Y$ is the *pretzel*, or sphere with two handles (see Figure A.4).

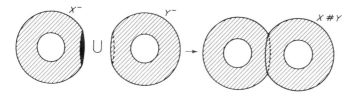

FIGURE A.4

(A.5) If X and Y are manifolds modelled on **E** and **F** respectively, then $X \times Y$ is a manifold modelled on $\mathbf{E} \times \mathbf{F}$. If ξ and η are charts on X and Y respectively, then $\xi \times \eta$ is a chart on $X \times Y$. For example, the torus T^2 as above defined may be identified with $S^1 \times S^1$ in the obvious way. Precisely, there is a homeomorphism taking $[(x, y)]$ to $([x], [y])$.

We say that a linear subspace **F** of **E** *splits* **E** if it is closed in **E**, and, for some closed linear subspace **G** of **E**, $\mathbf{E} = \mathbf{F} \oplus \mathbf{G}$. We call dim **G** the *codimension* of F; it may be ∞. Splitting is guaranteed to occur if **F** is finite dimensional. Let X be a manifold modelled on **E** and let Y be a connected subset of X. We say that Y is a *(locally flat) submanifold* of X if there is a subspace **F** of **E** that splits **E** and, for each $y \in Y$, a chart $\xi : U \to U'$ at y such that ξ maps $U \cap Y$ onto $U' \cap \mathbf{F}$ (see Figure A.6). The *codimension* of Y is the codimension of **F**. Then obviously the submanifold Y is (with the induced topology) a manifold modelled on **F**.

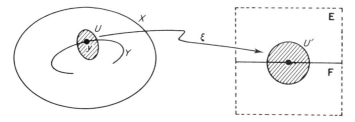

FIGURE A.6

(A.7) Submanifolds of $\mathbf{E}$ often arise in the following way. Suppose that V is an open subset of $\mathbf{E}$ and that $f: V \to \mathbf{H}$ is a C^r map $(r \geqslant 1)$ to a Banach space $\mathbf{H}$. Fix on some value $a \in \mathbf{H}$ and consider the level set $f^{-1}(a)$. If, for all $x \in f^{-1}(a)$, the differential $Df(x) \in L(\mathbf{E}, \mathbf{H})$ is surjective (*split surjective* in the infinite dimensional case, which means that ker $Df(x)$ splits $\mathbf{E}$) we say that a is a *regular value* of f, and, in this case, it is a consequence of the inverse mapping theorem that $f^{-1}(a)$ is a disjoint union of submanifolds of V (see Exercise C.13 of Appendix C). For example, if $\mathbf{E} = \mathbf{R}^{n+1}$, $\mathbf{H} = \mathbf{R}$ and $f(x_1, \ldots, x_{n+1}) = x_1^2 + \cdots + x_{n+1}^2$, then 1 is a regular value and $f^{-1}(1)$ is the sphere $x_1^2 + \cdots + x_{n+1}^2 = 1$, the *unit n-sphere* in $\mathbf{R}^{n+1}$, denoted S^n. If $n = 1$, we get the *unit circle* S^1, which is homeomorphic to the manifold defined in A.3.

(A.8) Real n-dimensional projective space $\mathbf{R}P^n$ is defined to be the set of lines through the origin in $\mathbf{R}^{n+1}$. The topology of the space is given by a metric, the distance between two such lines being the distance between their closest points of intersection with S^n. There is a map $p: S^n \to \mathbf{R}P^n$ which takes any point to the line joining it to the origin. Since this map is a (double) covering, $\mathbf{R}P^n$ is a manifold, by A.3.

Exercises

(A.9) Let Σ^n be the unit sphere $\{x \in \mathbf{R}^{n+1}: \|x\| = 1\}$ in $\mathbf{R}^{n+1}$ with respect to any norm $\| \ \|$. Show that Σ^n is a manifold homeomorphic to S^n.

(A.10) Show that when M and N are submanifolds of X neither $M \cup N$ nor $M \cap N$ need be a submanifold of X.

(A.11) Show that if X and Y are manifolds then, for all $y \in Y$, $X \times \{y\}$ is a submanifold of $X \times Y$.

II. SMOOTH MANIFOLDS AND MAPS

If U is an open subset of a Banach space $\mathbf{E}$, and $\mathbf{F}$ is another Banach space, one has the notion of a C^r ($= r$ times continuously differentiable) map $f: U \to \mathbf{F}$. We shall assume familiarity with the basic theory of such maps: some definitions appear in Appendix B below. We shall not be concerned much with C^ω ($=$ analytic) maps. The term *smooth* is used to mean C^r for some r with $1 \leqslant r \leqslant \infty$.

We wish to extend the theory of differentiable maps from the context of Banach spaces to the context of manifolds. That is to say, we have to extend the notion of differentiability of a map from the local, chart level to the whole manifold. We do this by restricting ourselves to charts which are smoothly related where they overlap. By doing so we get what is known as a *smooth structure* on the manifold.

Let $\mathcal{A} = \{\xi_i : i \in I\}$ be an atlas for a topological manifold X, where I is some indexing set and $\xi_i : U_i \to U_i' \subset \mathbf{E}$ are charts. We say that $\mathcal{A}$ is a C *atlas* if, for all $i, j \in I$, the coordinate change map $\xi_j \xi_i^{-1} : \xi_i(U_i \cap U_j) \to \xi_j(U_i \cap U_j)$ is C^r (see Figure A.12). A chart ξ on X is C^r-*compatible* with $\mathcal{A}$ if $\mathcal{A} \cup \{\xi\}$ is a C^r

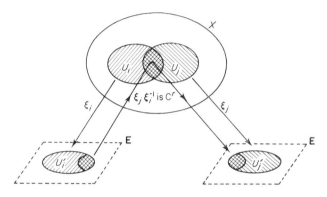

FIGURE A.12

atlas. The set of all charts C^r-compatible with $\mathcal{A}$ is the C^r *structure* generated by $\mathcal{A}$. A C^r *manifold* is a topological manifold X together with a C^r structure $\mathcal{A}$ on X. Any chart in $\mathcal{A}$ is said to be an *admissible* chart on the C^r manifold. *In the text of the book, and from now on in the appendices, "manifold" always means "C^r manifold for some $r \geq 1$" and "chart" means "admissible chart." We often use the symbol X for a C^r-manifold rather than the cumbersome $(X, \mathcal{A})$.*

All the above examples and constructions for topological manifolds may be modified to fit into this new setting. For example the inclusion of an open set U of $\mathbf{E}$ in $\mathbf{E}$ is, by itself, a C^∞ atlas for U, since we have no overlaps to worry about beyond the trivial one of U with itself. If $h : X \to Y$ is a homeomorphism from a C^r manifold X to a topological space Y then h induces a C^r structure on Y, with a chart $\xi h^{-1} : h(U) \to U'$ corresponding to each chart $\xi : U \to U'$ on X. This also works at a local level, so that any connected open subset of a C^r manifold has an induced C^r structure, and so does any space with covering a C^r manifold. We may define C^r *submanifold* by making the charts in the above definition of submanifold admissible charts of a C^r manifold. The level surfaces of the C^r map $f : V \to H$ of Example A.7 turn out to be unions of C^r submanifolds at regular values.

Thus, for example, S^n, and hence $\mathbf{R}P^n$, are C^∞ manifolds. (The C^∞ structure of S^n may be defined, equivalently, by the two charts $\xi: S^n\backslash\{N\} \to \mathbf{R}^n$ and $\eta: S^n\backslash\{S\} \to \mathbf{R}^n$, where N and S are the north and south poles and ξ and η are stereographic projection from those poles.) Products of C^r manifolds are C^r manifolds. The only construction that needs some care is the connected sum construction; it takes a little ingenuity to make $X \# Y$ smooth at the seam.

Now let X and Y be C^r manifolds and let $f: X \to Y$ be any continuous map. We say that f is C^r if, for all admissible charts $\xi: U \to U' \subset \mathbf{E}$ on X and $\eta: V \to V' \subset \mathbf{F}$ on Y, the *local representative*

$$\eta f \xi^{-1}: \xi(U \cap f^{-1}(V)) \to V'$$

is C^r (see Figure A.13). Obviously it is sufficient to check this for all charts in a pair of admissible atlases, one for X and one for Y. If f is C^r and has a C^r inverse $f^{-1}: Y \to X$, we call it a C^r *diffeomorphism* $(r \geqslant 1)$. Any C^r map $f: X \to \mathbf{R}$, where $\mathbf{R}$ has its standard C^∞ structure, is called a C^r *function* on X.

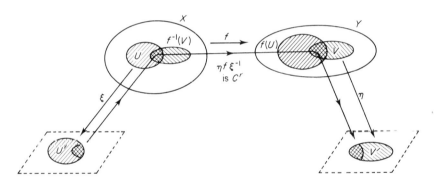

FIGURE A.13

Examples

(A.14) For any C^r manifold X the identity function $id: X \to X$ is a C^r diffeomorphism, and any constant map $c: X \to Y$ onto a point of a C^r manifold Y is a C^r map.

(A.15) Any C^r map $f: U \to V$ of open subsets of Banach spaces is a C^r map when U and V are given their standard C^∞ manifold structures (induced by inclusion).

(A.16) Let $X = Y \times Y$, where Y is a C^r manifold, and let $\delta: Y \to X$, $\tau: X \to X$ and $\pi: X \to Y$ be defined by $\delta(y) = (y, y)$, $\tau(y, z) = (z, y)$ and $\pi(y, z) = y$. Then δ and π are C^r maps, and τ is a C^r diffeomorphism.

(A.17) If Y has the C^r structure induced by a homeomorphism $h: X \to Y$ from a C^r structure on X, then h is a C^r diffeomorphism.

(A.18) Any admissible chart $\xi: U \to U'$ of a C^r manifold X is a C^r diffeomorphism, where U and U' inherit their structures from X and $\mathbf{E}$ via the inclusions.

Exercises

(A.19) Prove that the composite $gf: X \to Z$ of C^r maps $f: X \to Y$ and $g: Y \to Z$ is a C^r map.

(A.20) For any positive integer n, let X_n be the real line $\mathbf{R}$ with the C^∞ structure given by the chart $\xi: \mathbf{R} \to \mathbf{R}$ defined by $\xi(x) = x^{2n-1}$. Show that the map $f_{nm}: X_n \to X_m$ defined by $f_{nm}(x) = x^p$, where $p = (2n-1)/(2m-1)$, is a C^∞ diffeomorphism, but that the identity map $id: \mathbf{R} \to \mathbf{R}$ is a C^1 diffeomorphism if and only if $m = n$.

Any C^s atlas on a topological manifold is trivially a C^r atlas for all r with $r \leqslant s$, and so any C^s structure may be extended to a C^r structure, by adding all C^r charts C^r-compatible with it. Conversely, but non-trivially,

(A.21) Theorem. *Any C^r structure $(r \geqslant 1)$ on a finite dimensional topological manifold contains a C^s structure for all s with $r \leqslant s \leqslant \infty$. Any two such C^s structures are C^s-diffeomorphic.*

For a proof, see Theorem 2.2.9 of Hirsch [1]. Two $C^\sim$ structures $\mathcal{A}$ and $\mathcal{B}$ on a topological manifold X are C^s-*diffeomorphic* if there is a map $f: X \to X$ that is a C^s diffeomorphism from $(X, \mathcal{A})$ to $(X, \mathcal{B})$. Theorem A.21 encourages us *to restrict our attention to C^∞ manifolds, and we do this in the text of the book.* Note, however, that there are topological manifolds which admit no differentiable structure at all (Theorem A.20 needed $r \geqslant 1$). Kervaire [1] and Smale [1] discovered compact 8-dimensional examples.

As we have seen in Exercise A.20, it is not hard to find different smooth structures on the same topological manifold; to say there that $id: \mathbf{R} \to \mathbf{R}$ is not a diffeomorphism is equivalent to saying that the structures are different. However, it is very much harder, and correspondingly more interesting, to find non-diffeomorphic smooth structures on the same manifold. Nevertheless they do exist. For example there are 28 C^∞ structures on the topological 7-sphere S^7, no two of which are C^1-diffeomorphic (see Milnor [1]).

If $f: X \to Y$ is a C^s map of C^s manifolds, and we extend the C^s structures of the manifolds to C^r structures, for $r < s$, then, of course, we decrease the degree of smoothness of f to C^r. If, conversely, we have a C^r map $f: X \to Y$ of C^r manifolds, and we restrict the C^r structures to C^s structures (as we may by Theorem A.21) we do not usually increase the smoothness of f to C^s; it would be a fluke if we did so. The best we can say about C^r maps $f: X \to Y$ of C^s manifolds is that (in finite dimensions) they may be approximated arbitrarily C^s-closely by C^s maps. See Theorem 2.6 of Hirsch [1]; for an exact statement, one needs to discuss the topology of map spaces, as in Appendix B and Hirsch [1].

III. SMOOTH VECTOR BUNDLES

This section is designed to provide a simple framework for the description of the tangent bundle of a smooth manifold and other associated concepts. The definition follows a pattern that can be used to specify many important structures on a manifold. The essential idea is to consider special types of chart, and to restrict the coordinate change maps to some particular type. We have already done this in defining smooth structures on a manifold.

Let $\mathbf{E}$ and $\mathbf{F}$ be Banach spaces. Let U be open in $\mathbf{E}$ and let $\pi_1: U \times \mathbf{F} \to U$ be projection onto the first factor (i.e. $\pi_1(x, y) = x$). We call π_1 (or sometimes $U \times \mathbf{F}$) a *local vector bundle*. To describe the coordinate change maps that interest us, we need the notion of a C^r *local vector bundle map*. This is a C^r map $\tilde{f}: U \times \mathbf{F} \to V \times \mathbf{H}$ (for V open in $\mathbf{G}$, where $\mathbf{G}$ and $\mathbf{H}$ are Banach spaces) which

(i) covers a C^r map $f: U \to V$, in the sense that the diagram

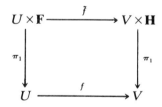

commutes, and

(ii) is a C^r-smoothly varying linear map of fibres. That is to say, for all $(x, y) \in U \times \mathbf{F}$,

$$\tilde{f}(x, y) = (f(x), T_x(y)),$$

where $T_x \in L(\mathbf{F}, \mathbf{H})$ and the map $x \mapsto T_x$ from U to $L(\mathbf{F}, \mathbf{H})$ is C^r.

Note that in finite dimensions $\tilde{f}$ C^r automatically implies that $x \mapsto T_x$ is C^r.

Now let X be a C^r manifold modelled on E, let B be a topological space and let $\pi: B \to X$ be a continuous surjection. A C^r *vector bundle chart* (or C^r

local trivialization) on π is, for some open subset U of X a homeomorphism $\tilde{\xi}$ from $\tilde{U} = \pi^{-1}(U)$ to a local vector bundle $U' \times \mathbf{F}$ that covers some admissible chart $\xi: U \to U'$ on X. That is to say the diagram

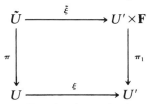

commutes. A C^r *vector bundle atlas* for π is, for some indexing set I, a set $\tilde{\mathcal{A}} = \{\tilde{\xi}_i: i \in I\}$ of C^r vector bundle charts $\tilde{\xi}_i: \tilde{U}_i = \pi^{-1}(U_i) \to U'_i \times \mathbf{F}$ satisfying

(i) $\{U_i: i \in I\}$ is a covering of X, and

(ii) for all $i, j \in I$ the map

$$\tilde{\xi}_j \tilde{\xi}_i^{-1}: \tilde{\xi}_i(\tilde{U}_i \cap \tilde{U}_j) \to \tilde{\xi}_j(\tilde{U}_i \cap \tilde{U}_j)$$

is a C^r local vector bundle map. We now proceed as when defining smooth structures on a manifold and say that a C^r vector bundle chart $\tilde{\xi}$ on π is C^r-compatible with $\tilde{\mathcal{A}}$ if $\tilde{\mathcal{A}} \cup \{\tilde{\xi}\}$ is a C^r vector bundle atlas. A C^r *vector bundle structure* on π is a maximal C^r vector bundle atlas on π, and a C^r *vector bundle* $(\pi, \tilde{\mathcal{A}})$ is π together with a given C^r *vector bundle structure* $\tilde{\mathcal{A}}$. As usual we abbreviate $(\pi, \tilde{\mathcal{A}})$ by π if the structure $\tilde{\mathcal{A}}$ is unambiguous, and we refer to the elements of $\tilde{\mathcal{A}}$ as (*admissible vector bundle*) *charts* on π. Moreover a further common abuse of language which we sometimes follow is to refer to "the vector bundle B" if the map $\pi: B \to X$ is unambiguous. We call π, B, X and $\mathbf{F}$ the *projection*, *total space*, *base* (*space*) and *fibre* of the vector bundle, and say that π is *over* X.

(A.22) Example. If X is any C^r manifold and $\mathbf{F}$ is any Banach space, then $\pi_1: X \times \mathbf{F} \to X$ is a C^r vector bundle with base X, fibre $\mathbf{F}$ and an atlas given by charts of the form $\xi \times id: U \times \mathbf{F} \to U' \times \mathbf{F}$ where $\xi \in$ an atlas of X. Such a bundle is called the *product* or *trivial* bundle.

(A.23) Example. We may regard any Banach space $\mathbf{F}$ as a smooth vector bundle with base $\{0\}$ and fibre $\mathbf{F}$.

(A.24) Example. The Möbius band $M = \mathbf{R}^2/\mathbf{Z}$ defined in Example A.3 (iii) as a topological 2-manifold is the total space of a non-trivial C^∞ vector bundle with base S^1 and fibre $\mathbf{R}$. The projection $\pi: M \to S^1$ takes $[(x, y)]$ to $[x]$, where $[\ ]$ denotes the equivalence class under the relations defining $M = \mathbf{R}^2/\mathbf{Z}$ and $S^1 = \mathbf{R}/\mathbf{Z}$. A C^∞ vector bundle atlas giving the structure consists of the two vector bundle charts $\tilde{\xi}: \tilde{U} \to U' \times \mathbf{R}$ and $\tilde{\eta}: \tilde{V} \to V' \times \mathbf{R}$, where $U' =]0, 1[$, $V' =]\frac{1}{2}, \frac{3}{2}[$, $\tilde{U} = \pi^{-1}(U) = [U' \times \mathbf{R}]$, $\tilde{V} = [V' \times \mathbf{R}]$, and $\tilde{\xi}$ and $\tilde{\eta}$ are the maps $[(s, t)] \to (s, t)$. The coordinate change map is $\theta : \tilde{\xi}(\tilde{W}) \to \tilde{\eta}(\tilde{W})$, where $\tilde{W} = \tilde{U} \cap \tilde{V}$, $\theta(s, t) = (s, t)$ for $s \in]\frac{1}{2}, 1[$ and $\theta(s, t) = (1 + s, -t)$ for $s \in]0, \frac{1}{2}[$. It is clear that the bundle is not trivial, since M is not orientable†.

(A.25) Remark. In the case $r = 0$, the condition that the first factor V of a local vector bundle is an open subset of a Banach space is inappropriate and unnecessarily restrictive. The definition works perfectly well with V any topological space. With this modification, we may define a C^0 vector bundle over any topological space as base. Such bundles occur naturally in the theory of dynamical systems (for example the tangent bundle over an exotic basic set X of a dynamical system on a manifold).

Any C^r vector bundle structure $\tilde{\mathcal{A}}$ for $\pi : B \to X$ is certainly a C^r manifold atlas on B, and thus determines a C^r manifold structure on B. (*As usual we restrict attention to B Hausdorff with a countable basis of open sets.*) We always regard B as furnished with this structure. With respect to it, π is a C^r map of manifolds. For all $\tilde{\xi}_i : \tilde{U}_i \to U_i' \times \mathbf{F}$ in $\tilde{\mathcal{A}}$, let Z_i denote the subset $\tilde{\xi}_i^{-1}(U' \times \{0\})$ of B. Since local vector bundle maps are linear on the second factor, it is clear that $Z_i \cap \tilde{U}_j = Z_j \cap \tilde{U}_i$ for all $i, j \in I$. Thus the union of all Z_i, $i \in I$, forms a C^r submanifold, Z say, of B which is termed, rather loosely, the *zero section* of B (see Figure A.26). Clearly π maps Z C^r diffeomorphically onto

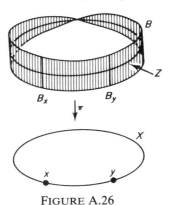

FIGURE A.26

the base X. For each point $x \in X$, the subset $B_x = \pi^{-1}(x)$ is a C^r submanifold of B, C^r diffeomorphic to $\mathbf{F}$, called the *fibre over x*. It inherits a linear structure from $\mathbf{F}$. That is to say, we may perform addition and multiplication by scalars in B_x by mapping to $\{x\} \times \mathbf{F}$ by some chart ξ, performing the corresponding operations in $\{x\} \times \mathbf{F}$ (identified with $\mathbf{F}$) and then mapping

† A finite dimensional manifold is *orientable* if it has an atlas all of whose coordinate change maps have differentials with positive determinant at all points (see Hirsch [1]).

back by ξ^{-1}. By definition of local vector bundle maps the net result does not depend on ξ. Thus B is indeed "a bundle of vector spaces", whence, of course, the name. Notice that although each fibre B_x also inherits via charts a topology from that of $\mathbf{F}$, it does *not* generally inherit a norm from $\mathbf{F}$, since a coordinate change map is not generally an isometry of fibres.

(A.27) Example. Let $\pi: B \to X$ and $\rho: C \to X$ be C^r vector bundles over a common base X, with fibres $\mathbf{F}$ and $\mathbf{G}$ respectively. As we have seen, for each $x \in X$ the fibres B_x and C_x inherit topologies, and so we may consider the topological vector space $L(B_x, C_x)$ of continuous linear maps from B_x to C_x. Let $L(B, C) = \bigcup_{x \in X} L(B_x, C_x)$. Then $L(B, C)$ is the total space of a C^r vector bundle, which we denote $L(\pi, \rho)$, with fibre $L(B_x, C_x)$ over x, defined as follows. Let $\tilde{\xi}: \tilde{U} \to U' \times \mathbf{F}$ and $\tilde{\eta}: \tilde{V} \to V' \times \mathbf{G}$ be charts on π and ρ covering the same chart $\xi: U \to U'$ (i.e. $\xi = \eta$, $U = V$, $U' = V'$). We have an associated chart $\tilde{\zeta}: L(\tilde{U}, \tilde{V}) \to L(U' \times \mathbf{F}, U' \times \mathbf{G}) = U' \times L(\mathbf{F}, \mathbf{G})$ defined, for all $T \in L(B_x, C_x)$ and for all $v \in B_x$, by $\tilde{\zeta}(T)(\tilde{\xi}(v)) = \tilde{\eta}(T(v))$. We may take C^r atlases $\{\tilde{\xi}_i : i \in I\}$ for π and $\{\tilde{\eta}_i : i \in I\}$ for ρ such that, for each $i \in I$, $\tilde{\xi}_i$ and $\tilde{\eta}_i$ have the above property. Tthen the associated atlas $\{\tilde{\zeta}_i : i \in I\}$ is a C^r atlas for $L(\pi, \rho)$. We call $L(\pi, \rho)$ the *linear map bundle* from π to ρ.

(A.28) Example. (*Products and sums of vector bundles*) Let $\pi: B \to X$ and $\rho: C \to Y$ be C^r vector bundles with fibres $\mathbf{E}$ and $\mathbf{F}$ respectively. Then the product $\pi \times \rho: B \times C \to X \times Y$ has a C^r vector bundle structure, with fibre $\mathbf{E} \times \mathbf{F}$, defined as follows. Let $\tilde{\xi}: \tilde{U} \to U' \times \mathbf{E}$ and $\tilde{\eta}: \tilde{V} \to V' \times \mathbf{F}$ be C^r vector bundle charts for π and ρ respectively. Then

$$\tilde{\xi} \times \tilde{\eta}: \tilde{U} \times \tilde{V} \to (U' \times \mathbf{E}) \times (V' \times \mathbf{F}) = (U' \times V') \times (\mathbf{E} \times \mathbf{F})$$

is a vector bundle chart on $\pi \times \rho$, and the set of all such product maps $\tilde{\xi} \times \tilde{\eta}$ forms a C^r vector bundle atlas for $\pi \times \rho$. If $X = Y$, and one restricts $\pi \times \rho$ to the fibres over the diagonal $\{(x, x): x \in X\}$, one has a bundle over X (identified with the diagonal by $x = (x, x)$) called the (*Whitney*) *sum* $\pi \oplus \rho$ of the bundles π and ρ. Its fibre is still, of course, $\mathbf{E} \times \mathbf{F}$, which is canonically isomorphic to $\mathbf{E} \oplus \mathbf{F}$.

Let $\pi: B \to X$ and $\rho: C \to Y$ be C^r vector bundles. A map $f: B \to C$ is a C^r *vector bundle map* if, for all admissible charts $\tilde{\xi}; \tilde{U} \to U' \times \mathbf{E}$ on π and $\tilde{\eta}: \tilde{V} \to V' \times \mathbf{F}$ on ρ, the induced map (or *local representative*)

$$\tilde{\eta} f \tilde{\xi}^{-1}: \tilde{\xi}(\tilde{f}^{-1}(\tilde{V}) \cap \tilde{U}) \to V' \times \mathbf{F}$$

is a C^r local vector bundle map. Thus $\tilde{f}$ is a C^r map of C^r manifolds which maps fibres of π linearly onto fibres of ρ. Note that $\tilde{f}$ maps the zero section Z of π into the zero section T of ρ, and thus induces (via the diffeomorphisms $\pi|Z: Z \to X$ and $\rho|T: T \to Y$) a C^r map $f: X \to Y$ such that the diagram over the page commutes. We say that $\tilde{f}$ *covers*, or is *over*, f.

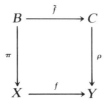

If $\tilde{f}: B \to C$ is a bijective C^r vector bundle map, and $\tilde{f}^{-1}: C \to B$ is also a C^r vector bundle map, then $\tilde{f}$ is said to be a C^r *vector bundle isomorphism*, and π is C^r *vector bundle isomorphic* to ρ.

(A.29) Exercise. Prove that the composite of two C^r vector bundle maps is another C^r vector bundle map.

(A.30) Exercise. Prove that there are, up to C^r vector bundle isomorphism, precisely two C^r vector bundles with base S^1 and fibre $\mathbf{R}^n$ ($n \geqslant 1$).

IV. THE TANGENT BUNDLE

Suppose that we are given a C^r manifold X ($r \geqslant 1$). We shall associate with each point x of X a vector space $T_x X$ which we may think of as the set of all possible velocities at x of a particle moving on X (see the section on vector fields in Chapter 3). This gives us a vector bundle, with base space X, which is called the *tangent bundle* of X. There is more than one way of constructing $T_x X$. We do it via charts. Any chart ξ at x takes the path of a moving particle on X to the path of a moving particle in the model space $\mathbf{E}$. We could define the velocity of the first particle at x to be the velocity of the second particle at $\xi(x)$, but this latter depends on the chart ξ. We get round this problem by the ingenious use of an equivalence relation.

Let $\mathscr{A} = \{\xi_i : U_i \to U_i' : i \in I\}$ be the C^r structure of X. Consider the subset A of $X \times \mathbf{E} \times I$ given by $A = \{(x, p, i): x \in U_i\}$ and define an equivalence relation $\sim$ on A putting $(x, p, i) \sim (y, q, j)$ if and only if $x = y$ and $q = D\chi(\xi_i(x))(p)$, where $\chi: \xi_i(U_i \cap U_j) \to \xi_j(U_i \cap U_j)$ is the coordinate change map, and $D\chi: \xi_i(U_i \cap U_j) \to L(\mathbf{E})$ is its derivative (see Appendix B), which is C^{r-1}. Let $TX = A/\sim$, and denote the $\sim$ class of (x, p, i) by $[x, p, i]$. Then there is a map $\pi_X: TX \to X$ given by $\pi_X([x, p, i]) = x$. Let $\tilde{U}_i = \pi_X^{-1}(U_i)$, and consider, for all $i \in I$, the set of maps $\tilde{\xi}_i: \tilde{U}_i \to U_i' \times \mathbf{E}$ given by $\tilde{\xi}_i([x, p, i)] = (\xi_i(x), p)$. It follows trivially from the definition of $\sim$ that $\{\tilde{\xi}_i : i \in I\}$ is a C^{r-1} vector bundle atlas for π_X. This determines a C^{r-1} vector bundle structure for π_X with fibre $\mathbf{E}$. With this structure, π_X (or TX) is called the *tangent bundle* of X.

The fibre $(TX)_x$ is called the *tangent space* to X at x. It is usually written as $T_x X$, or X_x (by abuse of notation). Its points are called *tangent vectors* (to X)

at x. It is sometimes useful to think of the base space X of π_X as identified with the zero section of π_X, the point x with the zero vector 0_x of T_xX.

(A.31) Example. If U is an open subset of $\mathbf{E}$, with its usual C^∞ manifold structure, then TU is C^∞ vector bundle isomorphic to the trivial bundle $U \times \mathbf{E}$. An explicit isomorphism is given as follows. Let $\xi_i : U \to \mathbf{E}$ be the inclusion, where $i \in I$. Then any point $v \in TU$ can be written uniquely as $[x, p, i]$ where $x \in U$ and $p \in \mathbf{E}$. Define $\tilde{f} : TU \to U \times \mathbf{E}$ by $f(v) = (x, p)$. Then $\tilde{f}$ is a C^∞ vector bundle isomorphism. It is very common to identify TU with $U \times \mathbf{E}$ by this isomorphism.

(A.32) Exercise. Construct a C^∞ vector bundle isomorphism from TS^1 to the trivial bundle $S^1 \times \mathbf{R}$.

(A.33) Exercise. Let X and Y be C^r manifolds modelled on $\mathbf{E}$ and $\mathbf{F}$ respectively. By Example A.28, $\pi_X \times \pi_Y$ is a C^{r-1} vector bundle over $X \times Y$. Construct a C^{r-1} vector bundle isomorphism from $\pi_X \times \pi_Y$ to $\pi_{X \times Y}$, the tangent bundle of the product manifold $X \times Y$.

(A.34) Example. (*Parallelizable manifolds*) A C^r manifold X modelled on a Banach space $\mathbf{E}$ is said to be *parallelizable* if there is a C^{r-1} vector bundle isomorphism from π_X to the trivial bundle $X \times \mathbf{E}$. Such an isomorphism is called a *trivialization*. Thus any open subset of $\mathbf{E}$ is parallelizable, and so is S^1. The product of two parallelizable manifolds is parallelizable, by Exercise A.33. As we commented in the introduction to the book in connection with the spherical pendulum, S^2 is not parallelizable.

The concept of *derivative*, or *linear approximation map*, is basic to differential calculus, and, when we work on smooth manifolds, it makes its appearance as the *tangent map*. Let X and Y be C^r manifolds modelled on $\mathbf{E}$ and $\mathbf{F}$ respectively, and let $f : X \to Y$ be a C^r map ($r \geq 1$). Let $v \in T_xX$ and let $\xi_i : U_i \to U_i'$ and $\eta_j : V_j \to V_j'$ be charts at x and $f(x)$ respectively. Then we have the local representative $\phi : W \to V_j'$, where $W = \xi_i(U_i \cap f^{-1}(V_j))$ and $\phi\xi_i(x) = \eta_j f(x)$ for all $x \in \xi_i^{-1}(W)$. If $v = [x, p, i]$, we define an element $Tf(v)$ of TY by $Tf(v) = [f(x), q, j]$, where $q = D\phi(\xi_i(x))(p)$. One must check that $Tf(v)$ is independent of choice of charts ξ_i and η_j. This is a routine exercise, and we leave it to the reader, together with the proof of the following result:

(A.35) Proposition. *The map* $Tf : TX \to TY$ *is a* C^{r-1} *bundle morphism. If* $g : Y \to Z$ *is another* C^r *map of manifolds, then* $T(gf) = (Tg)(Tf)$. *For any* X, $T(id_X) = id_{TX}$. □

In the language of category theory, we have constructed a *covariant functor* from the category of C^r manifolds and C^r maps to the category of C^{r-1} vector bundles and C^{r-1} vector bundle maps. We call T the tangent functor. We denote by T_xf the restriction $(Tf)_x : T_xX \to T_{f(x)}Y$ of Tf to a single fibre. It is a continuous linear map of topological vector spaces.

(A.36) Example. If X and Y are open subspaces of Banach spaces $\mathbf{E}$ and $\mathbf{F}$ then we may identify TX and TY with $X \times \mathbf{E}$ and $Y \times \mathbf{E}$ as explained in Example A.31. The derivative $Df: X \to L(\mathbf{E}, \mathbf{F})$ and the tangent map $Tf: TX \to TY$ are related by $Tf(x, p) = (f(x), Df(x)(p))$. The *double tangent map* $T^2f = T(Tf): T(TX) \to T(TY)$ is given by $T^2f((x, p), (u, v)) = ((f(x), Df(x)(p)), (Df(x)(u), D^2f(x)(p, u) + Df(x)(v)))$, where $T(TX)$ is identified with $(X \times \mathbf{E}) \times (\mathbf{E} \times \mathbf{E})$.

If U is an open subspace of a C^r manifold X and $\iota: U \to X$ is the inclusion, then the C^{r-1} vector bundle map $T\iota: TU \to TX$ maps TU bijectively onto the open subset $\pi_X^{-1}(U)$ of TX. One customarily identifies TU with its image under $T\iota$. If $\xi_i: U_i \to U_i'$ is any C^r admissible chart on X then the map $T\xi_i: TU_i \to TU_i' = U_i' \times \mathbf{E}$ is a C^{r-1} admissible chart on TX. Modulo the identifications, it is precisely the chart $\tilde{\xi}_i$ in the definition of TX.

(A.37) Example. If $I = [a, b]$ is a real interval and $\gamma: I \to X$ is a smooth map[†], then γ is called a *curve* on X. The tangent space TI is identified with $I \times \mathbf{R}$, and for all $t \in I$, 1_t denotes the element $(t, 1)$ of TI. The vector $T\gamma(1_t)$ in the tangent space $T_{\gamma(t)}X$ is called the *velocity* of γ at (time) t. We usually abbreviate $T\gamma(1_t)$ to $\gamma'(t)$. If $X = \mathbf{E}$ and we identify $T_{\gamma(t)}X$ with $\mathbf{E}$, this usage fits in with the standard notation of differential calculus.

(A.38) Example. (*The canonical involution*) If $\pi: B \to Y$ is a C^r vector bundle and if $f: B \to C$ is a bijection onto a set C, then f induces on C a C^r vector bundle structure such that f is a C^r vector bundle isomorphism. If C has from the outset its own C^r manifold structure (and in particular if $C = B$) then we are usually interested in maps f for which the induced C^r structure is the original one. Trivially this is so if and only if f is a C^r diffeomorphism with respect to the original structure on C.

An example of this occurs when $B = C = T^2X = T(TX)$, for any C^r manifold X ($r \geqslant 2$). In forming $T(TX)$ from X, we, in effect, twice take tangents to X; firstly in forming TX, secondly in taking tangents to TX (for we envisage X as embedded in TX as the zero section). Thus T^2X has a built in symmetry (to use the term rather loosely), and there is a C^{r-2} *involution* (= diffeomorphism of period 2) that exchanges the two "tangent spaces to X" at each point x. This map is called the *canonical involution* of T^2X. We now give it a precise description.

We continue with the notations in the definition of TX. Let $\mathcal{B} = \{\tilde{\xi}_i: i \in I\}$ be the atlas of TX corresponding to the atlas $\mathcal{A} = \{\xi_i: i \in I\}$ of X. Points of $T(TX)$ are of the form $[[x, p, i], (q, r), j]$, where $j \in I$, $(q, r) \in \mathbf{E}^2$ and $[x, p, i]$ is a point of TX in the domain of $\tilde{\xi}_j$. The outer brackets denote equivalence class with respect to $\sim$ in the construction of T^2X. We may write any such point as $[[x, p, i], (u, v), i]$ for some $(u, v) \in \mathbf{E}^2$, since certainly $[x, p, i] \in \hat{U}_i$.

[†] That is to say, γ can be extended to a smooth map of some open interval J with $I \subset J$.

The canonical involution f takes this point to $[[x, u, i], (p, v), i]$. Checking that f is well defined and a $\dot{C}^{r-2}$ diffeomorphism is a useful exercise. Now recall that the C^{r-1} map π_X takes $[x, p, i]$ to x. Thus we have two important maps from T^2X to TX, namely π_{TX} and $T(\pi_X)$. The first takes $[[x, p, i], (u, v), i]$ to $[x, p, i]$, and the second, as may easily be verified, takes it to $[x, u, i]$. Thus the diagram

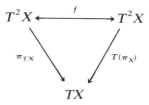

commutes.

V. IMMERSIONS, EMBEDDINGS AND SUBMERSIONS

Let $f: X \to Y$ be a C^r map of C^r manifolds $(r \geq 1)$. For each $x \in X$, the tangent map $T_x f: T_x X \to T_y Y$, where $y = f(x)$, is continuous linear. Its kernel ker $T_x f$ and its image im $T_x f$ are linear subspaces of $T_x X$ and $T_y Y$ respectively. Ker $T_x f$ is automatically a closed subspace; im $T_x f$ is not necessarily closed unless Y is finite dimensional. We say that f is *immersive* at x if $T_x f$ is injective and im $T_x f$ splits $T_y Y$. Dually, f is *submersive* at x if $T_x f$ is surjective and ker $T_x f$ splits $T_x X$. We say that f is an *immersion* if it is immersive at x for all $x \in X$, and a *submersion* if it is submersive at x for all $x \in X$.

(A.39) Note. If im $T_x f$ (resp. ker $T_x f$) has finite dimension or finite codimension in $T_y Y$ (resp. $T_x X$) then splitting is automatic. This is because, firstly, any finite dimensional subspace or closed finite codimensional subspace splits any Banach space and, secondly, any finite codimensional image of a continuous linear map is closed (see Lang [2]). In particular, one may omit the splitting condition from the above definitions when either X or Y is finite dimensional.

(A.40) Example. The C^∞ map $f: S^1 \to S^1$ given by $f([x]) = [nx]$ for all $[x] \in S^1 = \mathbf{R}^2/\mathbf{Z}$, where n is any integer, is an immersion (and also a submersion) if and only if $n \neq 0$.

(A.41) Example. For any C^r vector bundle $\pi: B \to X$, π is a C^r submersion. In particular, if X is a C^r manifold, π_X is a C^{r-1} submersion for $r \geq 2$.

The image of an injective C^r immersion is sometimes called a C^r *immersed submanifold*. This is an abuse of language, since it need not be a topological submanifold; think of the numeral 6 regarded as the image of a C^∞ immersion of $\mathbf{R}$ in $\mathbf{R}^2$. If $f: X \to Y$ is an injective immersion, we say that im $T_x f$ is the *tangent space to the immersed submanifold* at $f(x)$. An injective C^r immersion $f: X \to Y$ whose image is a C^r submanifold of Y is called a C^r *embedding*. Any C^r immersion is locally a C^r embedding, by Exercise C.12 of Appendix C.

(A.42) Example. In irrational flow on the torus (see Examples 1.25 and 2.9) the map: $\mathbf{R} \to T^2$ taking t to $([x + t], [y + \theta t])$ is an injective immersion but not an embedding.

(A.43) Example. The inclusion of a C^r submanifold of X in X is a C^r embedding.

(A.44) Example. If X and Y are C^r manifolds, and we define $i : X \to X \times Y$ by $i(x) = (x, y)$ for some given $y \in Y$, then i is an embedding. Similarly the diagonal map $x \mapsto (x, x)$ from X to $X \times X$ is an embedding.

(A.45) Exercise. Prove that $f: X \to Y$ is a C^r embedding ($r \geq 1$) if and only if it is a C^r immersion and a *topological embedding* (i.e. maps X homeomorphically onto $f(X)$).

(A.46) Example. (*Foliations and laminations*) Let X be a C^r manifold modelled on $\mathbf{E}$ and let $\mathbf{E} = \mathbf{F} \times \mathbf{G}$ be a splitting of $\mathbf{E}$. A C^r *foliation* is a disjoint decomposition of X into C^r injectively immersed submanifolds, called *leaves*, satisfying the following condition: There is an admissible atlas of charts of the form $\xi: U \to \mathbf{F} \times \mathbf{G}$, called *foliation boxes*, such that, for all $y \in \mathbf{G}$, $\xi^{-1}(\mathbf{F} \times \{y\})$ is contained in a single leaf, and is the image of an open set under the injective immersion giving the leaf (see Figure A.46). The

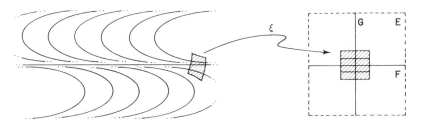

FIGURE A.46

dimensions of $\mathbf{F}$ and $\mathbf{G}$ are called respectively the *dimension* and *codimension* of the foliation. It is also possible to give an equivalent purely local definition of foliation by considering a maximal atlas of C^r admissible charts for which all coordinate change maps χ have the property $\chi_2(x, y) = \chi_2(x', y')$ if and only if $y = y'$, for all (x, y) and (x', y') in $\mathbf{F} \times \mathbf{G}$.

Rational and irrational flows both give 1-dimensional C^∞ foliations of the torus (the leaves being the orbits of the flow). As we have seen in Chapter 6, the stable manifold of a hyperbolic closed orbit of a C^r flow is C^r foliated by the stable manifolds of the individual points of the orbit. This situation is not typical of hyperbolic sets in general. Usually the stable manifolds of the individual points only C^r *laminate* the stable manifold of the set. To define this notion, we regard $\mathbf{E}$ as a trivial vector bundle with base $\mathbf{G}$ and fibre $\mathbf{F}$, and weaken the definition of C^r foliation by relaxing the condition that the foliation boxes ξ are C^r admissible charts and insisting only that they are homeomorphisms with F^r inverses (see Appendix B for F^r maps). Of course they must still form a topological atlas for X and satisfy the foliation box condition.

If $f : X \to Y$ is a C^r map of manifolds, and, for some $y \in Y$, f is submersive at every point of $f^{-1}(y)$, then $f^{-1}(y)$ is a disjoint union of C^r submanifolds of X, with dimension dim $X - \dim Y$ if this makes sense. This result generalizes the remarks of Example A.6 and is, again, a consequence of the inverse mapping theorem. More generally still, let W be a C^r-immersed submanifold of Y. We say that f is *transverse* to W, written $f \pitchfork W$, if for all $y \in f(X) \cap W$, $T_y(Y) = W_y + \operatorname{im} T_x f$, where $y = f(x)$ and W_y is the tangent space to W at y, and $(T_x f)^{-1}(W_y)$ splits $T_x X$. In this case, $f^{-1}(W)$ is a C^r submanifold of X whose codimension equals the codimension of W in Y. If V and W are two C^r immersed submanifolds of Y, we say that V is transverse to W, written $V \pitchfork W$, if some injective C^r immersion f with image V is transverse to W. It follows that $V \cap W$ is a C^r immersed submanifold.

VI. SECTIONS OF VECTOR BUNDLES

Let $\pi : B \to X$ be a C^r vector bundle. A map $\sigma : X \to B$ such that $\pi\sigma : B \to B$ is the identity on B is called a *section* of π. Thus σ is a section if and only if it maps every point x of X into the fibre B_x over x. Figure A.47 illustrates this

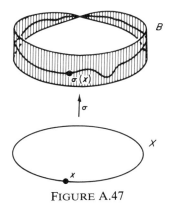

FIGURE A.47

idea in the case when B is the Möbius band (Example A.24). When π is the tangent bundle $\pi_X: TX \to X$, sections are called *vector fields* on X.

In the case of the trivial bundle $\pi_1: X \times \mathbf{F} \to X$, the sections of π_1 are precisely all maps $(id, f): X \to X \times \mathbf{F}$, and, of course, the section (id, f) is C^r if and only if the map $f: X \to \mathbf{F}$ is C^r. The map f is called the *principal part* of the section. For a general vector bundle $\pi: B \to X$ we may trivialize the situation locally, using an admissible chart. Let $\tilde{\xi}: \tilde{U} \to U' \times \mathbf{F}$ be such a chart, so that the diagram

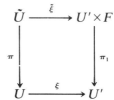

commutes. Then the section σ of π, when restricted to U, induces a section, τ say, of the trivial bundle π_1, defined by commutativity of the diagram

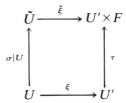

In this case τ is said to be a *local representative* of σ.

(A.48) Exercise. Prove that any C^r section of a vector bundle is a C^r embedding.

(A.49) Example. (*The spherical pendulum*) In the introduction to the book we discussed the spherical pendulum, and found that its motion could be modelled by a C^∞ vector field v on the C^∞ manifold $X = TS^2$ of dimension 4. If U is the complement in S^2 of a single meridian of longitude, then there is an admissible chart $\xi: U \to U'$ on S^2 given by $\xi(y) = (\theta, \phi)$ where θ and ϕ are the Euler angles. Correspondingly there are admissible charts $\tilde{\xi}: \tilde{U} \to U' \times \mathbf{R}^2$ for X and $\tilde{\tilde{\xi}}: \tilde{\tilde{U}} \to (U' \times \mathbf{R}^2) \times R^4$ for TX, where $\tilde{U} = \pi_{S^2}^{-1}(U)$ and $\tilde{\tilde{U}} = \pi_X^{-1}(\tilde{U})$. In terms of these local coordinates, the vector field $v: X \to TX$ has local representative

$$(\theta, \phi, \lambda, \mu) \to ((\theta, \phi, \lambda, \mu), p)$$

where $(\theta, \phi, \lambda, \mu) \in U' \times \mathbf{R}^2$ and $p \in \mathbf{R}^4$ is the point

$$(\lambda, \mu, \mu^2 \sin\theta \cos\theta + g \sin\theta, -2\lambda\mu \cot\theta).$$

(A.50) Example. (*Second order equations and sprays*) Notice that, for the vector field v of Example A.49, the first two coordinates of the principal part of $v(x)$ are the same as the last two coordinates of X, namely (λ, μ). This came about because we originally converted a system of second order equations on U' into a system of first order equations on $U' \times \mathbf{R}^2$ by the substitution $\theta' = \lambda$, $\phi' = \mu$. Since, as we commented in Note 3.17, this is a standard procedure, we ought to analyse the situation a bit further.

Let $X = TM$, where M is a C^r manifold ($r \geqslant 2$). *A first order ordinary differential equation* on M is just a vector field on M. *A second order ordinary differential equation* on M is a vector field v on X satisfying $T(\pi_M)v = id_X$. This is precisely the coordinate free generalization of the condition in the last paragraph. If $\xi: U \to U'$ is a chart on M, with corresponding charts $\tilde{\xi}: \tilde{U} \to U' \times \mathbf{E}$ on X and $\tilde{\tilde{\xi}}: \tilde{\tilde{U}} \to (U' \times \mathbf{E}) \times (\mathbf{E} \times \mathbf{E})$ on TX, and if $f: U' \times \mathbf{E} \to \mathbf{E} \times \mathbf{E}$ is the principal part of the local representative of v, then the condition says that the first coordinate of $f(x, y)$ is y. Now let $\gamma: I \to X$ be any integral curve of v, and let $\delta: I \to M$ be the projection of γ onto M by π_M. Thus $\delta = \pi_M \gamma$. We call δ a *solution of the second order equation*. Differentiating at $t \in I$, we deduce that

$$\delta'(t) = T\pi_M(\gamma'(t)) = T\pi_M(v\gamma(t)) = \gamma(t).$$

That is to say, the velocity of the curve δ at t is the value of the curve γ at t. See Figure A.50. There is no reason why δ should not have self intersections.

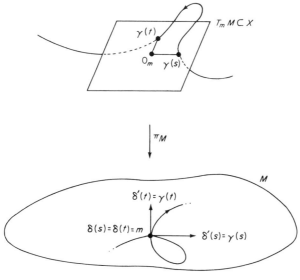

FIGURE A.50

Uniquenesss of the integral curve of v at a given point of X corresponds to uniqueness of the solution δ of the second order equation through a given point of M with a given velocity at $t = 0$.

Concentrating on a particular point $m \in M$ for a moment, we have infinitely many solutions δ_x starting at m at $t = 0$, one for each possible velocity x at m. In fact we have infinitely many starting off in a given direction with various speeds, and there is no reason why these should be in any way related (see Figure A.51(i)). However, given $x \in T_m M$ and $a \in \mathbf{R}$, there is a very natural way of obtaining from δ_x a curve with velocity ax at m at time $t = 0$, and that is by speeding δ_x up by a factor a. That is to say, the curve δ_{ax} defined by $\delta_{ax}(t) = \delta_x(at)$ has the required property (see Figure A.51(ii)). Note that for $a \neq 0$, δ_x and δ_{ax} have the same image. It is a particularly nice situation when the solution curves δ fit together in this way.

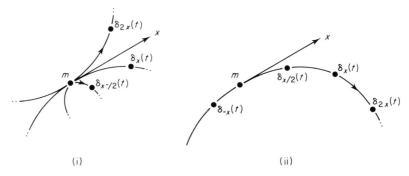

(i)   (ii)

FIGURE A.51

When they do so, we call the second order differential equation a *spray*, and the images of the solution curves *geodesics* of the spray.

Let us find the condition on v that makes it a spray. We require that if $\delta'_{ax}(0) = a\delta'_x(0) = ax$, then $\delta_{ax}(t) = \delta_x(at)$. Differentiating this latter relation gives $\delta'_{ax}(t) = a\delta'_x(t)$. Since v is a second order equation, we may restate this as $\gamma_{ax}(t) = a\gamma_x(at)$. To find the condition on v, we differentiate again. This not completely straightforward, because the outside a is scalar multiplication on the vector bundle TM; it affects the fibre direction but not the "zero section direction" (if we could make sense of this at points of TM which are not on the zero section). However, if we denote by $\mu_a: TM \to TM$ scalar multiplication by a on the fibres, then we can write $\gamma_{ax}(t) = \mu_a \gamma_x(at)$ and obtain $\gamma'_{ax}(t) = T\mu_a(a\gamma'_x(at))$. Putting $t = 0$, we get, for all $x \in TM$ and $a \in \mathbf{R}$,

(A.52) $v(ax) = aT\mu_a(v(x)).$

Conversely, if v is a C^1 second order differential equation on M satisfying (A.52) then, by the uniqueness theorem (Theorem 3.34) its integral curves satisfy $\gamma_{ax} = a\gamma_x(at)$, and hence the solution curves satisfy $\delta_{ax}(t) = \delta_x(at)$ as required.

Let v be a C^1 spray on M. By (A.52), 0_m is a zero of v, for all $m \in M$. Thus, by Theorem 3.22, there is some neighbourhood of 0_m in TM such that, for all x in it, integral curves of v at x are defined on $[0, 1]$. Thus, for some neighbourhood N of the zero section of TM, there is a map $\exp\colon N \to M$ defined by $\exp x = \delta_x(1)$. This is called the *exponential* map of the spray. It is C^r when v is C^r. For all $m \in M$, the restriction $\exp_m$ of $\exp$ to $N \cap T_mM$ maps rays through the origin onto geodesics at m, since $\exp_m(tx) = \delta_{tx}(1) = \delta_x(t)$. Moreover, since $d/dt(\exp_m(tx)) = \delta_x'(t) = x$ at $t = 0$, the derivative of $\exp_m$ at 0_m is the identity map on T_mM. More precisely, if we identify $T(T_mM)$ with $T_mM \times T_mM$ in the standard way, then $T \exp_m\colon T(T_mM) \to TM$ satisfies $T \exp_m(0_m, x) = x$ for all $x \in T_mM$. It follows from the inverse function theorem that $\exp_m$ maps some neighbourhood of 0_m in T_mM diffeomorphically onto a neighbourhood of m in M.

(A.53) Example. If U is an open subset of R^n then the *standard spray* $v\colon U \times \mathbf{R}^n = TU \to T^2U = (U \times \mathbf{R}^n) \times (\mathbf{R}^n \times \mathbf{R}^n)$ is given by $v(m, x) = (m, x, x, 0)$. The integral curve of v at (m, x) is $t \mapsto (m + tx, x)$ and the solution curve through m with velocity (m, x) is $t \mapsto m + tx$. The exponential map is given by $\exp(m, x) = m + x$.

(A.54) Example. The circle $S^1 = \mathbf{R}/\mathbf{Z}$ inherits its *standard spray* from the standard spray of $\mathbf{R}$ by the covering map $m \to [m]$. Thus $v([m], x) = ([m], x, x, 0)$, where we write TS^1 as $S^1 \times \mathbf{R}$. The exponential map is $\exp([m], x) = [m + x]$. In terms of the standard embedding of S^1 in $\mathbf{R}^2$, $\exp$ maps each tangent space round the circle, taking the tangent vector x at p to the point $p\, e^{2\pi i x}$. For higher dimensional spheres S^n, the geodesics are great circle arcs which may be derived from the above example by taking 2-plane sections through $0 \in \mathbf{R}^{n+1}$.

A *zero* of a section $\sigma\colon X \to B$ is a point $x \in X$ such that $\sigma(x)$ is the zero 0_x of the vector space B_x. The *zero section* is the section defined by $\sigma(x) = 0_x$ for all $x \in X$ (but, as we commented earlier, the phrase is sometimes used loosely for the image of the map). A section without zeros is said to be *nowhere zero*. For many vector bundles, nowhere zero sections do not exist. For example, the reader can soon verify experimentally that the Möbius band has no nowhere zero sections. Neither does the tangent bundle TS^{2n}, by the Poincaré–Hopf index theorem (Theorem 5.79).

The set $C^r(\pi)$ of all C^r sections of π has algebraic structures generated by the linear structure of the fibres. It is a real vector space, with structure defined pointwise by

$$(a\sigma + b\tau)(x) = a\sigma(x) + b\tau(x)$$

for $\sigma, \tau \in C^r(\pi)$, $a, b \in \mathbf{R}$ and $x \in X$. Moreover it is a module over the ring $\mathscr{F}^r(X)$ of C^r functions on X, when we put

$$(f \cdot \sigma + g \cdot \tau)(x) = f(x) \cdot \sigma(x) + g(x) \cdot \tau(x)$$

for $f, g \in \mathscr{F}^r(X)$. We shall discuss the space $C^r(\pi)$ further in Appendix B, and give it a natural Banach space structure in certain circumstances. When π is the tangent bundle of X, we denote $C^r(\pi)$ by $\Gamma^r(X)$.

VII. TENSOR BUNDLES

The total space of a C^k vector bundle $\pi: B \to X$ is partitioned into a set of fibres B_x. Each fibre B_x has an associated dual topological vector space $B_x^* = L(B_x, \mathbf{R})$. If we cobble these dual spaces together in a natural way determined by the structure of π, we get a new C^k vector bundle, called the dual π^* of π. Similarly we can, for each fibre B_x, form the space $L_r^s(B_x, \mathbf{R})$ of all real $(r + s)$-linear functions on $(B_x)^r \times (B_x^*)^s$, which we call *tensors of type* (r, s) and we can make these into a C^k vector bundle which we denote by π_r^s. In particular, π^* is π_1^0.

We now describe the structure of π_r^s. We first observe that any (topological) linear automorphism $f: \mathbf{F} \to \mathbf{G}$ of Banach spaces determines a linear automorphism $f_r^s: L_r^s(\mathbf{F}, \mathbf{R}) \to L_r^s(\mathbf{G}, \mathbf{R})$ by the formula

$$f_r^s(T)(p_1, \ldots, p_r, q_1, \ldots, q_s) = T(f^{-1}(p_1), \ldots, f^{-1}(p_r), q_1 f, \ldots, q_s f),$$

where $(p_1, \ldots, p_r, q_1, \ldots, q_s) \in \mathbf{G}^r \times (\mathbf{G}^*)^s$. For any subset U of X, let $L_r^s(\tilde{U}, \mathbf{R})$ denote the disjoint union of $L_r^s(B_x, \mathbf{R})$ for all fibres B_x of π with $x \in U$. Here, as usual, $\tilde{U} = \pi^{-1}(U)$. Let $\pi_r^s: L_r^s(B, \mathbf{R}) \to X$ send $L_r^s(B_x, \mathbf{R})$ to x. Any admissible chart $\tilde{\xi}: \tilde{U} \to U' \times \mathbf{F}$ on π has the form $\tilde{\xi}(u) = (\xi(x), f_x(u))$ where $\pi(u) = x$, and $f_x: B_x \to \mathbf{F}$ is a linear automorphism. We define a map $\tilde{\xi}_r^s: L_r^s(\tilde{U}, \mathbf{R}) \to U' \times L_r^s(\mathbf{F}, \mathbf{R})$ by $\tilde{\xi}_r^s(T) = (\xi(x), (f_x)_r^s(T))$, where $x = \pi_r^s(T)$. We may topologize $L_r^s(B, \mathbf{R})$ by insisting that for all admissible charts $\tilde{\xi}$ on π the maps $\tilde{\xi}_r^s$ are homeomorphisms, and it is easy to check that the $\tilde{\xi}_r^s$ then form a C^k vector bundle atlas for π_r^s. We call π_r^s, with the C^k vector bundle structure determined by this atlas, the *bundle of tensors on π of type (r, s)* (or *covariant of order r, contravariant of order s*). A section of π_r^s is called a *tensor field of type (r, s) on π*.

(A.55) Example. (*The derivative of a smooth function*) The dual π_X^* of the tangent bundle of a smooth manifold is called the *cotangent bundle* of X. The total space is denoted by T^*X, the fibre T^*_xX (or X^*_x) is called the *cotangent space* at x and its elements are called *cotangent vectors*. Sections of π^*_X are called 1-*forms* on X. If $f: X \to \mathbf{R}$ is a C^r map ($r \geqslant 1$), then T_xf maps T_xX linearly to $T_{f(x)}\mathbf{R} = \{f(x)\} \times \mathbf{R}$. Thus if we identify $\{f(x)\} \times \mathbf{R}$ with $\mathbf{R}$ in the obvious way, T_xf becomes a cotangent vector at x. The C^{r-1} map from X to T^*X taking x to T_xf is a 1-form on X, and we call it the *derivative Df* of f. If X is an open subset of $\mathbf{E}$, and we identify the tangent space $T_xX = \{x\} \times \mathbf{E}$ with $\mathbf{E}$, then Df is the derivative in the usual sense (see Example A.36).

VIII. RIEMANNIAN MANIFOLDS

As we commented earlier, a vector bundle does not usually come equipped with an inner product, or even a norm, on each individual fibre. However these may be given to the bundle as extra structure, and they then give rise to extra theory. In the case of the tangent bundle of a manifold, they enable us to define lengths of curves on the manifold and also to give the manifold a natural metric space structure.

A C^r *Riemannian structure*, or *Riemannian metric*, on a vector bundle $\pi: B \to X$ is a C^r section $\rho: X \to L_2(B, \mathbf{R})$ of the tensor bundle π^0_2 such that the bilinear form $\rho(x)$ on B_x is symmetric and positive definite for all $x \in X$. One usually writes $\rho(x)(p, q)$ as $\langle p, q \rangle_x$ when there is no doubt as to which metric ρ is in use. A C^r Riemannian metric gives rise to a C^r *Finsler*, which is a norm $|\ |_x$ on the fibre B_x that depends on x in a C^r fashion. Of course, $|\ |_x$ is defined by $|p|_x = \sqrt{\langle p, p \rangle_x}$.

(A.56) Example. Any trivial bundle $\pi_1: X \times \mathbf{H} \to X$, where X is a smooth manifold and $\mathbf{H}$ is a Hilbert space, has a Riemannian structure given by the inner product of $\mathbf{H}$. In particular if U is an open subset of $\mathbf{R}^n$, the tangent bundle of U has a Riemannian structure given by the standard inner product on $\mathbf{R}^n$. That is to say

$$\langle (x, y), (x, y') \rangle_x = y_1 y'_1 + \cdots + y_n y'_n$$

for all (x, y) and (x, y') in $TU = U \times \mathbf{R}^n$.

A Riemannian metric on the tangent bundle of a smooth manifold X is also said to be a *Riemannian metric* on X. A C^{r+1} *Riemannian manifold* is a C^{r+1} manifold X together with a C^r Riemannian metric on X. The metric of Example A.56 is the *standard Riemannian metric* on an open subset of $\mathbf{R}^n$. One may construct a C^r Riemannian metric on any finite dimensional C^{r+1} manifold X, or, indeed, on any C^{r+1} manifold X modelled on a Hilbert space provided that X admits C^{r+1} partitions of unity.

(A.57) Example. (*Gradient vector fields*) A C^r Riemannian metric on a manifold X enables us to convert C^r 1-forms on X to C^r vector fields on X and vice versa. The former operation is called *sharpening*, the latter *flatten-ing*. By the representation theorem for Hilbert spaces, given any element λ of T_x^*X, there exists a unique element p of T_xX such that $\lambda(q) = \langle p, q \rangle_x$ for all $q \in T_xX$. Conversely any $p \in T_xX$ gives rise to an element λ of T_x^*X defined by the given formula. We thus obtain inverse C^r vector bundle isomorphisms $\rho_\# : T^*X \to TX$ and $\rho_\flat : TX \to T^*X$, and, correspondingly, isomorphisms $\tilde{\rho}_\# : \Omega'X \to \Gamma'X$ and $\tilde{\rho}_\flat : \Gamma'X \to \Omega'X$, where $\Omega'X$ is the Banach space of C^r 1-forms on X, and

$$\tilde{\rho}_\#(w) = \rho_\# w, \qquad \tilde{\rho}_\flat(v) = \rho_\flat v.$$

In particular, if $f : X \to \mathbf{R}$ is a C^{r+1} function on X, then the derivative Df is a C^r 1-form on X (see Example A.55) and $\tilde{\rho}_\#(Df)$ is a C^r vector field on X called the *gradient* ∇f of f. For more detail, see Example 3.3 of the text.

(A.58) Exercise. Let $f : X \to \mathbf{R}$ be C^{r+1}. Prove that ∇f is orthogonal to the contours of f, in the sense that, for all $x \in X$ and $p \in \ker T_x f$, $\langle p, \nabla f(x) \rangle_x = 0$.

Let X be a C^{r+1} Riemannian manifold, for sufficiently large r ($r \geq 4$ covers all eventualities). The Riemannian structure on X gives rise to a C^{r-1} spray σ on X, called the *Riemannian spray*, defined as follows. Let $\xi : U \to V$ be an admissible chart on X, so that $T\xi : TU \to TV = V \times \mathbf{E}$ is an admissible chart on TX. We transfer the Riemannian structure to V by the chart. That is to say, we define a function $K : V \times \mathbf{E} \times \mathbf{E} \to \mathbf{R}$ by

(A.59) $K(y, p, q) = \langle T\xi^{-1}(y, p), T\xi^{-1}(y, q) \rangle_{\xi^{-1}(y)}$

so that, for fixed $y \in V$, $K(y, ., .)$ is a symmetric positive definite bilinear function on $\mathbf{E} \times \mathbf{E}$. We wish to define a local representative τ of σ with respect to the chart $T^2\xi : T^2U \to T^2V = (V \times \mathbf{E}) \times (\mathbf{E} \times \mathbf{E})$ on T^2X. We do so by the formula

$$\tau(y, p) = (y, p, p, v),$$

where $v \in \mathbf{E}$ satisfies, for all $u \in \mathbf{E}$, the formula

(A.60) $K(y, u, v) = (\tfrac{1}{2}TK(y, p, p, u, 0, 0) - TK(y, p, u, p, 0, 0))_2,$

the subscript denoting the second coordinate in $T\mathbf{R} = \mathbf{R} \times \mathbf{R}$. This, by the representation theorem for Hilbert spaces, gives rise to a uniquely defined element $v \in \mathbf{E}$, and it is clear from the formula that the second order equation τ satisfies the condition (A.52) for a spray, using the bilinearity of $K(y, ., .)$. Obviously the left-hand side of (A.60) is connected with the derivative of $\langle , \rangle_x$ with respect to x, but the geometrical motivation for its detailed expression is less clear. A very natural, but rather subtle, explana-tion of it appears in Lang [1]. However, this requires more of the machinery of differential forms than we are prepared to introduce here. We content ourselves by verifying that σ is well defined, in that it does not depend on the choice of chart ξ.

Let $\xi': U' \to V'$ be another chart on X, and let $\chi: \xi(U \cap U') \to \xi'(U \cap U')$ be the coordinate change map. The map K of (A.59) and the corresponding map $K': V' \times \mathbf{E} \times \mathbf{E} \to \mathbf{R}$ induced by ξ' are related on the overlap by

(A.61) $K(y, p, q) = K'(y', p', q')$

where $y' = \chi(y)$, $p' = D\chi(y)(p)$ and $q' = D\chi(y)(q)$. We have to show that $T^2\chi$ takes $\tau(y, p)$ to (y', p', p', v') where v' satisfies (A.60)', which is (A.60) with K and all the variables dashed. Now, by Example A.36,

$$T^2\chi(y, p, p, v) = (y', p', p', D^2\chi(y)(p, p) + D\chi(y)(v)).$$

Moreover, differentiating (A.61),

$$TK'(y', p', q', u', 0, 0) = TK(y, p, q, u, -D\chi(y)^{-1}D^2\chi(y)(p, u),$$
$$-D\chi(y)^{-1}D^2\chi(y)(q, u)),$$

where $u' = D\chi(y)(u)$. Thus

$$\tfrac{1}{2}TK'(y', p', p', u', 0, 0) - TK'(y', p', u', p', 0, 0)$$
$$= \tfrac{1}{2}TK(y, p, p, u, -D\chi(y)^{-1}D^2\chi(y)(p, u), -D\chi(y)^{-1}D^2\chi(y)(p, u))$$
$$- TK(y, p, u, p, -D\chi(y)^{-1}D^2\chi(y)(p, p), -D\chi(y)^{-1}D^2\chi(y)(u, p)).$$

By (A.60) and the bilinearity of $K(y, ., .)$, the second coordinate of this expression reduces to

$$K(y, u, v) - \tfrac{1}{2}K(y, p, D\chi(y)^{-1}D^2\chi(y)(p, u))$$
$$- \tfrac{1}{2}K(y, D\chi(y)^{-1}D^2\chi(y)(p, u), p)$$
$$+ K(y, p, D\chi(y)^{-1}D^2\chi(y)(u, p)) + K(y, D\chi(y)^{-1}D^2\chi(y)(p, p), u)$$
$$= K(y, u, v + D\chi(y)^{-1}D^2\chi(y)(p, p))$$
$$= K'(y', u', v')$$

as required.

By the theory of sprays (Example A.50) we now have a C^{r-1} exponential map $\exp: N \to X$ where N is some neighbourhood of the zero section in TX. We call this the *Riemannian exponential map*.

The Riemannian metric on X gives rise to a metric on X in the sense of metric space theory. We denote this by d, and always call it the *Riemannian distance function*, to avoid confusion. We define $d(x, x')$ for $x, x' \in X$ as follows. Since X is connected, we may join x to x' by *piecewise C^1 curve* $\gamma: I \to X$ (that is to say, I is a real interval $[a, b]$, γ is a continuous map and is C^1 on each subinterval $[a_r, a_{r+1}]$ of some subdivision

$$a = a_0 < a_1 < \cdots < a_n = b).$$

Since $|\gamma'(t)|_{\gamma(t)}$ has at worst a finite number of jump discontinuities on $[a, b]$, the length of γ, $\int_a^b |\gamma'(t)|_{\gamma(t)} \, dt$, exists finite, and we define $d(x, x')$ to be the infimum of the lengths of γ as γ ranges over all piecewise C^1 curves joining x to x'. Trivially d is symmetric and satisfies the triangle inequality, and $d(x, x') = 0$. To complete a proof that (X, d) is a metric space, we need to show that $d(x, x) = 0$ implies that $x = x'$. This is an easy consequence of our final theorem, Theorem A.62 below, which says that, near x, distances from x in X correspond under the exponential map $\exp_x$ to distances from 0_x in $T_x X$, and shortest paths from x correspond to rays through 0_x (and hence are geodesics through x). Moreover, since small open balls with centre 0_x correspond to open balls with centre x with respect to d, and since $\exp_x$ is, near 0_x, a homeomorphism with respect to the original topology on X, we deduce that the metric topology of d is the original topology.

Theorem A.62 has other important corollaries. One, which we used in the proof of the generalized stable manifold theorem (Theorem 6.21), is that if Λ is a compact subset of X then we may choose a number $a > 0$ such that, for all $x \in \Lambda$, $\exp_x$ maps the ball in $T_x X$ with centre 0_x and radius a with respect to $| \quad |_x$ onto the ball in X with centre x and radius a with respect to d.

(A.62) Theorem. *Let $x \in X$, and let B be any open ball with centre 0_x in $T_x X$ small enough for $\exp_x$ to map B diffeomorphically onto its image. Then, for all $y \in B$, $d(x, \exp y) = |y|_x$. Moreover $|y|_x$ is the length of the curve $\gamma : [0, 1] \to X$ defined by $\gamma(t) = \exp ty$.*

Proof. We first comment that any C^3 diffeomorphism $h : X \to Y$ from X to a smooth manifold Y induces a Riemannian structure on Y, defined for all $p, q \in T_y Y$ by $\langle p, q \rangle_y = \langle Th^{-1}(p), Th^{-1}(q) \rangle_{h^{-1}(y)}$. Correspondingly, the Riemannian spray $\sigma : TX \to T^2 X$ induces a C^1 Riemannian spray $\tau : TY \to T^2 Y$, and these are related by the commutative diagram

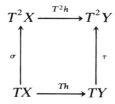

Each solution curve ε of τ at $y \in Y$ is of the form $h\delta$ where δ is a solution curve of σ at $h^{-1}(y)$. The length of a curve γ in X is equal to the length of the curve $h\gamma$ in Y, and thus the distance from x to x' in X equals the distance from $h(x)$ to $h(x')$ in Y.

We apply this idea with $X = \exp_x B$, $Y = B$ and h the inverse of the restriction $\exp_x : B \to \exp_x B$. To prove the theorem, we have to show that, with respect to the distance function of the new Riemannian metric induced on B by h, the distance from 0_x to y is $|y|_x$, for all $y \in B$, and that this is the length of the straight line segment $\gamma : [0, 1] \to B$ joining 0_x to y (i.e. $\gamma(t) = ty$). We denote the new Riemannian metric on B by $\langle\!\langle \ \rangle\!\rangle_y$ and the associated Finsler by $\| \ \|_y$. Since T_xX is a Banach space, we identify TB with $B \times T_xX$ in the usual way, and write $\langle\!\langle p, q \rangle\!\rangle_y$ or $K(y, p, q)$ for $\langle\!\langle (y, p), (y, q) \rangle\!\rangle_y$ and $\|p\|_y$ for $\|(y, p)\|_y$. Thus K is a real function on $B \times T_xX \times T_xX$. Perhaps we ought to emphasize that B already has a Riemannian metric induced from the inner product $\langle \ \rangle_x$ on T_xX. We denote this by $\langle \ \rangle_x$ as well; thus $\langle (y, p), (y, q) \rangle_x = \langle p, q \rangle_x$. The associated Finsler is denoted by $| \ |_x$.

Now T_xX may be identified with $\mathbf{E}$ (for example, by the isomorphism $T_x\xi$ for any chart ξ at x) and so we may think of h as being a chart. (In the literature h is usually called a *normal* chart at x.) Thus the Riemannian spray $\tau : B \times \mathbf{E} \to (B \times \mathbf{E}) \times (\mathbf{E} \times \mathbf{E})$ is defined by the formulae for the local representative of the Riemannian spray σ on X (see (A.60)). We know that solution curves of τ come under h from solution curves of σ, and thus are curves of the form $\delta_p(t) = tp$. Differentiating each curve δ_p gives an integral curve of τ of the form $\gamma_p(t) = (tp, p)$. Differentiating again, we deduce that, for all $p \in B$ and for all $t \in \mathbf{R}$ such that $tp \in B$, $\tau(tp, p) = (tp, p, p, 0)$. Thus, by (A.60).

(A.63) $(\tfrac{1}{2}TK(tp, p, p, u, 0, 0) - TK(tp, p, u, p, 0, 0))_2 = 0$

for all $u \in \mathbf{E}$, and in particular $(TK(tp, p, p, p, 0, 0))_2 = 0$. This says that $\|p\|_z$ is constant for z on the line joining 0_x and p, and hence that $\|p\|_{tp} = \|p\|_{0_x} = |p|_x$. It is now clear that the line segment γ joining 0_x to y has length $|y|_x$ with respect to $\| \ \|$. Note that we have, in fact, shown that the curve $t \mapsto \exp ty$ in X has constant speed $|y|_x$ at every point.

The above property of $\|p\|_x$ holds, more generally, for $\langle\!\langle p, q \rangle\!\rangle_x$, for all $q \in \mathbf{E}$. We assume this for the time being:

(A.64) Lemma. *For all $p, q \in \mathbf{E}$ and for all $t \in \mathbf{R}$ with $tp \in B$,*

$$K(tp, p, q) = \langle p, q \rangle_x,$$

and complete the proof of the theorem. Let $\delta : [a, b] \to B$ be any piecewise C^1 curve joining 0_x to y. We wish to show that δ has length $\geqslant |y|_x$. We may assume that $\delta(t) \neq 0_x$ for $t > a$, for otherwise we may shorten δ. Then, for all but finitely many values of t, the Schwarz inequality gives

$$\|\delta'(t)\|_{\delta(t)} \geqslant |\langle\!\langle \delta(t), \delta'(t) \rangle\!\rangle_{\delta(t)}| / \|\delta(t)\|_{\delta(t)}$$

and by Lemma A.64 the right-hand side equals $|\langle \delta(t), \delta'(t) \rangle_x| / |\delta(t)|_x$. Now consider the curve $\varepsilon : [a, b] \to B$ defined by $\varepsilon(t) = |\delta(t)|_x y / |y|_x$. This parametrizes the line segment $[0_x, y]$, possibly covering it more than once in places, and so its length is at least $|y|_x$. But the formula for its length is $\int_a^b \|\varepsilon'(t)\|_{\varepsilon(t)}\, dt$, which reduces to $\int_a^b (|\langle \delta(t), \delta'(t)\rangle_x|/|\delta(t)|_x)\, dt$. Thus the length of δ is $\geq$ the length of ε.   □

Proof of Lemma A.64. We first show that, for all $u \in \mathbf{E}$ with $\langle p, u\rangle_x = 0$, $K(tp, p, u) = 0$. We may assume that p and u have unit length with respect to $|\ |_x$, so that, for all $s \in \mathbf{R}$, $tp + su$ has length $\sqrt{t^2 + s^2}$. For all $t > 0$, we differentiate the relation

$$K(tp + su, (tp + su)/\sqrt{t^2 + s^2}, (tp + su)/\sqrt{t^2 + s^2}) = 1$$

with respect to s at $s = 0$, and obtain

$$(TK(tp, p, p, u, u/t, u/t))_2 = 0$$

or, equivalently,

$$(TK(tp, p, p, u, 0, 0))_2 + (2/t)K(tp, p, u) = 0,$$

using the symmetry and bilinearity of $K(y, ., .)$. But, by (A.63),

$$(TK(tp, p, p, u, 0, 0))_2 = 2(TK(tp, p, u, p, 0, 0))_2.$$

Thus if the real function ψ is defined by $\psi(t) = K(tp, p, u)$ then it satisfies the differential equation $t\psi'(t) = -\psi(t)$ for $t > 0$. Hence $t\psi(t) =$ constant. Since ψ is continuous and $\psi(0) = 0$, the constant is zero, and hence $K(tp, p, u) = 0$, as required. Finally, for all $q \in \mathbf{E}$,

$$\langle p, q - \langle p, q\rangle_x p/\langle p, p\rangle_x\rangle_x = 0$$

and hence

$$\begin{aligned}
0 &= K(tp, p, q - \langle p, q\rangle_x p/\langle p, p\rangle_x)\\
&= K(tp, p, q) - (\langle p, q\rangle_x/\langle p, p\rangle_x)K(tp, p, p)\\
&= K(tp, p, q) - \langle p, q\rangle_x.
\end{aligned}$$

          □

Map Spaces

In proving theorems about dynamical systems in the text of the book we have often applied the contraction mapping theorem to map spaces. Typically, we have had a collection of contractions indexed by some parameter. We have deduced smoothness results for the original dynamical systems by showing that the fixed point of the contraction depends smoothly on the parameter. This technique requires familiarity with the differential calculus in map spaces, and we outline some of the theory below. The proofs are rather straightforward and uninteresting, the main problem being how to organize the results in the least indigestible way. We give one approach; the reader may prefer a more economical treatment that appears in Franks [1].

The basic theorem is to the effect that if $g: Y \to Z$ is a C^{r+s} map then the map $g_* : C^r(X, Y) \to C^r(X, Z)$ defined by $g_*(f) = gf$ is C^s, where $C^r(X, Y)$ is the space of C^r maps from X to Y. For more intricate applications we need results of the form that the map comp$: C^r(X, Y) \times C^{r+s}(Y, Z) \to C^r(X, Z)$ sending (f, g) to gf is C^s. Unfortunately, it does not seem to be practicable to develop the theory in a sufficiently general context to cover all applications simultaneously. One needs, in the proofs, a certain element of uniformity in some of the maps concerned, and this can be introduced in various ways. For example, one can make certain spaces compact, or others finite dimensional, or one can restrict one's attention to spaces of uniform maps. Trying to cope with all these tactics simultaneously would further complicate what is already a not particularly appealing piece of theory. We prefer to stick to one approach in our exposition and to indicate possible modifications, some of which are actually applied in the text.

We first deal with maps between Banach spaces, and later come on to the more general theory of sections of vector bundles. In the final section we define topologies for spaces of dynamical systems on a compact manifold. As we have previously commented, these spaces can actually be given Banach manifold structures, but we do not do so here.

Throughout **E**, **F**, **G**, etc. are real Banach spaces, with norm written $|\ |$.

I. SPACES OF SMOOTH MAPS

Let X be any set. The set $C_b(X, \mathbf{F})$ of all bounded maps from X to $\mathbf{F}$ inherits a vector space structure from $\mathbf{F}$. That is to say, we define $f + g$ and αf by

$$(f + g)(x) = f(x) + g(x), \qquad (\alpha f)(x) = \alpha f(x)$$

where $f, g \in C_b(X, \mathbf{F})$, $x \in X$ and $\alpha \in \mathbf{R}$. We define a norm $| \ |_0$ on $C_b(X, F)$ by

$$|f|_0 = \sup \{|f(x)|: x \in X\},$$

and this makes $C_b(X, \mathbf{F})$ into a Banach space. If X is a topological space, the subspace $C^0(X, \mathbf{F})$ of all bounded continuous maps is closed in $C_b(X, \mathbf{F})$ and hence is itself a Banach space.

The vector space $L(\mathbf{E}, \mathbf{F})$ of all continuous linear maps from $\mathbf{E}$ to $\mathbf{F}$ has a Banach space structure with norm $| \ |$ defined by

$$|T| = \sup \{|T(x)|: |x| \leq 1\} = \sup \{|T(x)|/|x|: x \neq 0\}.$$

If X is an open subset of $\mathbf{E}$, we say that $f : X \to \mathbf{F}$ is (*Frechet*) *differentiable* at $x \in X$ if, for some map $T \in L(\mathbf{E}, \mathbf{F})$

$$|f(x + h) - f(x) - T(h)| = o(|h|)$$

as $h \to 0$. If T exists it is unique, and we call it $Df(x)$, the *differential* of f at x. If $Df(x)$ exists for all $x \in X$, we say that f is *differentiable*. The map $Df : X \to L(\mathbf{E}, \mathbf{F})$ is called the *derivative* of f. Higher derivatives are defined inductively by $D^r f = D(D^{r-1} f)$. We say that f is C^0 if f is continuous, C^r if $D^r f$ is continuous and C^∞ if f is C^r for all $r \geq 0$. Strictly speaking $D^r f$ is a map from X to $L(\mathbf{E}, L(\mathbf{E}, \ldots L(\mathbf{E}, \mathbf{F}) \ldots))$, but, as usual, we identify the latter space with the space $L_r(\mathbf{E}, \mathbf{F})$ of all r-linear maps from $\mathbf{E}^r$ to $\mathbf{F}$, putting $S = T$ when $S(x_1)(x_2) \ldots (x_r) = T(x_1, x_2, \ldots, x_r)$ for all $(x_1, x_2, \ldots, x_r) \in \mathbf{E}^r$. Thus $L_r(\mathbf{E}, \mathbf{F})$ has norm $| \ |$ given by

$$|T| = \sup \{|T(x_1, \ldots, x_r)|: |x_i| \leq 1, 1 \leq i \leq r\}.$$

If f is C^r, $D^r f(x)$ is a symmetric r-linear map. Note that continuous multilinear maps are themselves C^∞.

If X is any subset of $\mathbf{E}$ with $X \subset \overline{\text{int } X}$, we shall say that $f : X \to \mathbf{F}$ is C^r ($0 \leq r \leq \infty$) if it has a C^r extension $\bar{f}$ to some open neighbourhood of X in $\mathbf{E}$. In this case the differentials $D^i f(x)$, $1 \leq i \leq r$ at points x of the frontier ∂X of X are independent of the choice of the extension $\bar{f}$, and we define $D^i f(x)$ to be $D^i \bar{f}(x)$ for such x. *We make the standing assumption that whenever a map* $f : X \to Y$ *is said or implied to be differentiable, its domain X always satisfies* $X \subset \overline{\text{int } X}$.

If X is a subset of $\mathbf{R}$, one commonly writes $Df(x)$ or $f'(x)$ instead of $Df(x)(1)$. The context decides whether $Df(x)$ is an element of $\mathbf{F}$ or $L(\mathbf{R}, \mathbf{F})$.

This remark also applies to partial derivatives $D_j f(x)(=\partial f/\partial x_i)$ when $\mathbf{E}$ is a product of Banach spaces $\mathbf{E}_1 \times \cdots \times \mathbf{E}_n$ and the ith factor $\mathbf{E}_i$ is $\mathbf{R}$. Recall that the jth *partial derivative* of f at $a = (a_1, \ldots, a_n) \in X$ is defined to be the differential of the partial map

$$x_j \mapsto f(a_1, \ldots, a_{j-1}, x_j, a_{j+1}, \ldots, a_n)$$

at a_j. It is an element of $L(\mathbf{E}_j, \mathbf{F})$.

The C^r map $f : X \to \mathbf{F}$ is C^r-*bounded* if the number

$$|f|_r = \sup \{|D^i f(x)| : x \in X, 0 \leq i \leq r\}$$

exists and is finite.

(B.1) Exercise. Prove that if $f : X \to \mathbf{F}$ and $g : \mathbf{F} \to \mathbf{G}$ are maps with Df and Dg C^r-bounded then $D(gf)$ is C^r-bounded. Prove that if, in addition g is C^0-bounded then gf is C^{r+1}-bounded.

The set $C^r(X, \mathbf{F})$ of all C^r-bounded maps from X to $\mathbf{F}$ has the vector space structure of $C^0(X, \mathbf{F})$, and we give it the norm $| \ |_r$. In some cases, $C^r(X, \mathbf{F})$ inherits completeness from $\mathbf{F}$; we need to know this in the following cases:

(B.2) Exercise. Prove that if X is open in E then $C^r(X, \mathbf{F})$ is a Banach space. Prove that if I is a compact real interval then $C^1(I, \mathbf{F})$ is a Banach space. (This is basically a well known theorem on uniform convergence; see, for example (8.6.4) of Dieudonne [1]).

When Y is a subset of $\mathbf{F}$, $C^r(X, Y)$ is (identified with) the subset of $C^r(X, \mathbf{F})$ consisting of maps taking values in Y. If Y is closed in $\mathbf{F}$ then $C^r(X, Y)$ is closed in $C^r(X, \mathbf{F})$. If X is compact and Y is open in F, then $C^r(X, Y)$ is open in $C^r(X, \mathbf{F})$. If $X = Y$, we abbreviate $C^r(X, X)$ to $C^r(X)$.

If Z is a subset of X, we say that $f : X \to \mathbf{F}$ is *uniformly* C^r at Z $(r \geq 0)$ if given $\varepsilon > 0$ there exists $\delta > 0$ such that, for all $x \in X$ and for all $z \in Z$ with $|x - z| < \delta$, $\sup \{|D^i f(x) - D^i f(z)| : 0 \leq i \leq r\} < \varepsilon$. We say that f is uniformly C^r if it is uniformly C^r at X. We denote by $UC^r(X, \mathbf{F})$ the closed subspace of $C^r(X, \mathbf{F})$ consisting of all uniformly C^r maps.

(B.3) Exercise. Let $f : X \to Y$ and $g : Y \to \mathbf{G}$ be uniformly C^r maps. Prove that if Df and Dg are C^{r-1}-bounded then the composite gf is uniformly C^r.

II. COMPOSITION THEOREMS

Most of the following theorems are proved by induction. When the integer inducted upon is the degree of smoothness r of the map space $C^r(,)$ concerned, the inductive step always depends upon the trivial relation

$|f|_{k+1} = \max \{|f|_k, |Df|_k\}$ for $0 \leqslant k < r$. Since these proofs run to a pattern, we tend to cut down on the detail. Throughout Y is a subspace of $\mathbf{F}$, and X is a topological space if $r = 0$ and a subset of $\mathbf{E}$ if $r > 0$. The first two lemmas are special cases of a general result about left composition with a continuous multilinear map.

(B.4) Lemma. *If $T \in L(\mathbf{F}, \mathbf{G})$ then, for all $f \in C^r(X, \mathbf{F})$, $|Tf|_r \leqslant |T||f|_r$. Thus $T_*(f) = Tf$ defines a continuous linear map $T_* : C^r(X, \mathbf{F}) \to C^r(X, \mathbf{G})$.*

Proof. By induction on r.

$r = 0$. $|Tf|_0 \leqslant |T|\|f|_0$, trivially.

Inductive step. Assume the inequality holds for $r = k$. Let $\tau : L(\mathbf{E}, \mathbf{F}) \to L(\mathbf{E}, \mathbf{G})$ be the continuous linear map $S \mapsto TS$. Then $D(Tf) = \tau Df$, and so, by hypothesis, $|D(Tf)|_k \leqslant |\tau||Df|_k = |T||Df|_k$. Combined with $|Tf|_k \leqslant |T|\|f|_k$, this gives $|Tf|_{k+1} \leqslant |T|\|f|_{k+1}$.   □

(B.5) Lemma. *If $B : \mathbf{F} \times \mathbf{G} \mapsto \mathbf{H}$ is continuous bilinear, then for all $f \in C^r(X, \mathbf{F})$ and $g \in C^r(X, \mathbf{G})$ the map $B_*(f, g) \in C^r(X, \mathbf{H})$ defined by $B_*(f, g)(x) = B(f(x), g(x))$ satisfies*

$$|B_*(f, g)|_r \leqslant 2^r |B|\|f|_r|g|_r.$$

Thus $B_* : C^r(X, \mathbf{F}) \times C^r(X, \mathbf{G}) \to C^r(X, \mathbf{H})$ is continuous bilinear.

Proof. By induction on r.

Inductive step. This uses $|D(B_*(f, g)|_k = |\beta_*(f, Dg) + \gamma_*(Df, g)|_k$ where $\beta : \mathbf{F} \times L(\mathbf{E}, \mathbf{G}) \to L(\mathbf{E}, \mathbf{H})$ and $\gamma : L(\mathbf{E}, \mathbf{F}) \times \mathbf{G} \to L(\mathbf{E}, \mathbf{H})$ are the continuous bilinear maps $\beta(y, T) = (x \mapsto B(y, T(x)))$ and $\gamma(S, z) = (x \mapsto B(S(x), z))$. Note that $|\beta| = |\gamma| = |B|$.   □

We are particularly concerned with the case when B is the composition map $B : L(\mathbf{F}, \mathbf{G}) \times L(\mathbf{G}, \mathbf{H}) \to L(\mathbf{F}, \mathbf{H})$ taking (S, T) to TS. Note that $|B| = 1$. In this case we simplify notation by writing $g \cdot f$ for the so-called *compositional product* $B_*(f, g)$ of $f \in C^r(X, L(\mathbf{F}, \mathbf{G}))$ and $g \in C^r(X, L(\mathbf{G}, \mathbf{H}))$. Thus Lemma B.5 becomes

(B.6) $|g \cdot f|_r \leqslant 2^r |f|_r|g|_r.$

The space $L(\mathbf{R}, \mathbf{G})$ is commonly identified with $\mathbf{G}$, equating maps with their value at 1, and in this case B becomes the *evaluation map* $(x, g) \mapsto g(x)$.

(B.7) Lemma. *For all $r \geqslant 0$, there is a constant A (independent of X, Y, $\mathbf{G}$) such that, for all $f : X \to Y$ with Df C^{r-1}-bounded and for all $g \in C^r(Y, \mathbf{G})$,*

$$|gf|_r \leqslant A|g|_r M_r(f),$$

where $M_r f = \max \{1, (|Df|_{r-1})^r\}$.

Proof. By induction on r.

$r = 0$. $|gf|_0 \leqslant |g|_0$.

Inductive step. $|D(gf)|_k = |(Dg)f \cdot Df|_k \leqslant 2^k |(Dg)f|_k |Df|_k$ by (B.6).   □

If X is compact and $g: Y \to \mathbf{G}$ is C^r then g induces the *left composition map* $g_*: C^r(X, Y) \to C^r(X, \mathbf{G})$ defined by $g_*(f) = gf$. This map may also exist in other circumstances, for example if g is C^r-bounded (by Lemma B.7).

(B.8) Lemma. *Let X be compact and $g: Y \to \mathbf{G}$ be C^r. Then $g_*: C^r(X, Y) \to C^r(X, \mathbf{G})$ is continuous. If g is uniformly continuous then $g_*: C^0(X, Y) \to C^0(X, \mathbf{G})$ is uniformly continuous.*

Proof. By induction on r.

$r = 0$. Let $f_0 \in C^0(X, Y)$. Then g is uniformly continuous at the compact subset $f_0(X)$. Thus $|g_*(f) - g_*(f_0)|_0$ is small for f C^0-near f_0. Moreover if g is uniformly continuous, $|g_*(f) - g_*(f_0)|_0$ is uniformly small (as f_0 varies).

Inductive step. For all $f, f_0 \in C^{k+1}(X, Y)$, where $0 \leqslant k \leqslant r$,

$$|D(g_*(f) - g_*(f_0))|_k = |(Dg)_*(f) \cdot Df - (Dg)_*(f_0) \cdot Df_0|_k$$
$$\leqslant 2^k (|(Dg)_*(f) - (Dg)_*(f_0)|_k |Df|_k$$
$$+ |(Dg)_*(f_0)|_k |Df - Df_0|_k).   □$$

(B.9) Exercise. Prove that if g is uniformly C^r and Dg is C^{r-1}-bounded then $g_*: C^r(X, Y) \to C^r(X, \mathbf{G})$ exists and is uniformly continuous at subsets $\mathscr{F}$ of $C^r(X, Y)$ such that $\sup\{|Df|_{r-1} : f \in \mathscr{F}\} < \infty$.

We now come to a result dealing with smoothness of the map g_*. In the applications the domain $C^r(X, Y)$ of g_* is usually open (e.g. if $Y = \mathbf{F}$, or if X is compact and Y open). However, occasionally it is not. In order to have $C^r(X, Y) \subset \overline{\text{int } C^r(X, Y)}$ (to fit in with our standing assumption) we need some other set of conditions on X and Y. The only one that need concern us here is X compact and Y a closed ball in $\mathbf{F}$.

(B.10) Theorem. *Let X be compact, Y be open or a closed ball, and $g: Y \to \mathbf{G}$ be C^{r+s}. Then $g_*: C^r(X, Y) \to C^r(X, \mathbf{G})$ is C^s, with Dg_* given, for all $f \in C^r(X, Y)$ and $\eta \in C^r(X, \mathbf{F})$ by*

$$Dg_*(f)(\eta) = (Dg)f \cdot \eta.$$

If g is uniformly C^s, and $r = 0$, then g_* is uniformly C^s.

Proof. We first prove that if $s \geqslant 1$ g_* is differentiable, by induction on r. We may assume that Y is open (otherwise g has a C^{r+1} extension to an open neighbourhood Y' of Y, inducing an extension of g_* to $C^r(X, Y')$).

$r = 0$. Let $f_0 \in C^0(X, Y)$. For all sufficiently C^0-small $\eta \in C^0(X, \mathbf{F})$ (i.e. such that, for all $x \in X$, the line segment $L_x = [f_0(x), f_0(x) + \eta(x)]$ is in Y),

$$|g_*(f_0 + \eta) - g_*(f_0) - (Dg)f_0 \cdot \eta|_0$$
$$= \sup \{|g(f_0(x) + \eta(x)) - gf_0(x) - D_g(f_0(x))(\eta(x))| : x \in X\}$$
$$\leqslant |\eta|_0 \sup \{|Dg(y) - Dg(f_0(x))| : y \in L_x, x \in X\}$$

by the mean value theorem (Corollary 2 in § 4 of Chapter 5 of Lang [2]). Since Dg is uniformly continuous at $f_0(X)$, the right-hand side is $0(|\eta|_0)$ as $|\eta|_0 \to 0$.

Inductive step. Let $f_0 \in C^{k+1}(X, Y)$, for $0 \leqslant k < r$. For all C^{k+1}-small $\eta \in C^{k+1}(X, \mathbf{F})$,

$$|D(g_*(f_0 + \eta) - g_*(f_0) - (Dg)f_0 \cdot \eta)|_k$$
$$= |Dg(f_0 + \eta) \cdot (Df_0 + D\eta) - (Dg)f_0 \cdot Df_0$$
$$- (D^2g)f_0 \cdot \eta \cdot Df_0 - (Dg)f_0 \cdot D\eta|_k$$
$$\leqslant 2^k (|(Dg)_*(f_0 + \eta) - (Dg)_*(f_0) - (D^2g)f_0 \cdot \eta|_k |Df_0 + D\eta|_k$$
$$+ 2^{2k} |D^2g)f_0|_k |\eta|_k |D\eta|_k).$$

The right-hand side is $0(|\eta|_{k+1})$ as $|\eta|_{k+1} \to 0$.

The remaining results follow by induction on s. The case $s = 0$ is Lemma B.8. If the theorem holds for $s = k$ and g is C^{r+k+1} then Dg_* is clearly C^k since it is $(Dg)_*$ followed by the continuous linear map λ from $C^r(X, L(\mathbf{F}, \mathbf{G}))$ to $L(C^r(X, \mathbf{F}), C^r(X, \mathbf{G}))$ that takes ζ to $(\eta \mapsto \zeta \cdot \eta)$. Similarly for the uniformity result. □

(B.11) Corollary. *Given $r \geqslant 0$ and $s \geqslant 1$, there exists a constant A (independent of X, Y and $\mathbf{G}$) such that if X and Y are as above, $f \in C^r(X, Y)$ and g is c^{r+s} with Dg C^{r+s-1}-bounded, then*

$$|D^s g_*(f)| \leqslant A |Dg|_{r+s-1} M_r(f),$$

where, again, $M_r(f) = \max \{1, (|Df|_{r-1})^r\}$.

Proof. By induction on s.

$s = 1$. $|Dg_*(f)| = \sup \{|(Dg)f \cdot \eta|_r / |\eta|_r : \eta \neq 0 \in C^r(X, \mathbf{F})\}$

$$\leqslant 2^r B |Dg|_r M_r(f) \qquad \text{by Lemma B.7.}$$

Inductive step. $|D^{k+1} g_*(f)| = |D^k (\lambda (Dg)_*)(f)| \qquad (\lambda \text{ as in Theorem B.10})$

$$\leqslant |\lambda| |D^k (Dg)_*(f)|$$
$$\leqslant 2^k |D^k (Dg)_*(f)|. \qquad \square$$

(B.12) Corollary. *Let X and Y be as above. For all C^s maps g with Dg C^{s-1}-bound the map $g_*: C^0(X, Y) \to C^0(X, \mathbf{G})$ is C^s-bounded. Moreover the map from $C^s(Y, \mathbf{G})$ to $C^s(C^0(X, Y), C^0(X, \mathbf{G}))$ taking g to g_* is continuous linear.*   □

The above theory is sufficient for most applications. However, we now move on to smoothness of the map $\mathrm{comp}: C^r(X, Y) \times C^{r+s}(Y, \mathbf{G}) \to C^r(X, \mathbf{G})$ defined by $\mathrm{comp}(f, g) = gf$, and we approach this via its partial derivatives. We have already dealt with differentiability of the partial map g_*. We also have to consider *right composition* maps of the form $f^*: C^{r+s}(Y, \mathbf{G}) \to C^r(X, \mathbf{G})$, where $f \in C^r(X, Y)$ and $f^*(g) = gf$. In fact, such maps are defined for all C^r maps f such that Df is C^{r-1}-bounded, by Lemma B.7. Since they are trivially linear, Lemma B.7 also gives:

(B.13) Lemma. *The maps f^* are continuous linear.*   □

However, this remark is not sufficient for our purposes. We now know that $Df^*: C^{r+s}(Y, \mathbf{G}) \to L(C^{r+s}(Y, \mathbf{G}), C^r(X, \mathbf{G}))$ is the constant map with value $(\zeta \mapsto \zeta f)$. We need to know how smoothly this value depends on f. The necessary result is:

(B.14) Lemma. *Let X be compact and Y be open or a closed ball. Then for $s \geq 1$, the map $\theta: C^r(X, Y) \to L(C^{r+s}(Y, \mathbf{G}), C^r(X, \mathbf{G}))$ defined by $\theta(f) = (\zeta \mapsto \zeta f)$ is C^{s-1}.*

Proof. By induction on s. We may assume that Y is open in $\mathbf{F}$, and so $C^r(X, Y)$ is open.

$s = 1$. Let $f_0 \in C^r(X, Y)$. For all sufficiently C^r-small $\eta \in C^r(X, \mathbf{F})$,

$$|\theta(f_0 + \eta) - \theta(f_0)| = |\zeta \mapsto (\zeta(f_0 + \eta) - \zeta f_0)|$$

$$= \sup \{|\zeta_*(f_0 + \eta) - \zeta_*(f_0)|_r / |\zeta|_{r+1}: \zeta \neq 0 \in C^{r+1}(Y, \mathbf{G})\}$$

$$\leq \sup \{|D\zeta_*(f_0 + t\eta)||\eta|_r / |\zeta|_{r+1}: 0 \leq t \leq 1, \zeta \neq 0\}$$

we have by the mean value theorem applied to ζ_*. By Corollary B.11, $|D\zeta_*(f_0 + t\eta)| \leq A |\zeta|_{r+1} M_r(f_0 + t\eta)$, and since $M_r(f_0 + t\eta)$ is bounded for C^r-small η, $\theta(f_0 + \eta) \to \theta(f_0)$ as $\eta \to 0$.

Inductive step. We assume that the theorem holds for $s = k \geq 1$ and prove it for $s = k+1$. Let $f_0 \in C^r(X, Y)$ and $\eta \in C^r(X, \mathbf{F})$. We assert that θ is differentiable with derivative given by $D\theta(f_0)(\eta) = (\zeta \mapsto (D\zeta)f_0 \cdot \eta)$. This is because, for sufficiently C^r-small η,

$$|\theta(f_0 + \eta) - \theta(f_0) - (\zeta \mapsto (D\zeta)f_0 \cdot \eta)|$$

$$= |\zeta \mapsto (\zeta_*(f_0 + \eta) - \zeta_*(f_0) - D\zeta_*(f_0)(\eta))|$$

$$= \sup \{|\zeta_*(f_0+\eta) - \zeta_*(f_0) - D\zeta_*(f_0)(\eta)|_r / |\zeta|_{r+k+1} : \zeta \neq 0\}$$

$$\leq \sup \{|D\zeta_*(f_0+t\eta) - D\zeta_*(f_0)\|\eta|_r / |\zeta|_{r+k+1} : 0 \leq t \leq 1, \zeta \neq 0\}$$

$$\leq \sup \{|D^2\zeta_*(f_0+ut\eta)|t(|\eta|_r)^2 / |\zeta|_{r+k+1} : 0 \leq t \leq 1, 0 \leq u \leq 1, \zeta \neq 0\}$$

by two applications of the mean value theorem, and this is $o(|\eta|_r)$ as $|\eta|_r \to 0$, using Corollary B.11 again. Thus $D\theta$ is the composite of $\theta: C^r(X, Y) \to L((C^{r+k}(Y, L(\mathbf{F}, \mathbf{G})), C^r(X, L(\mathbf{F}, \mathbf{G}))$, which is C^{k-1} by hypothesis, and the continuous linear map ϕ from the target of θ to

$$L(C^r(X, \mathbf{F}), L(C^{r+k+1}(Y, \mathbf{G}), C^r(X, \mathbf{G})))$$

defined by

$$\phi(T) = (\eta \mapsto (\zeta \mapsto T(D\zeta) \cdot \eta)). \qquad \square$$

We can now prove:

(B.15) Theorem. *Let X be compact and Y be open or a closed ball. Then* comp: $C^r(X, Y) \times C^{r+s}(Y, \mathbf{G}) \to C^r(X, \mathbf{G})$ *is* C^s.

Proof. By induction on s.

$s = 0$. This uses the inequality

$$|\text{comp}(f, g) - \text{comp}(f_0, g_0)|_r \leq |(g - g_0)f|_r + |g_0 f - g_0 f_0|_r$$

$$\leq A|g - g_0|_r M_r(f) + |g_{0*}(f) - g_{0*}(f_0)|$$

and Lemma B.8.

Induction step. We know that for $s = k + 1 \geq 1$ comp has partial derivatives given by

$$D_1 \text{comp}(f, g) = (\eta \mapsto (Dg)f \cdot \eta) \in L(C^r(X, \mathbf{F}), C^r(X, \mathbf{G}))$$

$$D_2 \text{comp}(f, g) = (\zeta \mapsto \zeta f) \in L(C^{r+s}(Y, \mathbf{G}), C^r(X, \mathbf{G})).$$

Thus D_2 Comp is C^k by Lemma B.14. We break down D_1 Comp as a composite of three maps

$$(f, g) \overset{1}{\mapsto} (f, Dg) \overset{2}{\mapsto} (Dg)f \overset{3}{\mapsto} (\eta \mapsto (Dg)f \cdot \eta).$$

The first and third are continuous linear, and the second is comp: $C^r(X, Y) \times C^{r+k}(Y, L(\mathbf{F}, \mathbf{G}) \to C^r(X, L(\mathbf{F}, \mathbf{G}))$, which is C^k by inductive hypothesis. $\qquad \square$

By putting $X = $ a singleton $\{x\}$ and identifying $C^0(X, Y)$ with $Y(f$ with $f(x))$ we have:

(B.16) Corollary. *If Y is open or a closed ball, then the evaluation map* ev: $Y \times C^s(Y, \mathbf{G}) \to \mathbf{G}$ *defined by* $ev(y, g) = g(y)$ *is* C^2.

We also need the following result, obtained similarly from Lemma B.14.

(B.17) Corollary. *If Y is open or a closed ball, the map $ev^{\cdot}$ from Y to $L(C^s(Y, \mathbf{G})), \mathbf{G})$ defined by $ev^{\cdot}(y) = ev^y = (g \mapsto g(y))$ is C^{s-1}.* $\square$

This is as far as we wish to take the theory with X compact. Let us just recall how that hypothesis was used. Firstly, we needed it in the case $s \geq 1$, together with Y open or a closed ball, in order to ensure $C^r(X, Y) \subset$ int $C^r(X, Y)$. Secondly, we used it in Lemma B.8 and Theorem B.10 to infer that the continuous maps g and Dg were uniformly continuous at $f_0(X)$. How can we dispense with compactness of X? If $s \geq 1$, we can put $Y = \mathbf{F}$ so that $C^r(X, Y)$ becomes the whole space $C^r(X, \mathbf{F})$. If $\mathbf{F}$ is finite dimensional, we can enclose $f_0(X)$ in a compact set K (since f_0 is C^0-bounded) and the proof proceeds as before. Actually, if $r > 0$, we also need $\mathbf{E}$ finite dimensional, so that $L(\mathbf{E}, \mathbf{F})$ is finite dimensional, or the induction will not work. We leave it to the reader to reformulate the main theorems in this finite dimensional case.

Another possible course of action is to build in the uniformity by restricting attention to maps g that, together with their derivatives, are uniformly continuous. We have used theorems of this sort in the text, so we state them below, leaving proofs to the reader. We are able to consider a slightly more general situation than before. Let $f_0 : X \to \mathbf{F}$ be a fixed C^r map with C^{r-1}-bounded derivative Df_0.

(B.18) Theorem. *Let $g \in UC^{r+s}(\mathbf{F}, \mathbf{G})$. The map from $C^r(X, \mathbf{F})$ to $C^r(X, \mathbf{G})$ taking f to $g(f + f_0)$ is C^s (and uniformly continuous if $s = 0$). For $s > 0$ its differential at f is $(\eta \mapsto Dg(f + f_0) \cdot \eta)$.* $\square$

(B.19) Theorem. *The map from $C^r(X, \mathbf{F}) \times UC^{r+s}(\mathbf{F}, \mathbf{G})$ to $C^r(X, \mathbf{G})$ taking (f, g) to $g(f + f_0)$ is C^s.* $\square$

In the text, in the course of proving the stable manifold theorem, we made use of spaces of sequences. These may be interpreted as map spaces, and so the above theory may be invoked. Perhaps we should be more explicit. Let $\mathbf{N} = \{0, 1, 2, 3, \ldots\}$ and let $Y^{\mathbf{N}}$ be the set of all sequences in Y. As before Y is a subset of the Banach space $\mathbf{F}$, so $Y^{\mathbf{N}}$ is a subset of the vector space $\mathbf{F}^{\mathbf{N}}$. We denote by $\mathscr{B}(\mathbf{F})$ the Banach space $C^0(\mathbf{N}, \mathbf{F})$ of bounded sequences in $\mathbf{F}$, and by $\mathscr{S}(\mathbf{F})$ the closed subspace consisting of convergent sequences. Notice that $\mathscr{S}(\mathbf{F})$ may be identified with $C^0(\bar{\mathbf{N}}, \mathbf{F})$, where $\bar{\mathbf{N}}$ is the one point compactification of $\mathbf{N}$. Let $\mathscr{B}(Y) = C^0(\mathbf{N}, Y)$ and $\mathscr{S}(Y) = C^0(\bar{\mathbf{N}}, Y)$.

Any bounded map (or any Lipschitz map) $g : Y \to \mathbf{G}$ induces a map $g_* : \mathscr{B}(Y) \to \mathscr{B}(\mathbf{G})$ taking γ to $g\gamma$. Similarly, any continuous map $g : Y \to \mathbf{G}$ induces $g_* : \mathscr{S}(Y) \to \mathscr{S}(\mathbf{G})$. We obtain results about these spaces by putting $r = 0$, $X = \bar{\mathbf{N}}$ or, when possible, $\mathbf{N}$ in the above composition theorems. In the text, it was convenient to use $\mathscr{S}_0(\mathbf{G})$, which is the closed subspace of $\mathscr{S}(\mathbf{G})$

consisting of series converging to 0 in **G**. The relevant theorems are:

(B.20) Theorem. *If Y is open or a closed ball and if $g: Y \to \mathbf{G}$ is C^s, with $g(0) = 0$, then $g_*: \mathscr{S}_0(Y) \to \mathscr{S}_0(\mathbf{G})$ is C^s. If, further, g is uniformly C^s then so is g_*.* □

(B.21) Theorem. *If Y is open or a closed ball then the map* comp *from $\mathscr{S}_0(Y) \times C_0^s(Y, \mathbf{G})$ to $\mathscr{S}_0(\mathbf{G})$ is C^s, where* comp $(f, g) = gf$ *and $C_0^s(Y, \mathbf{G})$ is the subspace of $C^s(Y, \mathbf{G})$ consisting of maps g with $g(0) = 0$.* □

Finally in Exercise 6.11 of the text, and its applications to hyperbolic closed orbits, we need the following generalization of the previous pair of theorems:

(B.22) Exercise. Let α be a positive real number, and let $\tilde{\alpha}: \mathbf{F}^{\mathbf{N}} \to \mathbf{F}^{\mathbf{N}}$ be the isomorphism taking γ to $(n \mapsto \alpha^n \gamma(n))$. Let $_\alpha| \ |$ denote the norm induced by $\tilde{\alpha}$ on $\tilde{\alpha}(\mathscr{B}(\mathbf{F}))$ from that of $\mathscr{B}(\mathbf{F})$ (so $_\alpha|\delta| = \sup \{\alpha^{-n}|\delta(n)|: n \in \mathbf{N}\}$). Suppose that Y is a ball with centre 0 in $\mathbf{F}$ and that $g: Y \to \mathbf{G}$ is a Lipschitz map with $g(0) = 0$. Prove that g induces a Lipschitz map $g_*: \tilde{\alpha}(\mathscr{S}_0(Y)) \to \tilde{\alpha}(\mathscr{S}_0(\mathbf{G}))$ (defined by $g_*(\gamma) = g\gamma$).

Now suppose that $\alpha \leq 1$. Show that g_* is C^s when g is C^s $(s \geq 1)$ and uniformly C^s when g is uniformly C^s. Prove also that the map comp: $\tilde{\alpha}(\mathscr{S}_0(Y)) \times C^s(Y, \mathbf{G}) \to \tilde{\alpha}(\mathscr{S}_0(\mathbf{G}))$ is C^s. Investigate the situation when α is greater than 1.

III. SPACES OF SECTIONS

Let $\pi: B \to X$ be a C^r vector bundle with fibre $\mathbf{F}$. *We suppose throughout for simplicity that X is compact.* Of course for $r > 0$ X is a C^r manifold. Recall (Appendix A) that $C^r(\pi)$ (or, sometimes, $C^r(B)$) denotes the vector space of all C^r sections of π. We wish to give $C^r(\pi)$ a norm. We say that an admissible chart $\tilde{\xi}: \tilde{U} \to U' \times \mathbf{F}$ on π is a *norm chart* if it is the restriction of an admissible chart $\tilde{\eta}: \tilde{V} \to V' \times \mathbf{F}$ where $U' \subset K \subset V'$ for some compact K. A *norm atlas* is a finite atlas of norm charts. Let $\mathscr{A} = \{\tilde{\xi}_i: 1 \leq i \leq n\}$ be such an atlas for π, and let σ be a C^r section of π. Then, for each i, we have a local representative $(id, \sigma_i): U_i' \to U_i' \times \mathbf{F}$ for σ. The definition of norm chart ensures that $\sigma_i: U_i' \to \mathbf{F}$ is C^r-bounded. That is to say, $|\sigma_i|_r$ exists, as defined in the last section. We define a norm $| \ |_{\mathscr{A},r}$ on $C^r(\pi)$ by

$$|\sigma|_{\mathscr{A},r} = \max \{|\sigma_i|_r: 1 \leq i \leq n\}.$$

It is not hard to prove that, with this norm, $C^r(\pi)$ is a Banach space.

Moreover if $\mathscr{B}$ is another norm atlas for π then $|\ |_{\mathscr{A},r}$ and $|\ |_{\mathscr{B},r}$ are equivalent norms. We suppose from now on that some atlas $\mathscr{A}$ has been chosen, and we abbreviate $|\ |_{\mathscr{A},r}$ to $|\ |_r$.

(B.23) Exercise. Show that norm atlases exist, and fill in the details of the previous paragraph.

Notice that there is a canonical norm-preserving isomorphism between $C^r(U_i', \mathbf{F})$ and $C^r(\pi_i)$, where $\pi_i : U_i' \times \mathbf{F} \to U_i'$ is the trivial bundle. This points to the way in which spaces of sections generalize the map spaces of the preceding sections. The generalization is a substantial one for, whereas above we composed elements of the map space with C^{r+s} maps of $\mathbf{F}$, we now compose with maps of the total space of the vector bundle satisfying a weaker differentiability condition. Clearly we have to introduce some new condition, because it does not usually make sense to talk of a C^{r+s} map of the total space of a C^r vector bundle. The condition that we introduce occurs naturally because of the fundamental distinction between base and fibre coordinates in vector bundles.

Let $\pi : B \to X$ and $\rho : C \to X$ be C^r vector bundles, and let $g : B \to C$ be a fibre preserving map (over the identity). Thus g maps each fibre B_x to C_x, not necessarily linearly. Recall (Example A.27 of Appendix A) that the linear map bundle $L(\pi, \rho)$ is a vector bundle with base space X and fibre $L(B_x, C_x)$ over x. Its total space is denoted by $L(B, C)$. We define inductively $L^n(\pi, \rho) = L(\pi, (L^{n-1}(\pi, \rho)))$. Suppose that, for all $x \in X$, the restriction $g_x : B_x \to C_x$ of g is n times differentiable. We say that g is n *times fibre differentiable* and define its nth *fibre derivative* $F^n g : B \to L^n(B, C)$ by

$$F^n g(v) = D^n g_x(v)$$

where $v \in B_x$. We say that g is of class $^r F^s$ if $F^n g$ exists and is C^r for $0 \leqslant n \leqslant s$. We write F^s for $^0 F^s$. Thus $^r F^s$ is a stronger condition than C^r but weaker than C^{r+s}. In terms of local coordinates, $^r F^s$ means that all partial derivatives up to order s in the fibre direction exist and are C^r (as functions of all coordinates together, both fibre and base). We may also use this local criterion to define what we mean by a map $g : B \to Y$ being $^r F^s$, where B is the total space of a vector bundle and Y is any smooth manifold.

In the text we used the following results, which may be given proofs closely resembling those of analogous theorems in the last section (see Eliasson [1] and Foster [1] for details).

(B.24) Theorem. *Let $\pi : B \to X$ and $\rho : C \to X$ be C^r vector bundles and let $g : B \to C$ be $^r F^s$. Then $g_* : C^r(\pi) \to C^r(\rho)$ defined by $g_*(\sigma) = g\sigma$ is C^s. For*

$s \geqslant 1$ *its derivative* Dg_* *is defined by*

(B.25) $Dg_*(\sigma)(\tau) = (Fg)\sigma . \tau$

for all σ and τ in $C'(\pi)$, where, as in the last section, . denotes compositional product.   □

(B.26) Remarks. The above theorem may be modified in various ways. For example, the domain of g may be an open neighbourhood N of the image of some given section of π, rather than the whole of B. Of course g_* is then only defined on sections taking values in N. Again, we may (and in fact, in the proof of the generalized stable manifold theorem, do) replace $C^0(\pi)$ by $C_b(\pi)$, the space of bounded but not necessarily continuous sections. We may in this case prove that $g F^s$ implies $g_* C^s$, with Dg_* as in (B.25), provided that the fibres of π and σ are finite dimensional (this condition being required for uniformity arguments in the proof). Finally, if one gives the space of $'F^s$ maps from B to C a natural Banach space structure, one may prove an analogue of Theorem B.15 above (see Theorem 15 of Foster [1]).

Theorem B.24 has a partial converse. This is very useful when, as in the proof of the generalized stable manifold theorem, one attempts to establish results about a subset of a manifold by applying differential calculus to the space of sections of the tangent bundle of the manifold over the subset. In this situation, one needs a tool for transferring results from the space of sections back down to the manifold again, and the following theorem (due to Foster [2]) serves this purpose:

(B.27) Theorem. *Let $\pi : B \to X$ and $\rho : C \to X$ be C^0 vector bundles and let $g : B \to C$ be a fibre preserving map. If $g_* : C^0(\pi) \to C^0(\rho)$ is C^s then g is C^s.*

Before proving this theorem, we establish two preliminary results. The first tells us that small open sets in the total space of a vector bundle may be embedded nicely in the space of sections.

(B.28) Lemma. *Let $\pi : B \to X$ be a C^0 vector bundle, and let $x_0 \in X$. There exists an open neighbourhood U of x_0 in X and a continuous map $\sigma . \tilde{U} \to C^0(\pi)$ (where $\tilde{U} = \pi^{-1}(U)$) with value at p denoted σ_p such that*

(i) *for all $x \in U$, $\sigma . |B_x$ is continuous linear,*

(ii) *for all $p \in \tilde{U}$, σ_p takes the value p at $\pi(p)$,*

(iii) *for all $p \in \tilde{U}, |\sigma_p|_0 = |p|$ where $|$ $|$ is the norm induced on the fibres of $\tilde{U}$ by some chart of $\mathscr{A}$.*

Proof. Let $\tilde{\xi} : \tilde{W} \to W' \times \mathbf{F}$ be a norm chart in $\mathscr{A}$ with $x_0 \in W$. Choose open neighbourhoods U and V of x_0 in X with $\bar{U} \subset V$ and $\bar{V} \subset W$. Let $\lambda : W \to [0, 1]$ be a continuous function which takes the value 1 on U and 0 on $W \backslash V$

(λ exists by Urysohn's theorem). For all $p \in \tilde{U}$, define $\sigma_p \in C^0(\pi)$ by

$$\sigma_p(x) = \begin{cases} 0_x & \text{if } x \in X \backslash V \\ \tilde{\xi}^{-1}(\xi(x), \lambda(x)\tilde{\xi}(p)_2) & \text{if } x \in V. \end{cases}$$

Then $\sigma. = (p \mapsto \sigma_p)$ has the required properties. $\square$

(B.29) Lemma. *The formula $\psi(\gamma)(\sigma) = \gamma . \sigma$ defines a continuous linear map $\psi: C^0(L(\pi, \rho)) \to L(C^0(\pi), C^0(\rho))$ which is injective and has a closed image.*

Proof. We take local representatives with respect to norm charts for π and ρ and the associated chart for $L(\pi, \rho)$ (see Example A.27). In terms of these charts, ψ sends (id, δ) to $((id, \tau) \to (id, \delta . \tau))$, where $\delta \in C^0(U', L(\mathbf{E}, \mathbf{F}))$, $\tau \in C^0(U', \mathbf{E})$ and $\mathbf{E}$, $\mathbf{F}$ are the fibres of π, ρ. It is now clear that ψ is well defined and continuous linear. For injectivity we need to show that if $\gamma . \sigma = 0 \in C^0(\rho)$ for all $\sigma \in C^0(\pi)$ then $\gamma = 0 \in C^0(L(\pi, \rho))$. This follows providing that for any $p \in B$ there is a C^0 section $\sigma \in C^0(\pi)$ taking the value p at $\pi(p)$, and this is so by Lemma B.28. Similarly, closure of Im ψ amounts to showing that if $\sup \{|\gamma_m . \sigma - \gamma_n . \sigma|_0 : |\sigma|_0 \leqslant 1\}$ is arbitrarily small for sufficiently large m, n then (γ_n) is a Cauchy sequence, and this again uses Lemma B.28. $\square$

Proof of Theorem B.27. We prove the theorem by induction on s. Let $\sigma. : \tilde{U} \to C^0(\pi)$ be as in Lemma B.28. We write g on $\tilde{U}$ as the composite

$$\tilde{U} \xrightarrow{(\pi, \sigma.)} U \times C^0(\pi) \xrightarrow{id \times g_*} U \times C^0(\rho) \xrightarrow{ev} C.$$

Thus continuity of g_* implies continuity of g (the evaluation map ev is continuous, essentially by Corollary B.16). If we restrict g to a single fibre B_x of U we have the composite $ev_x g_* \sigma.$, so, since ev_x and $\sigma.$ are continuous linear, differentiability of g_* implies the existence of Fg. From the definition of derivatives, we deduce that Fg satisfies the relation $Dg_* = \psi(Fg)_*$, where ψ is the map of Lemma B.29. Now assume that the theorem holds for $s = t \geqslant 0$ and that g_* is C^{t+1}. Then Dg_* is C^t, and hence, by the above relation and Lemma B.29, $(Fg)_*$ is C^t. Thus by the inductive hypothesis Fg is F^t. We deduce that g is F^{t+1}. $\square$

IV. SPACES OF DYNAMICAL SYSTEMS

Let X be a compact differentiable manifold. We wish to describe a topology for the set of all C^r dynamical systems on X which takes into account all derivatives of the system up to order $r (r \geqslant 1)$. For vector fields on

X we already have such a topology. A C^r vector field on X is just a C^r section of the tangent bundle π_X of X, and we have seen in the last section how to give the space $C^r(\pi_X)$ a C^r norm $|\ |_r$. As we commented above, the norms corresponding to any two norm atlases are equivalent, so we may sensibly talk of the C^r *topology* on the space. We usually write $\Gamma^r(X)$ for $C^r(\pi_X)$.

The situation for the set $\text{Diff}^r(X)$ of all C^r diffeomorphisms of X is rather more complicated. The easiest way to link it with what we have done so far is to note that if X has a Riemannian metric then right composition with the exponential map $\exp: TX \to X$ takes $0 \in \Gamma^r(X)$ to the identity map $id_X \in \text{Diff}^r(X)$, and induces a bijection from a small enough neighbourhood U of 0 in $\Gamma^r(X)$ to a subset V of $\text{Diff}^r(X)$ containing id_X. We define a basic system of neighbourhoods of id_X in $\text{Diff}^r(X)$ to be the subset V as U ranges over all discs in $\Gamma^r(X)$ with centre 0 and with radius (small) $\varepsilon > 0$. We may now obtain a basic system of neighbourhoods at any $f \in \text{Diff}^r(X)$ by composition (either right or left) with f.

An equivalent and more straightforward way to define a basic system of neighbourhoods at f is as follows: Take any pair of finite atlases $\mathcal{A} = \{\xi_i : U_i \to U_i'\}$ and $\mathcal{B} = \{\eta_j : V_j \to V_j'\}$ with the property that, for all i, $f(U_i) \subset V_{j(i)}$ for some $j(i)$, and define $W_\varepsilon(f)$ to be the set of all $g \in \text{Diff}^r(X)$ such that, for all i,

$$\overline{g(U_i)} \subset V_{j(i)}$$

and

$$\sup_i |\eta_{j(i)} g \xi_i^{-1} - \eta_{j(i)} f \xi_i^{-1}|_r < \varepsilon.$$

Then $\{W_\varepsilon(f): \varepsilon > 0\}$ is a basic system of neighbourhoods of f in the C^r-topology. Yet another description of the C^r topology is in terms of jet bundles (see, for example, § 2.1 of Hirsch [1]).

We may easily define topologies for $\Gamma^\infty(X)$ and $\text{Diff}^\infty(X)$ by taking as a basis the C^r topologies for all finite r. For example U is open in $\text{Diff}^\infty(X)$ if and only if for all $f \in U$ there is an open subset V of $\text{Diff}^r(X)$, for some r, with $f \in V \subset U$.

One may now consider the case when X is finite dimensional but non-compact. There is more than one candidate for the C^r topology. See § 2.1 of Hirsch [1] for a good discussion of the *weak* and *strong* (*Whitney*) topologies. One may also extend these ideas to X with boundary, and even to infinite dimensional X.

The Contraction Mapping Theorem

Let X and Y be non-empty metric spaces, with distance function denoted by d. Let κ be any positive number. A map $f: X \to Y$ is *Lipschitz* (*with constant* κ) if, for all x and $x' \in X$,

$$d(f(x), f(x')) \leq \kappa d(x, x').$$

The chords of the graph of f have slope $\leq \kappa$ (see Figure C.1). Clearly any

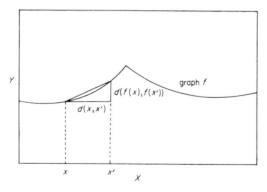

FIGURE C.1

Lipschitz map is continuous (in fact, uniformly continuous). An invertible Lipschitz map with Lipschitz inverse is sometimes called a *Lipeomorphism*. A map f is *locally Lipschitz* if every $x \in X$ has a neighbourhood on which f is Lipschitz.

(C.2) Proposition. *Let X and Y be subsets of Banach spaces and let $f: X \to Y$ be a map. If f is C^1 with $|Df(x)|$ bounded by κ and if X is convex, then f is Lipschitz with constant κ. In particular, any C^1 map is locally Lipschitz.*

Conversely, if f is Lipschitz with constant κ and f is differentiable at x then $|Df(x)| \le \kappa$.

Proof. These are immediate consequences of the mean value theorem (see § 4 of Chapter 5 of Lang [2]) and the definition of differentiability. □

(C.3) Exercise. Which of the following maps are Lipschitz?

    (i) $f: \mathbf{R} \to \mathbf{R}$ defined by $f(x) = \sin^2 x$,

    (ii) $f: \mathbf{R} \to \mathbf{R}$ defined by $f(x) = x^{1/3}$,

    (iii) $f: \mathbf{R}^2 \to \mathbf{R}$ defined by $f(x, y) = x^2 + y^2$,

    (iv) $f: \mathbf{E} \to \mathbf{R}$ defined by $f(x) = |x|$, for any norm $|\ \ |$ on a vector space $\mathbf{E}$.

We say that f is a (*metric*) contraction if it is Lipschitz with constant $\kappa < 1$. If f is invertible and f^{-1} is a contraction we call f an *expansion*. When $X \cap Y$ is non-empty, a *fixed point* of f is any $x \in X \cap Y$ such that $f(x) = x$. One of the simplest and yet most widely used of all fixed point theorems is due to Banach and Cacciopoli. The idea is as follows. Suppose that $\chi: X \to X$ is a contraction, with Lipschitz constant $\kappa < 1$. Let $x_0 \in X$, and choose a number r with $r(1 - \kappa) > d(x_0, \chi(x_0))$. If $B_r(x)$ denotes the closed ball with centre x and radius r in X, then $B_{\kappa r}(\chi(x_0))$ is contained in $B_r(x_0)$ (see Figure C.4). Since χ

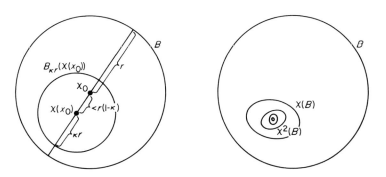

FIGURE C.4

maps $B = B_r(x_0)$ into $B_{\kappa r}(\chi(x_0))$, the iterates $\chi^n(B)$ for $n = 0, 1, 2, \ldots$ form a nested sequence, as in Figure C.4. Since χ decreases diameters by a factor κ, it is intuitively obvious that there is single point at the core of the sequence, and this must be a fixed point of f.

It is almost as quick to give a proper proof:

(C.5) Theorem. (*Contraction mapping theorem*) *A contraction* $\chi: X \to Y$ *has at most one fixed point. If* $X = Y$ *and* X *is complete then* χ *has a fixed point.*

Proof. Let x and x' be fixed points of χ. Then

$$d(x, x') = d(\chi(x), \chi(x')) \leqslant \kappa d(x, x')$$

where $\kappa < 1$ is the Lipschitz constant of χ. Thus $d(x, x') = 0$, and so $x = x'$.

Suppose that $X = Y$ is complete. Let $x \in X$. Consider the sequence $(\chi^n(x))_{n \geqslant 0}$. For all integers $n \geqslant m \geqslant 0$

$$d(\chi^n(x), \chi^m(x)) \leqslant \kappa^m d(\chi^{n-m}(x), x)$$

$$\leqslant \kappa^m \sum_{r=0}^{n-m-1} d(\chi^{r+1}(x), \chi^r(x))$$

$$\leqslant \kappa^m \sum_{r=0}^{n-m-1} \kappa^r d(\chi(x), x)$$

$$\leqslant \kappa^m (1-\kappa)^{-1} d(\chi(x), x),$$

which tends to 0 as $m \to \infty$. Thus the sequence is a Cauchy sequence. Since X is complete, the sequence converges to some limit, l say, in x. By the continuity of χ,

$$\chi(l) = \chi\left(\lim_{n \to \infty} \chi^n(x)\right) = \lim_{n \to \infty} \chi(\chi^n(x)) = l. \qquad \square$$

We often find in applications that there is a variable parameter present, and that we need to know how the fixed point depends on this parameter. Let us be more precise. A map $\chi : X \times Y \to Z$ is *uniformly Lipschitz on the first factor* if for some constant $\kappa > 0$ and all $y \in Y$ the map $\chi_y : X \to Z$ taking x to $\chi(x, y)$ is Lipschitz with constant κ. Similarly for the second factor. Clearly χ is Lipschitz if and only if it is uniformly Lipschitz on both factors. It is a *uniform contraction* on either factor if it is uniformly Lipschitz on that factor with constant <1. When χ is a uniform contraction on the second factor, say, and $Y = Z$ is complete, each map χ^x has a unique fixed point, which we denote by $g(x)$. This defines the *fixed point map* $g : X \to Y$ of χ. Such a map, satisfying for all $x \in X$

(C.6) $$\chi(x, g(x)) = g(x)$$

may, of course, exist even when the above Lipschitz conditions do not hold. We now investigate the extent to which properties of χ influence properties of g.

(C.7) Theorem. *Let $\chi : X \times Y \to Z$ be a uniform contraction on the second factor, and let $g : X \to Y$ satisfy (C.6). If χ is continuous then g is continuous. If χ is Lipschitz then g is Lipschitz. If, further, X is a subset of a Banach space $\mathbf{E}$, Y and Z are subsets of a Banach space $\mathbf{F}$ and χ is C^r then g is C^r ($r \geqslant 1$). If, further, $D\chi$ is C^{r-1}-bounded then Dg is C^{r-1}-bounded.*

Proof. We denote the distance from p to p' by $|p - p'|$. Let $\kappa > 1$ be a Lipschitz constant for χ on the second factor. Then, for all x and $x' \in X$,

$$|g(x) - g(x')| = |\chi(x, g(x)) - \chi(x', g(x'))|$$
$$\leq |\chi(x, g(x)) - \chi(x', g(x))|$$
$$+ |\chi(x', g(x)) - \chi(x', g(x'))|$$
$$\leq |\chi(x, g(x)) - \chi(x', g(x))| + \kappa|g(x) - g(x')|,$$

and so

$$|g(x) - g(x')| \leq (1 - \kappa)^{-1}|\chi(x, g(x)) - \chi(x', g(x))|.$$

Thus g is continuous when χ is continuous, and Lipschitz when χ is Lipschitz.

Now suppose $X \subset \mathbf{E}$, $Y \cup Z \subset \mathbf{F}$ and that χ is Lipschitz and C^r $(r \geq 1)$. Then for all $(x, y) \in X \times Y, |D_2\chi(x, y)| \leq \kappa$, and thus $id - D_2\chi(x, y)$ is a linear homeomorphism of $\mathbf{F}$. We first show that g is differentiable at $x \in X$, with

(C.8) $$Dg(x) = T(x)D_1\chi(x, g(x))$$

where $T(x) = (id - D_2\chi(x, g(x)))^{-1}$. For all sufficiently small $\xi \in \mathbf{E}$,

$$|g(x + \xi) - g(x) - T(x)D_1\chi(x, g(x))(\xi)|$$
$$\leq |T(x)\|g(x + \xi) - g(x) - D_2\chi(x, g(x))(g(x + \xi) - g(x))$$
$$- D_1\chi(x, g(x))(\xi)|$$
$$= |T(x)\|\chi(x + \xi, g(x + \xi)) - \chi(x, g(x))$$
$$- D\chi(x, g(x))((x + \xi, g(x + \xi)) - (x, g(x)))|.$$

By the differentiability of χ, this expression is $o(|(\xi, g(x + \xi) - g(x))|)$ as $|(\xi, g(x + \xi) - g(x))| \to 0$, whence $o(|\xi|)$ as $|\xi| \to 0$ (since g is Lipschitz). This gives differentiability of g.

The proof that g is C^r is by induction on $r(\geq 0)$. The case $r = 0$ is trivial, since g is Lipschitz. The inductive step is clear, since (C.8) expresses Dg as a composite

(C.9) $$X \xrightarrow{(id, g)} X \times Y \xrightarrow{(D_1\chi, D_2\chi)} L(\mathbf{E}, \mathbf{F}) \times B \xrightarrow{id \times \rho}$$

$$\xrightarrow{id \times \rho} L(\mathbf{E}, \mathbf{F}) \times L(\mathbf{F}, \mathbf{F}) \xrightarrow{comp} L(\mathbf{E}, \mathbf{F}),$$

where B is the ball with centre 0 and radius κ in $L(\mathbf{F}, \mathbf{F})$, and $\rho: B \to L(\mathbf{F}, \mathbf{F})$ is the C^∞-bounded uniformly C^∞ map sending T to $(id - T)^{-1}$. Note that comp is here continuous bilinear.

The last part comes, similarly, by induction using Lemma B.7. $\square$

Notice that the proof of continuity of g works in principle when X is merely a topological space. Note also that continuity of χ is implied by continuity of the maps χ_y for $y \in Y$.

We now take the theory one stage further. Our attitude is that results in the text such as Theorem 3.45 (relating a change in a vector field to the corresponding change in its integral curves) should be immediate applications of theorems in this section. To achieve this, we introduce a further parameter, taking values in a topological space A. The spaces X, Y and Z are as in Theorem C.7. We are now, however, given a map $\chi: A \times X \times Y \to Z$ such that, for each $a \in A$, $\chi^a: X \times Y \to Z$ is a uniform contraction on the second factor with constant $\kappa < 1$. We also have, for each $a \in A$, a fixed point map $g^a: X \to Y$ satisfying $g^a(x) = \chi^a(x, g^a(x))$ for all $x \in X$.

(C.10) Theorem. *Let $a_0 \in A$. Suppose that, for all $a \in A$, χ^a is C^r $(r \geq 0)$ and, if $r > 0$, Lipschitz. Suppose also that $D\chi^a$ is C^{r-1}-bounded and that $\chi^a - \chi^{a_0}$ is C^0-bounded. Then $g^a - g^{a_0}$ is C^r-bounded. If, further, $D\chi^{a_0}$ is uniformly C^{r-1} and the map $\alpha: A \to C^r(X \times Y, Z)$ taking a to $\chi^a - \chi^{a_0}$ is continuous at a_0, then the map $\beta: A \to C^r(X, Y)$ taking a to $g^a - g^{a_0}$ is continuous at a_0.*

Proof. By Theorem C.7 Dg^a is C^{r-1}-bounded for all $a \in A$. Also, for all $a \in A$ and $x \in X$,

$$|g^a(x) - g^{a_0}(x)| = |\chi^a(x, g^a(x)) - \chi^{a_0}(x, g^{a_0}(x))|$$
$$\leq |\chi^a(x, g^a(x)) - \chi^{a_0}(x, g^a(x))|$$
$$+ |\chi^{a_0}(x, g^a(x)) - \chi^{a_0}(x, g^{a_0}(x))|$$
$$\leq |\chi^a - \chi^{a_0}|_0 + \kappa |g^a(x) - g^{a_0}(x)|,$$

and

$$|g^a(x) - g^{a_0}(x)| \leq (1 - \kappa)^{-1} |\chi^a - \chi^{a_0}|_0.$$

This completes the proof that $g^a - g^{a_0}$ is C^r-bounded. It also gives continuity of β at a_0, when α is continuous at a_0, in the $r = 0$ case. We complete the proof by induction. Suppose that $\beta: A \to C^k(X, Y)$ is continuous at a_0. To perform the inductive step, we show that the map $\gamma: A \to C^k(X, L(\mathbf{E}, \mathbf{F}))$ taking a to Dg^a is continuous at a_0.

First note that, by hypothesis, the map from A to $C^k(X, X \times Y)$ taking a to $(0, g^a - g^{a_0})$ is continuous at a_0. So is the map $(a \mapsto D\chi^a)$ from A to $C^k(X \times Y, L(\mathbf{E} \times \mathbf{F}, \mathbf{F}))$. Now (id, g^{a_0}) has a C^{k-1}-bounded derivative, and $D\chi^{a_0}$ is uniformly C^k. We may apply Theorem B.18 and the $s = 0$ argument from Theorem B.15 to show that the composition map from $C^k(X, X \times Y) \times C^k(X \times Y, L(\mathbf{E} \times \mathbf{F}, \mathbf{F}))$ to $C^k(X, L(\mathbf{E} \times \mathbf{F}, \mathbf{F}))$ taking (θ, ϕ) to $(\phi(\theta + (id, g^{a_0}))$ is continuous at $((0, 0), D\chi^{a_0})$. Thus the map λ from A to

$C^k(X, L(\mathbf{E} \times \mathbf{F}, \mathbf{F}))$ taking a to $D\chi^a(id, g^a)$ is continuous at a_0. We identify $C^k(X, L(\mathbf{E} \times \mathbf{F}, \mathbf{F}))$ with $C^k(X, L(\mathbf{E}, \mathbf{F})) \times C^k(X, L(\mathbf{F}, \mathbf{F}))$ by the canonical isomorphism. The second component of λ takes values in $C^k(X, B)$, where B is as in the proof of Theorem C.7. We now describe a decomposition of the map γ. One first applies λ. Then one operates on the second factor by ρ_*, where ρ is as in the proof of Theorem C.7. Finally one takes the compositional product of the two factors (continuous bilinear, by Lemma B.5). Since λ is continuous at a_0, and the maps that follow are continuous, γ is continuous at a_0. $\square$

(C.11) Exercise. (*Lipschitz inverse mapping theorem*) Let B be the closed ball with centre 0 and radius b (possibly $b = \infty$) in a Banach space $\mathbf{E}$. Let $T: \mathbf{E} \to \mathbf{E}$ be a (topological) linear automorphism, and let $\eta: B \to \mathbf{E}$ be Lipschitz with constant $\kappa < |T^{-1}|^{-1}$ and such that $\eta(0) = 0$. Let C be the closed ball with centre 0 and radius $b(|T^{-1}|^{-1} - \kappa)$ in $\mathbf{E}$. Prove that, for all $y \in C$, there is a unique $x \in B$ such that $(T + \eta)(x) = y$. (*Hint:* rewrite this as $x = T^{-1}(y - \eta(x))$.) Hence, if $D = \text{int } C$ and we write $x = g(y)$, then $g(D)$ is an open neighbourhood of 0 in B and the map $g: D \to g(D)$ is inverse to the restriction $T + \eta: g(D) \to D$. Prove that g is Lipschitz, and C^r ($r \geq 1$) when η is C^r. Deduce the following local form:

   If f is a C^r ($r \geq 1$) map of some open subset of $\mathbf{E}$ into $\mathbf{E}$ and if $Df(x_0)$ is an automorphism then there exist open neighbourhoods U of x_0 and V of $f(x_0)$ such that the restriction $f: U \to V$ is a C^r diffeomorphism.

(C.12) Exercise. (*Immersive mapping theorem*) Prove that if $f: X \to Y$ is a C^r map of manifolds ($r \geq 1$) and f is immersive at x_0 then f restricts to a C^r embedding of some neighbourhood of x_0. (*Hint:* Assume that X and Y are open in Banach spaces $\mathbf{E}$ and $\mathbf{F}$, $x_0 = f(x_0) = 0$, $\mathbf{F} = \mathbf{E} \times \mathbf{G}$ and $Df(0) = (id, 0)$. Apply the inverse mapping theorem to the map $\phi: X \times \mathbf{G} \to \mathbf{F}$ defined by $\phi(x, z) = f(x) + (0, z)$.)

(C.13) Exercise. (*Submersive mapping theorem*) Prove that if $f: X \to Y$ is a C^r map of manifolds ($r \geq 1$) and f is submersive at x_0 then some neighbourhood of x_0 in $f^{-1}(f(x_0))$ is a C^r submanifold of X modelled on $\ker Df(x_0)$. (*Hint:* Assume that X and Y are open in Banach spaces $\mathbf{E}$ and $\mathbf{F}$, $x_0 = f(x_0) = 0$, $\mathbf{E} = \mathbf{F} \times \ker Df(0)$ and $Df(0)$ is projection to the first factor. Apply the inverse mapping theorem to the map $\phi: X \to \mathbf{E}$ defined by $\phi(x) = \phi(x_1, x_2) = (f(x), x_2)$.)

(C.14) Exercise. (*Implicit mapping theorem*) The implicit mapping theorem is, basically, concerned with solving the equation

(C.15) $$T(y) + \eta(x, y) = 0$$

for y in terms of x, where T is an automorphism of a Banach space $\mathbf{F}$, x takes values in a topological space X and η is Lipschitz on the second factor. The theorem is usually presented in a local form, where we are given a single solution $y = b$ when $x = a$, and have to show the existence of a unique continuous map $x \mapsto g(x)$ defined on some neighbourhood of a in X such that $g(a) = b$ and

(C.16) $$T(g(x)) + \eta(x, g(x)) = 0$$

for all x in the neighbourhood. We can always modify η so that $a = b = 0$.

Let B be the closed ball in F with centre 0 and radius b (possibly $b = \infty$). Suppose that $x = 0$, $y = 0$ satisfies (C.15) and let $\eta : X \times B \to \mathbf{F}$ be uniformly Lipschitz on the second factor with constant $\kappa < |T^{-1}|^{-1}$. Suppose that, for all $x \in X$, $|\eta(x, 0)| \leqslant |T^{-1}|^{-1} - \kappa$. Prove that there is a unique map $g : X \to B$ satisfying (C.16) for all $x \in X$. (*Hint*: Rewrite (C.15) as $y = -T^{-1}\eta(x, y) = 0$.) Prove that g is continuous if η is continuous. Prove that if X is open in a Banach space $\mathbf{E}$ then g is Lipschitz if η is Lipschitz, and C^r ($r \geqslant 1$) if η is C^r. Deduce the following local form:

Let X and Y be open subsets of Banach spaces $\mathbf{E}$ and $\mathbf{F}$ respectively, and let $f : X \times Y \to \mathbf{F}$ be C^r ($r \geqslant 1$) with $f(a, b) = 0$ and $D_2 f(a, b)$ an automorphism, for some $(a, b) \in X \times Y$. Prove that there exist neighbourhoods U of a in X and V of b in Y such that there is a unique map $g : U \to V$ satisfying $f(x, g(x)) = 0$ for all $x \in U$. Moreover the map g is C^r.

BIBLIOGRAPHY

As a reading of Chapter 7 should make clear, many recent advances in dynamical systems research stem from ideas of Stephen Smale, and his 1967 survey article (Smale [5]) has been profoundly influential. Some other surveys that may help to guide further reading on the subject are Nemitskii [1], Palis [2], Robbin [3], Shub [2, 3, 4, 5] and Smale [7, 8]. In the list of references that follows, two conference proceedings appear so often that I have abbreviated them. *"Global Analysis"* denotes *"Global Analysis*, Proc. Symp. in Pure Math. **14** (Providence, Rhode Island: American Math. Soc., 1970)", and *"Dynamical Systems"* denotes *"Dynamical Systems* (Ed. Peixoto, M. M.) (New York: Academic Press, 1973).

ABRAHAM, R. (assisted by MARSDEN, J. E.)
  1. "Foundations of Mechanics", Benjamin, New York, 1967.
ABRAHAM, R. and ROBBIN, J.
  1. "Transversal Mappings and Flows". Benjamin, New York, 1967.
ABRAHAM, R. and SMALE, S.
  1. Non-genericity of Ω-stability. *Global Analysis*, 5–8.
ANDRONOV, A. and PONTRJAGIN, L.
  1. Systèmes grossiers. *Dokl. Akad. Nauk SSSR* **14** (1937), 247–251.
ANOSOV, D.
  1. Roughness of geodesic flows on compact Riemannian manifolds of negative curvature. *Dokl. Akad. Nauk SSSR* **145** (1962), 707–709 (English: *Soviet Math. Dokl.* **3** (1962), 1068–1070.)
  2. Geodesic flows on compact Riemannian manifolds of negative curvature. *Trudy Mat. Inst. Steklov* **90** (1967) (English: *Amer. Math. Soc.* translation (1969).)
ANTOSIEWICZ, H. A.
  1. A survey of Lyapunov's second method. "Contributions to the Theory of Non-Linear Oscillations". (Ed. Lefschetz, S.), Vol. 4, pp. 141–166. Annals of Math. Study **41**, Princeton University Press, Princeton, 1958.
ARNOLD, V. I.
  1. "Ordinary Differential Equations". MIT Press, Cambridge, Massachusetts, 1973.
BIRKHOFF, G. D.
  1. "Dynamical Systems". Amer. Math. Soc. Colloquium Pub. **9** American Math. Soc., New York, 1927.
BOURBAKI, N.
  1. "General Topology, Part II". Herman and Addison-Wesley, Reading, Mass., 1966.
BRICKELL, F. and CLARK, R. S.
  1. "Differentiable Manifolds". Van Nostrand Reinhold, London, 1970.

CHILLINGWORTH, D. R. J.
1. "Differential Topology with a View to Applications". Pitman, London, 1976.
CODDINGTON, E. A. and LEVINSON, N.
1. "Theory of Ordinary Differential Equations". McGraw-Hill, New York, 1955.
DANKNER, A.
1. On Smale's Axiom A dynamical systems. *Astérisque* **49** (1977), 19–22.
DE BAGGIS, H. F.
1. Dynamical systems with stable structures. "Contributions to the Theory of Non-Linear Oscillations". (Ed. Lefschetz, S), Vol. 2, pp. 37–59. Annals of Math. Study **29**, Princeton University Press, Princeton, 1952.
DENJOY, A.
1. Sur les courbes définies par les équations différentielles à la surface du tore. *J. de Math. Pure et Appl.* **11** (1932), 333–375.
DE OLIVIERA, M.
1. C^0-density of structurally stable vector fields. *Bull. Amer. Math. Soc.* **82** (1976).
DIEUDONNE, J.
1. "Foundations of Modern Analysis". Academic Press, New York and London, 1960.
DUNFORD, N. and SCHWARTZ, J. T.
1. "Linear Operators" Part 1. Interscience, New York, 1958.
ELIASSON, H. I.
1. Geometry of manifolds of maps. *J. Differential Geometry* **1** (1967), 169–194.
FOSTER, M. J.
1. Calculus on vector bundles. *J. London Math. Soc.* (2) **11** (1975), 65–73.
2. Fibre derivatives and stable manifolds: a note. *Bull. London Math. Soc.* **8** (1976), 286–288.
FRANKS, J.
1. "Manifolds of C^r-mappings".
2. Differentiably Ω-stable diffeomorphisms. *Topology* **11** (1972), 107–114.
3. Absolutely structurally stable diffeomorphisms. *Proc. Amer. Math. Soc.* **37** (1973), 293–296.
4. Constructing structurally stable diffeomorphisms. *Annals of Math.* **105** (1977), 343–359.
GANTMACHER, F. R.
1. "The Theory of Matrices" Vol. 1. Chelsea, New York, 1959.
GREENBERG, M. J.
1. "Lectures on Algebraic Topology". Benjamin, New York, 1967.
GROBMAN, D. M.
1. Homeomorphisms of systems of differential equations. *Dokl. Akad. Nauk SSSR* **128** (1959), 880–881.
2. Topological classification of the neighbourhood of a singular point in n-dimensional space. *Mat. Sb.* (*N.S*) **56** (98) (1962), 77–94.
GUCKENHEIMER, J.
1. Absolutely Ω-stable diffeomorphisms. *Topology* **11** (1972), 195–197.
2. Bifurcation and catastrophe. *Dynamical Systems*, 95–110.
3. One parameter families of vector fields on two-manifolds: another non-density theorem. *Dynamical Systems*, 111–127.

GUILLEMIN, V. and POLLACK, A.
1. "Differential topology". Prentice-Hall, Englewood Cliffs, New Jersey, 1974.
GUTIERREZ, C.
1. Structural stability for flows on the torus with a cross-cap. *Trans. Amer. Math. Soc.* **241** (1978), 311–320.
HALE, J. K.
1. "Ordinary Differential Equations". Wiley–Interscience, New York, 1969.
HARTMAN, P.
1. "Ordinary Differential Equations". Wiley, New York, 1964.
HELGASON, S.
1. "Differential Geometry and Symmetric Spaces". Academic Press, New York and London, 1953.
HIRSCH, M. W.
1. "Differential Topology". Springer-Verlag, New York, 1976.
HIRSCH, M.. W. and PUGH, C. C.
1. Stable manifolds and hyperbolic sets. *Global Analysis*, 133–163.
HIRSCH, M. W. and SMALE, S.
1. "Differential Equations, Dynamical Systems, and Linear Algebra". Academic Press, New York and London, 1974.
HIRSCH, M. W., PUGH, C. C. and SHUB, M.
1. "Invariant Manifolds". Springer Lecture Notes No. 583 (1977).
HOLMES, R. B.
1. A formula for the spectral radius of an operator. *Amer. Math. Monthly* **75** (1968), 163–166.
HOPF, E.
1. Anzweigung einer periodischen lösung von einer stationären lösung eines differentialsystems. *Ber. Verh. Sächs, Akad. Wiss. Leipzig Math. Phys.* **95** (1943), 3–22.
HU, S.-T.
1. "Introduction to general topology". Holden-Day, San Francisco, 1966.
2. "Homology theory". Holden-Day, San Francisco, 1966.
HUREWICZ, W.
1. "Lectures on Ordinary Differential Equations". MIT Press, Cambridge, Mass., 1958.
IRWIN, M. C.
1. Transformation groups with a common orbit. *Bull. London Math. Soc.* **5** (1973), 164–168.
2. A new proof of the pseudo-stable manifold theorem. *J. London Math. Soc.* (to appear).
KERVAIRE, M.
1. A manifold which does not admit any differentiable structure. *Comm. Math. Helv.* **34** (1960), 304–312.
KUIPER, N. H.
1. The homotopy type of the unitary group of Hilbert space. *Topology* **3** (1965), 19–30.
2. The topology of the solutions of a linear differential equation on R^n. *Manifolds-Tokyo 1973*, pp. 195–203. Univ. Tokyo Press, Tokyo, 1975.
KUIPER, N. H. and ROBBIN, J. W.
1. Topological classification of linear endomorphisms. *Invent. Math.* **19** (1973), 83–106.

KUPKA, I.
 1. Contribution à la théorie des champs génériques. "Contributions to Differential Equations" Vol. **2** (1963), pp. 457–484: Vol. **3** (1964), pp. 411–420. Wiley–Interscience, New York.
 2. On two notions of structural stability. *J. Differential Geometry* **9** (1974), 639–644.
LANG, S.
 1. "Differential manifolds". Addison-Wesley, Reading, Mass., 1972.
 2. "Analysis II". Addison-Wesley, Reading, Mass., 1969.
LESLIE, J.
 1. On a differential structure for the group of diffeomorphisms. *Topology* **6** (1967), 263–271.
LEFSCHETZ, S.
 1. "Differential Equations, Geometric Theory". Wiley–Interscience, New York, 1957.
LIAPUNOV, A. M.
 1. "Problème général de la stabilité du mouvement." Annals of Math. Study **17**. Princeton University Press, Princeton, 1947.
MANNING, A.
 1. There are no new Anosov diffeomorphisms on tori. *Amer. Jour. Math.* **96** (1974), 422–429.
MAUNDER, C. R. F.
 1. "Algebraic Topology". Van Nostrand Reinhold, London, 1970.
MARKLEY, N.
 1. Homeomorphisms of the circle without periodic points. *Proc. London Math. Soc.* (3) **20** (1970), 688–698.
MATHER, J.
 1. Characterization of Anosov diffeomorphisms. *Nederl. Akad. van Wetensch. Proc. Ser. A, Amsterdam* **71** = *Indag. Math.* **30** (1968), 479–483.
MAZUR, B.
 1. *Pub. Math. I.H.E.S.* **15** (1963) (also **22** (1964), 81–92).
MILNOR, J.
 1. On manifolds homeomorphic to the 7-sphere. *Annals of Math.* **64** (1956), 399–405.
 2. Topology from the Differentiable Viewpoint". University of Virginia Press, Charlottesville, 1966.
 3. "Morse Theory". Annals of Math. Study **51** Princeton University Press, Princeton, 1963.
MOSER, J.
 1. On a theorem of Anosov. *J. Differential Equations* **5** (1969), 411–440.
MUNKRES, J.
 1. "Elementary differential topology". Annals of Math Study **54**, Princeton University Press, Princeton, 1963.
NELSON, E.
 1. "Topics in Dynamics I: Flows". Princeton University Press, Princeton, 1969.
NEMITSKII, V. V.
 1. Some modern problems in the qualitative theory of ordinary differential equations. *Russian Math. Surveys* **20** (1965), 1–34.
NEMITSKII, V. V. and STEPANOV, V. V.
 1. "Qualitative Theory of Ordinary Differential. Equations". Princeton University Press, Princeton, 1960.

NEWHOUSE, S. E.
1. Nondensity of Axiom A(a) on S^2. *Global Analysis*, 191–202.
2. Hyperbolic limit sets. *Trans. Amer. Math. Soc.* **167** (1972), 125–150.
3. On simple arcs between structurally stable flows. "Dynamical Systems—Warwick 1974". Springer Lecture Notes No. 468, pp. 209–233.

NEWHOUSE, S. E. and PALIS, J.
1. Hyperbolic nonwandering sets on two-dimension manifolds. *Dynamical Systems*, 293–301.
2. Bifurcations of Morse–Smale dynamical systems. *Dynamical Systems*, 303–366.
3. Cycles and bifurcation theory. *Astérisque* **31** (1976), 43–140.

NEWHOUSE, S. E. and PEIXOTO, M.
1. There is a simple arc joining any two Morse–Smale flows. *Astérisque* **31** (1976), 15–41.

NITECKI, Z.
1. "Differentiable dynamics". MIT Press, Cambridge, Mass., 1971.

PALIS, J.
1. A note on Ω-stability. *Global Analysis*, 221–222.
2. Some developments on stability and bifurcation of dynamical systems. "Geometry and Topology". Springer Lecture Notes No. 597, pp. 495–509 (1977).

PALIS, J. and SMALE, S.
1. Structural stability theorems. *Global Analysis*, 223–231.

PEIXOTO, M. M.
1. On an approximation theorem of Kupka and Smale. *J. Differential Equations* **3** (1966), 214–227.
2. Structural stability on two dimensional manifolds. *Topology* **1** (1962), 101–120.
3. On the classification of flows on 2-manifolds. *Dynamical Systems*, 389–420.

PEIXOTO, M. and PUGH, C. C.
1. Structurally stable systems on open manifolds are never dense. *Annals of Math.* **87** (1968), 423–430.

POINCARÉ, H.
1. Mémoire sur les courbes définies par une équation différentielle. *J. de Math.* **7** (1881) 375–422 and **8** (1882), 251–296. Sur les courbes définies par les équations différentielles. *J. de Math. Pure et Appl.* **1** (1885) 167–244. (These are all in Vol. 1 of "Oeuvres de Henri Poincaré". Gauthier–Villars, Paris, 1951.

PUGH, C. C.
1. The closing lemma. *Amer. J. Math.* **89** (1967), 956–1009.
2. An improved closing lemma and a general density theorem. *Amer. J. Math.* **89** (1967), 1010–1021.
3. On a theorem of P. Hartman. *Amer. J. Math.* **91** (1969), 363–367.

PUGH, C. C. and SHUB, M.
1. Ω-stability for flows. *Inventiones Math.* **11** (1970), 150–158.

RENZ, P. L.
1. Equivalent flows on Banach manifolds. *Indiana Univ. Math. J.* **20** (1971), 695–698.

ROBBIN, J. W.
1. A structural stability theorem. *Annals of Math.* **94** (1971), 447–493.

2. Topological conjugacy and structural stability for discrete dynamical systems. *Bull. Amer. Math. Soc.* **78** (1972), 923–952.

ROBINSON, C.
1. Generic properties of conservative systems. *Amer. J. Math.* **92** (1970), 562–603 and 897–906.
2. Structural stability of C^1-flows. "Dynamical Systems—Warwick 1974". Springer Lecture Notes No. 468, pp. 262–277.
3. Structural stability of C^1-diffeomorphisms. *J. Differential Equations*, **22** (1976), 28–73.
4. The geometry of the structural stability proof using unstable discs. *Bol. Soc. Bras. Mat.* **6** (1975), 129–144.

ROBINSON, C. and WILLIAMS, R. F.
1. Finite stability is not generic. *Dynamical Systems*, 451–462.

ROSENBERG, H.
1. A generalization of Morse–Smale inequalities. *Bull. Amer. Math. Soc.* **70** (1974), 422–427.

SHUB, M.
1. Structurally stable systems are dense. *Bull. Amer. Math. Soc.* **78** (1972), 817–818.
2. Stability and genericity for diffeomorphisms. *Dynamical Systems*, 493–514.
3. Dynamical systems, filtrations and entropy. *Bull. Amer. Math. Soc.* **80** (1974), 27–41.
4. Stabilité globale des systèmes dynamiques. *Astérisque* **56** (1978).
5. The Lefschetz fixed point formula: smoothness and stability. "Dynamical Systems" (Ed. Cesaro, L., Hale, J. K. and LaSalle, J. P.) pp. 13–28. Academic Press, New York, 1976.

SHUB, M. and SMALE, S.
1. Beyond hyperbolicity. *Annals of Math.* **96** (1972), 587–591.

SHUB, M. and WILLIAMS, R. F.
1. Future stability is not generic. *Proc. Amer. Math. Soc.* **22** (1969), 483–484.

SIMMONS, G. F.
1. "Differential Equations with Applications and Historical Notes". McGraw-Hill, New York, 1972.

SMALE, S.
1. On gradient dynamical systems. *Annals of Math.* **74** (1961), 199–206.
2. Generalized Poincaré's conjecture in dimensions greater than four. *Annals of Math.* **74** (1961), 391–406.
3. Stable manifolds for differential equations and diffeomorphisms. *Ann. Scuola Normale Superiore Pisa* **18** (1963), 97–116.
4. Structurally stable systems are not dense. *Amer. J. Math.* **88** (1966), 491–496.
5. Differentiable dynamical systems. *Bull. Amer. Math. Soc.* **73** (1967), 747–817.
6. The Ω-stability theorem. *Global Analysis*, 289–297.
7. Notes on differentiable dynamical systems. *Global Analysis*, 277–287.
8. Stability and genericity in dynamical systems. "Seminaire Bourbaki 1969–1970". Springer Lecture Notes No. 180, pp. 177–185 (1971).
9. Stability and isotopy in discrete dynamical systems. *Dynamical Systems*, 527–531.
10. On the structure of manifolds. *Amer. J. Math.* **84** (1962), 387–399.

SOTOMAYOR, J.
1. Structural stability and bifurcation theory. *Dynamical Systems*, 549–560.
2. Generic bifurcations of dynamical systems. *Dynamical Systems*, 561–582.
3. Generic one-parameter families of vector fields on two-dimensional manifolds. *Publ. Math. IHES* **43** (1974), 5–46.

SPANIER, E. H.
1. "Algebraic topology". McGraw-Hill, New York, 1966.

SULLIVAN, D. and WILLIAMS, R. F.
1. On the homology of attractors. *Topology* **15** (1976), 259–262.

STERNBERG, S.
1. Local contractions and a theorem of Poincaré. *Amer. J. Math.* **79** (1957), 809–824.
2. On the structure of local homeomorphisms of euclidean n-space: II. *Amer. J. Math.* **80** (1958), 623–631.

TAKENS, F.
1. On Zeeman's tolerance stability conjecture. "Manifolds Amsterdam 1970". Springer Lecture Notes No. 197, pp. 209–219 (1971).
2. Tolerance stability. "Dynamical Systems—Warwick 1974". Springer Lecture Notes No. 468, pp. 293–304 (1975).
3. Integral curves near mildly degenerate singular points of vector fields. *Dynamical Systems*, 599–617.
4. Singularities of vector fields. *Publ. Math. IHES* **43** (1974), 47–100.

THOM, R.
1. "Stabilité Structurelle et Morphogénèse". Addison-Wesley, Reading, Mass., 1973. (English translation by D. H. Fowler, 1975).

WARNER, F. W.
1. "Foundations of Differentiable Manifolds and Lie Groups". Scott, Foresman, Glenview, Ill., 1971.

WELLS, J. C.
1. Invariant manifolds of nonlinear operators. *Pacific J. Math.* **62** (1976), 285–293.

WHITE, W.
1. On the tolerance stability conjecture. *Dynamical Systems*, 663–665.

WILLIAMS, R. F.
1. One dimensional non wandering sets. *Topology* **6** (1967), 473–487.
2. The "DA" maps of Smale and structural stability. *Global Analysis*, 329–339.
3. Expanding attractors. *Publ. Math. IHES* **43** (1974), 169–203.

ZEEMAN, E. C.
1. Morse inequalities for diffeomorphisms with shoes and flows with solenoids. "Dynamical Systems—Warwick 1974". Springer Lecture Notes No. 468, pp. 44–47 (1975).

Index

253